COMPLEX VARIABLES AND APPLICATIONS

COMPLEX VARIABLES AND APPLICATIONS

Fourth Edition

Ruel V. Churchill

Professor Emeritus of Mathematics
The University of Michigan

James Ward Brown

Professor of Mathematics
The University of Michigan — Dearborn

McGraw-Hill Book Company

New York St. Louis San Francisco Auckland Bogotá Hamburg
London Madrid Mexico Montreal New Delhi
Panama Paris São Paulo Singapore Sydney Tokyo Toronto

COMPLEX VARIABLES AND APPLICATIONS

7890HALHAL89

ISBN 0-07-010873-0

This book was set in Times Roman by Beacon Graphics Corporation.
The editors were John J. Corrigan, Peter R. Devine, and James W. Bradley;
the production supervisor was Diane Renda.
New drawings were done by J & R Services, Inc.
The cover was designed by Mark Wieboldt.
Halliday Lithograph Corporation was printer and binder.

Library of Congress Cataloging in Publication Data

Churchill, Ruel Vance, date
    Complex variables and applications.

    Bibliography: p.
    Includes index.
    1. Functions of complex variables.   I. Brown,
James Ward.   II. Title.
QA331.C524   1984   515.9   83-9384
ISBN 0-07-010873-0

CONTENTS

PREFACE

This book is a revision of the third edition, published in 1974. That edition has served, just as the first two editions did, as a textbook for a one-term introductory course in the theory and applications of functions of a complex variable. This revision preserves the basic content and style of the earlier editions, the first two of which were written by Ruel V. Churchill alone.

In this edition the authors have improved the exposition by making the examples more prominent in the text and by redrawing and adding a number of figures. The material on integrals and residues and their applications is now reached earlier, and the chapter on mapping by elementary functions leads more directly into the chapters on conformal mapping and its applications. To mention some other improvements, the sections on finding roots of complex numbers and calculating residues of functions at isolated singular points have been completely rewritten with special attention paid to the understanding of concepts and less reliance on formulas.

As was the case with the earlier editions, the first objective of this edition is to develop in a rigorous and self-contained manner those parts of the theory which are prominent in the applications of the subject. The second objective is to furnish an introduction to applications of residues and conformal mapping. Special emphasis is given to the use of conformal mapping in solving boundary value problems which arise in studies of heat conduction, electrostatic potential, and fluid flow. Hence the book may be considered as a companion volume to the authors' "Fourier Series and Boundary Value Problems" and Ruel V. Churchill's "Operational Mathematics," in which other classical methods for solving boundary value problems are treated. The latter book also contains applications of residues in connection with Laplace transforms.

The first nine chapters of this book, with various substitutions from the remaining chapters, have for many years formed the content of a 3-hour course given each term at The University of Michigan. The classes have consisted mainly of seniors and graduate students majoring in mathematics, engineering, or one of the physical sciences. The students have usually completed one term of advanced calculus. Some of the material is not covered in the lectures and is left for students to read on their own. If mapping by elementary functions and applications of conformal mapping are desired earlier in the course, one can skip to Chapters 7, 8, and 9 immediately after Chapter 3 on elementary functions.

Most of the basic results are stated as theorems, followed by examples and exercises which illustrate those results. A bibliography of other and, in many cases, more advanced books is provided in Appendix 1. A table of conformal transformations useful in applications appears in Appendix 2.

In preparing this revision, the authors have taken advantage of suggestions from a variety of people, especially Douglas G. Dickson, who suggested improvements in the treatment of antiderivatives. The authors are also indebted to their editors, John J. Corrigan, who provided a number of anonymous reviews of both the last edition and the present one in manuscript form, and Peter R. Devine, who saw this edition through its final stages.

Ruel V. Churchill
James Ward Brown

COMPLEX VARIABLES AND APPLICATIONS

COMPLEX NUMBERS

In this chapter we survey the algebraic and geometric structure of the complex number system. We assume various corresponding properties of real numbers to be known.

1. Definition

Complex numbers z can be defined as ordered pairs

$$(1) \qquad\qquad z = (x, y)$$

of real numbers x and $y,$ with operations of addition and multiplication to be specified below. It is customary to identify the pairs $(x, 0)$ with the real numbers x. The set of complex numbers thus includes the real numbers as a subset. Complex numbers of the form $(0, y)$ are called *pure imaginary numbers*. The real numbers x and y in expression (1) are known as the *real and imaginary parts* of z, respectively; and we write

$$(2) \qquad\qquad \operatorname{Re} z = x, \qquad \operatorname{Im} z = y.$$

Two complex numbers (x_1, y_1) and (x_2, y_2) are said to be *equal* whenever they have the same real parts and the same imaginary parts. That is,

$$(3) \qquad (x_1, y_1) = (x_2, y_2) \qquad \text{if and only if} \qquad x_1 = x_2 \text{ and } y_1 = y_2.$$

The *sum* $z_1 + z_2$ and *product* $z_1 z_2$ of two complex numbers $z_1 = (x_1, y_1)$ and $z_2 = (x_2, y_2)$ are defined by the equations

$$(4) \qquad (x_1, y_1) + (x_2, y_2) = (x_1 + x_2, y_1 + y_2),$$

$$(5) \qquad (x_1, y_1)(x_2, y_2) = (x_1 x_2 - y_1 y_2, y_1 x_2 + x_1 y_2).$$

In particular, $(x, 0) + (0, y) = (x, y)$ and $(0, 1)(y, 0) = (0, y)$; hence

$$(6) \qquad\qquad (x, y) = (x, 0) + (0, 1)(y, 0).$$

Note that the operations defined by equations (4) and (5) become the usual operations of addition and multiplication when restricted to the real numbers:

$$(x_1, 0) + (x_2, 0) = (x_1 + x_2, 0),$$
$$(x_1, 0)(x_2, 0) = (x_1 x_2, 0).$$

The complex number system is thus a natural extension of the real number system.

Thinking of a real number as either x or $(x, 0)$ and *letting i denote the pure imaginary number* $(0, 1)$, we can rewrite equation (6) as

(7) $$(x, y) = x + iy.*$$

Also, with the convention $z^2 = zz$, $z^3 = z^2 z$, etc., we find that

$$i^2 = (0, 1)(0, 1) = (-1, 0);$$

that is, $$i^2 = -1.$$

In view of expression (7), equations (4) and (5) become

(8) $$(x_1 + iy_1) + (x_2 + iy_2) = (x_1 + x_2) + i(y_1 + y_2),$$

(9) $$(x_1 + iy_1)(x_2 + iy_2) = (x_1 x_2 - y_1 y_2) + i(y_1 x_2 + x_1 y_2).$$

Observe that the right-hand sides of these equations can be obtained by formally manipulating the terms on the left as if they involved only real numbers and by replacing i^2 by -1 when it occurs.

2. Algebraic Properties

Various properties of addition and multiplication of complex numbers are the same as for real numbers. We list here the more basic of these algebraic properties and verify a few of them.

The commutative laws

(1) $$z_1 + z_2 = z_2 + z_1, \qquad z_1 z_2 = z_2 z_1$$

and the associative laws

(2) $$(z_1 + z_2) + z_3 = z_1 + (z_2 + z_3), \qquad (z_1 z_2)z_3 = z_1(z_2 z_3)$$

follow easily from the definitions of addition and multiplication of complex numbers and the fact that real numbers obey these laws. For example, if

$$z_1 = (x_1, y_1) \qquad \text{and} \qquad z_2 = (x_2, y_2),$$

then

$$z_1 + z_2 = (x_1, y_1) + (x_2, y_2) = (x_1 + x_2, y_1 + y_2) = (x_2 + x_1, y_2 + y_1)$$
$$= (x_2, y_2) + (x_1, y_1) = z_2 + z_1.$$

Verification of the rest of the above laws, as well as the distributive law

(3) $$z_1(z_2 + z_3) = z_1 z_2 + z_1 z_3,$$

is left to the exercises.

*In electrical engineering the symbol j is used instead of i.

According to the commutative law for multiplication, $iy = yi$; hence it is permissible to write either

$$z = x + iy \qquad \text{or} \qquad z = x + yi.$$

Also, because of the associative laws, a sum $z_1 + z_2 + z_3$ or a product $z_1 z_2 z_3$ is well defined without parentheses, just as it is with real numbers.

The additive identity $0 = (0, 0)$ and the multiplicative identity $1 = (1, 0)$ for real numbers carry over to the entire complex number system. That is,

$$(4) \qquad\qquad z + 0 = z \qquad \text{and} \qquad z \cdot 1 = z$$

for every complex number z. Furthermore, 0 and 1 are the only complex numbers with such properties. To establish the uniqueness of 0, we suppose that (u, v) is an additive identity, and we write

$$(x, y) + (u, v) = (x, y)$$

where (x, y) is any complex number. It follows that

$$x + u = x \qquad \text{and} \qquad y + v = y;$$

that is, $u = 0$ and $v = 0$. The complex number $0 = (0, 0)$ is therefore the only additive identity. A similar method can be used to show that 1 is unique as a multiplicative identity.

There is associated with each complex number $z = (x, y)$ an additive inverse

$$(5) \qquad\qquad -z = (-x, -y)$$

which satisfies the equation $z + (-z) = 0$. Moreover, there is only one additive inverse for any given z. Additive inverses are used to define subtraction:

$$(6) \qquad\qquad z_1 - z_2 = z_1 + (-z_2).$$

So if $z_1 = (x_1, y_1)$ and $z_2 = (x_2, y_2)$, then

$$(7) \qquad z_1 - z_2 = (x_1 - x_2, y_1 - y_2) = (x_1 - x_2) + i(y_1 - y_2).$$

Likewise, for any *nonzero* complex number $z = (x, y)$, there is a number z^{-1} such that $zz^{-1} = 1$. This multiplicative inverse is less obvious than the additive one. To find it, we seek real numbers u and v, expressed in terms of x and y, such that

$$(x, y)(u, v) = (1, 0).$$

According to equation (5), Sec. 1, which defines the product of two complex numbers, u and v must satisfy the pair

$$xu - yv = 1, \qquad yu + xv = 0$$

of linear simultaneous equations; and simple computation yields the unique solution

$$u = \frac{x}{x^2 + y^2}, \qquad v = \frac{-y}{x^2 + y^2}.$$

The multiplicative inverse of $z = (x, y)$ is, then,

$$(8) \qquad z^{-1} = \left(\frac{x}{x^2 + y^2}, \frac{-y}{x^2 + y^2} \right) \qquad (z \neq 0).$$

The existence of multiplicative inverses enables us to show that *if a product $z_1 z_2$ is zero, then so is at least one of the factors z_1 and z_2.* For suppose that $z_1 z_2 = 0$ and $z_1 \neq 0$. The inverse z_1^{-1} exists; and, according to the definition of multiplication, any complex number times zero is zero. Hence

$$(9) \qquad z_2 = 1 \cdot z_2 = (z_1^{-1} z_1) z_2 = z_1^{-1} (z_1 z_2) = z_1^{-1} \cdot 0 = 0.$$

That is, if $z_1 z_2 = 0$, either $z_1 = 0$ or $z_2 = 0$; or possibly both z_1 and z_2 equal zero. Another way to state this result is that *if two complex numbers z_1 and z_2 are nonzero, then so is their product $z_1 z_2$.*

Division by a nonzero complex number is defined:

$$(10) \qquad \frac{z_1}{z_2} = z_1 z_2^{-1} \qquad (z_2 \neq 0).$$

If $z_1 = (x_1, y_1)$ and $z_2 = (x_2, y_2)$, equations (8) and (10) show that

$$(11) \qquad \frac{z_1}{z_2} = \left(\frac{x_1 x_2 + y_1 y_2}{x_2^2 + y_2^2}, \frac{y_1 x_2 - x_1 y_2}{x_2^2 + y_2^2} \right) = \left(\frac{x_1 x_2 + y_1 y_2}{x_2^2 + y_2^2} \right) + i \left(\frac{y_1 x_2 - x_1 y_2}{x_2^2 + y_2^2} \right)$$

$$(z_2 \neq 0).$$

The quotient z_1 / z_2 is not defined when $z_2 = 0$; note that $z_2 = 0$ means that $x_2^2 + y_2^2 = 0$, and this is not permitted in expressions (11).

Finally, we mention some useful identities involving quotients. They are based on the relation

$$(12) \qquad \frac{1}{z_2} = z_2^{-1} \qquad (z_2 \neq 0),$$

which is equation (10) when $z_1 = 1$ and which allows us to write that equation in the form

$$(13) \qquad \frac{z_1}{z_2} = z_1 \left(\frac{1}{z_2} \right) \qquad (z_2 \neq 0).$$

Noticing that (see Exercise 11)

$$(z_1 z_2)(z_1^{-1} z_2^{-1}) = (z_1 z_1^{-1})(z_2 z_2^{-1}) = 1 \qquad (z_1 \neq 0, z_2 \neq 0),$$

and hence that $(z_1 z_2)^{-1} = z_1^{-1} z_2^{-1}$, one can use relation (12) to verify the identity

$$(14) \qquad \frac{1}{z_1 z_2} = \left(\frac{1}{z_1} \right) \left(\frac{1}{z_2} \right) \qquad (z_1 \neq 0, z_2 \neq 0).$$

With the aid of equations (13) and (14), it is then easy to show that

$$(15) \qquad \frac{z_1 + z_2}{z_3} = \frac{z_1}{z_3} + \frac{z_2}{z_3}, \qquad \frac{z_1 z_2}{z_3 z_4} = \left(\frac{z_1}{z_3} \right) \left(\frac{z_2}{z_4} \right) \qquad (z_3 \neq 0, z_4 \neq 0).$$

Example. Computations such as the following are now justified:

$$\left(\frac{1}{2-3i}\right)\left(\frac{1}{1+i}\right) = \frac{1}{5-i} = \left(\frac{1}{5-i}\right)\left(\frac{5+i}{5+i}\right) = \frac{5+i}{26} = \frac{5}{26} + \frac{1}{26}i.$$

EXERCISES

1. Verify that

   (a) $(\sqrt{2} - i) - i(1 - \sqrt{2}\,i) = -2i$; (b) $(2, -3)(-2, 1) = (-1, 8)$;

   (c) $(3, 1)(3, -1)\left(\dfrac{1}{5}, \dfrac{1}{10}\right) = (2, 1)$; (d) $\dfrac{1 + 2i}{3 - 4i} + \dfrac{2 - i}{5i} = -\dfrac{2}{5}$;

   (e) $\dfrac{5}{(1-i)(2-i)(3-i)} = \dfrac{1}{2}i$; (f) $(1-i)^4 = -4$.

2. Verify that each of the two numbers $z = 1 \pm i$ satisfies the equation $z^2 - 2z + 2 = 0$.

3. Solve the equation $z^2 + z + 1 = 0$ for $z = (x, y)$ by writing

   $$(x, y)(x, y) + (x, y) + (1, 0) = (0, 0)$$

   and then solving a pair of simultaneous equations in x and y.
   Suggestion: Note that $y \neq 0$ since no real number x satisfies the equation

   $$x^2 + x + 1 = 0.$$

   $$Ans. \quad z = \left(-\frac{1}{2}, \pm\frac{\sqrt{3}}{2}\right).$$

4. Prove that multiplication is commutative, as stated in the second of equations (1), Sec. 2.

5. Verify the associative laws (2), Sec. 2.

6. Verify the distributive law (3), Sec. 2.

7. Apply laws established in Exercises 5 and 6 to show that

   $$z(z_1 + z_2 + z_3) = zz_1 + zz_2 + zz_3.$$

8. Show that the complex number $1 = (1, 0)$ is the only multiplicative identity.

9. Show that $-z = (-x, -y)$ is the only additive inverse of a given complex number $z = (x, y)$.

10. Prove that

   (a) $\text{Im}(iz) = \text{Re } z$; (b) $\text{Re}(iz) = -\text{Im } z$; (c) $1/(1/z) = z$ $(z \neq 0)$;

   (d) $(-1)z = -z$.

11. Use the associative and commutative laws for multiplication to show that

   $$(z_1 z_2)(z_3 z_4) = (z_1 z_3)(z_2 z_4).$$

12. Prove that if $z_1 z_2 z_3 = 0$, then at least one of the three factors is zero.

13. Verify identity (14), Sec. 2.

14. Establish the first of identities (15), Sec. 2.

15. Establish the second of identities (15), Sec. 2, and use it to prove the cancellation law

$$\frac{zz_1}{zz_2} = \frac{z_1}{z_2} \qquad (z \neq 0, z_2 \neq 0).$$

16. Show that $(1 + z)^2 = 1 + 2z + z^2$.

17. Use mathematical induction to establish the binomial formula

$$(z_1 + z_2)^n = z_1^n + \frac{n}{1!}z_1^{n-1}z_2 + \frac{n(n-1)}{2!}z_1^{n-2}z_2^2 + \cdots$$

$$+ \frac{n(n-1)(n-2)\cdots(n-k+1)}{k!}z_1^{n-k}z_2^k + \cdots + z_2^n,$$

where z_1 and z_2 are any two complex numbers and n is a positive integer $(n = 1, 2, \ldots)$.

3. Geometric Interpretation

It is natural to associate the complex number $z = x + iy$ with a point in the plane whose cartesian coordinates are x and y. Each complex number corresponds to just one point, and conversely. The number $-2 + i$, for instance, is represented by the point $(-2, 1)$ in Fig. 1. The number z can also be thought of as the directed line

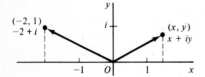

Figure 1

segment, or vector, from the origin to the point (x, y). In fact, we often refer to a complex number z as the point z or the vector z. When used for the purpose of displaying the numbers $z = x + iy$ geometrically, the xy plane is called the *complex plane,* or the z *plane.* The x axis is called the *real axis,* and the y axis is known as the *imaginary axis.*

According to the definition of the sum of two complex numbers $z_1 = x_1 + iy_1$ and $z_2 = x_2 + iy_2$, the number $z_1 + z_2$ corresponds to the point $(x_1 + x_2, y_1 + y_2)$. It also corresponds to a vector with those coordinates as its components. Hence $z_1 + z_2$ may be obtained vectorially as shown in Fig. 2. The difference $z_1 - z_2 = z_1 + (-z_2)$ corresponds to the sum of the vectors for z_1 and $-z_2$ (Fig. 3). Note that the number $z_1 - z_2$ can also be interpreted as the directed line segment from the point (x_2, y_2) to the point (x_1, y_1).

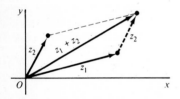

Figure 2

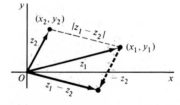

Figure 3

Although the product of two complex numbers z_1 and z_2 is itself a complex number represented by a vector, that vector lies in the same plane as the vectors for z_1 and z_2. Evidently, then, this product is neither the scalar nor the vector product used in ordinary vector analysis. The geometric interpretation of the product of z_1 and z_2 is discussed in Sec. 5.

The *modulus,* or *absolute value,* of a complex number $z = x + iy$ is defined as the nonnegative real number $\sqrt{x^2 + y^2}$ and is denoted by $|z|$; that is,

(1) $$|z| = \sqrt{x^2 + y^2}.$$

Geometrically, the number $|z|$ is the distance between the point (x, y) and the origin, or the length of the vector representing z. It reduces to the usual absolute value in the real number system when $y = 0$. Note that while *the inequality $z_1 < z_2$ is meaningless unless both z_1 and z_2 are real,* the statement $|z_1| < |z_2|$ means that the point z_1 is closer to the origin than the point z_2 is.

Example 1. Since $|-3 + 2i| = \sqrt{13}$ and $|1 + 4i| = \sqrt{17}$, the point $-3 + 2i$ is closer to the origin than $1 + 4i$ is.

The distance between two points $z_1 = x_1 + iy_1$ and $z_2 = x_2 + iy_2$ is $|z_1 - z_2|$. This is clear from Fig. 3, since $|z_1 - z_2|$ is the length of the vector representing $z_1 - z_2$. Alternatively, it follows from definition (1) and the expression

$$z_1 - z_2 = (x_1 - x_2) + i(y_1 - y_2)$$

that

$$|z_1 - z_2| = \sqrt{(x_1 - x_2)^2 + (y_1 - y_2)^2}.$$

The complex numbers z corresponding to the points lying on the circle with center z_0 and radius R thus satisfy the equation $|z - z_0| = R$, and conversely. We refer to this set of points simply as the circle $|z - z_0| = R$.

Example 2. The equation $|z - 1 + 3i| = 2$ represents the circle whose center is $z_0 = (1, -3)$ and whose radius is $R = 2$.

It also follows from definition (1) that the real numbers $|z|$, Re $z = x$, and Im $z = y$ are related by the equation

(2) $$|z|^2 = (\text{Re } z)^2 + (\text{Im } z)^2,$$

as well as the inequalities

(3) $$|z| \geq |\text{Re } z| \geq \text{Re } z, \qquad |z| \geq |\text{Im } z| \geq \text{Im } z.$$

The *complex conjugate,* or simply the conjugate, of a complex number $z = x + iy$ is defined as the complex number $x - iy$ and is denoted by $\bar{z}$; that is,

(4) $$\bar{z} = x - iy.$$

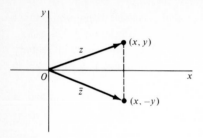

Figure 4

The number $\bar{z}$ is represented by the point $(x, -y)$, which is the reflection in the real axis of the point (x, y) representing z (Fig. 4). Note that $\bar{\bar{z}} = z$ and $|\bar{z}| = |z|$ for all z.

If $z_1 = x_1 + iy_1$ and $z_2 = x_2 + iy_2$, then

$$\overline{z_1 + z_2} = (x_1 + x_2) - i(y_1 + y_2) = (x_1 - iy_1) + (x_2 - iy_2).$$

So the conjugate of the sum is the sum of the conjugates:

(5) $$\overline{z_1 + z_2} = \bar{z}_1 + \bar{z}_2.$$

In like manner, it is easy to show that

(6) $$\overline{z_1 - z_2} = \bar{z}_1 - \bar{z}_2,$$

(7) $$\overline{z_1 z_2} = \bar{z}_1 \bar{z}_2,$$

(8) $$\overline{\left(\frac{z_1}{z_2}\right)} = \frac{\bar{z}_1}{\bar{z}_2} \qquad (z_2 \neq 0).$$

The sum $z + \bar{z}$ of a complex number $z = x + iy$ and its conjugate $\bar{z} = x - iy$ is the real number $2x$, and the difference $z - \bar{z}$ is the pure imaginary number $2iy$. Hence we have the identities

(9) $$\text{Re } z = \frac{z + \bar{z}}{2}, \qquad \text{Im } z = \frac{z - \bar{z}}{2i}.$$

An important identity relating the conjugate of a complex number $z = x + iy$ to its modulus is

(10) $$z\bar{z} = |z|^2,$$

where each side is equal to $x^2 + y^2$. It provides another way of determining the quotient z_1/z_2 in expressions (11), Sec. 2. The procedure is to multiply both numerator and denominator by $\bar{z}_2$ so that the denominator becomes the real number $|z_2|^2$.

Example 3. As an illustration,

$$\frac{-1 + 3i}{2 - i} = \frac{(-1 + 3i)(2 + i)}{(2 - i)(2 + i)} = \frac{-5 + 5i}{|2 - i|^2} = \frac{-5 + 5i}{5} = -1 + i.$$

Also, see the example at the end of Sec. 2.

4. Further Properties of Moduli

With the aid of identity (10) in Sec. 3, various other properties of moduli are easily obtained from properties of conjugates noted there.

We mention that

(1) $$|z_1 z_2| = |z_1|\,|z_2|,$$

(2) $$\left|\frac{z_1}{z_2}\right| = \frac{|z_1|}{|z_2|} \qquad (z_2 \neq 0).$$

To establish property (1), write

$$|z_1 z_2|^2 = (z_1 z_2)\overline{(z_1 z_2)} = (z_1\bar{z}_1)(z_2\bar{z}_2) = |z_1|^2|z_2|^2 = (|z_1|\,|z_2|)^2$$

and recall that a modulus is never negative. Property (2) can be verified in a similar way.

Using this same general technique, we include here an algebraic derivation of the *triangle inequality,*

(3) $$|z_1 + z_2| \leq |z_1| + |z_2|,$$

which is geometrically evident in Fig. 2 of Sec. 3. Indeed, it is merely a statement that the length of one side of a triangle is less than or equal to the sum of the lengths of the other two sides.

We start the proof by writing

$$|z_1 + z_2|^2 = (z_1 + z_2)\overline{(z_1 + z_2)} = (z_1 + z_2)(\bar{z}_1 + \bar{z}_2).$$

Multiplying out the right-hand side here shows that

$$|z_1 + z_2|^2 = z_1\bar{z}_1 + (z_1\bar{z}_2 + \overline{z_1\bar{z}_2}) + z_2\bar{z}_2.$$

But

$$z_1\bar{z}_2 + \overline{z_1\bar{z}_2} = 2\,\mathrm{Re}(z_1\bar{z}_2) \leq 2|z_1\bar{z}_2| = 2|z_1|\,|z_2|;$$

and so

$$|z_1 + z_2|^2 \leq |z_1|^2 + 2|z_1|\,|z_2| + |z_2|^2,$$

or

$$|z_1 + z_2|^2 \leq (|z_1| + |z_2|)^2.$$

Since moduli are nonnegative, inequality (3) now follows.

The triangle inequality applies to sums involving any finite number of terms. That is, one can write

$$|z_1 + z_2 + z_3| \leq |z_1 + z_2| + |z_3| \leq |z_1| + |z_2| + |z_3|,$$

which is generalized by means of mathematical induction to

(4) $$|z_1 + z_2 + \cdots + z_n| \leq |z_1| + |z_2| + \cdots + |z_n| \qquad (n = 2, 3, \ldots).$$

It also follows from inequality (3) that

(5) $$\left||z_1| - |z_2|\right| \leq |z_1 + z_2|.$$

To derive inequality (5), we write

$$|z_1| = |(z_1 + z_2) + (-z_2)| \leq |z_1 + z_2| + |-z_2|,$$

which means that

(6) $$|z_1| - |z_2| \leq |z_1 + z_2|.$$

This is inequality (5) when $|z_1| \geq |z_2|$. If $|z_1| < |z_2|$, we need only interchange z_1 and z_2 in inequality (6) to get

$$-(|z_1| - |z_2|) \leq |z_1 + z_2|,$$

which is desired result.

Inequality (5) and the triangle inequality (3) combine to become

(7) $$\left||z_1| - |z_2|\right| \leq |z_1 + z_2| \leq |z_1| + |z_2|.$$

Replacing z_2 by $-z_2$ here shows that

(8) $$\left||z_1| - |z_2|\right| \leq |z_1 - z_2| \leq |z_1| + |z_2|.$$

Example. If a point z lies on the unit circle $|z| = 1$ about the origin, then

$$|z^2 - z + 1| \leq |z|^2 + |z| + 1 = 3$$

and

$$|z^3 - 2| \geq \left||z|^3 - 2\right| = 1.$$

EXERCISES

1. Locate the numbers $z_1 + z_2$ and $z_1 - z_2$ vectorially when

   (a) $z_1 = 2i$, $z_2 = \dfrac{2}{3} - i$; (b) $z_1 = (-\sqrt{3}, 1)$, $z_2 = (\sqrt{3}, 0)$;

   (c) $z_1 = (-3, 1)$, $z_2 = (1, 4)$; (d) $z_1 = x_1 + iy_1$, $z_2 = x_1 - iy_1$.

2. Show that

   (a) $\overline{\overline{z} + 3i} = z - 3i$; (b) $\overline{iz} = -i\overline{z}$; (c) $\overline{(2 + i)^2} = 3 - 4i$;

   (d) $|(2\overline{z} + 5)(\sqrt{2} - i)| = \sqrt{3}|2z + 5|$.

3. Verify inequalities (3), Sec. 3.

4. Prove that $\sqrt{2}|z| \geq |\text{Re } z| + |\text{Im } z|$.

5. Verify properties (6) and (7) of $\overline{z}$ in Sec. 3.

6. Prove that (a) z is real if and only if $\overline{z} = z$; (b) z is either real or pure imaginary if and only if $(\overline{z})^2 = z^2$.

7. Use the property $\overline{z_1 z_2} = \overline{z}_1 \overline{z}_2$ to show that (a) $\overline{z_1 z_2 z_3} = \overline{z}_1 \overline{z}_2 \overline{z}_3$; (b) $\overline{(z^4)} = (\overline{z})^4$.

8. Verify property (2) of moduli in Sec. 4.

9. Use results in Secs. 3 and 4 to show that when z_2 and z_3 are nonzero,

(a) $\overline{\left(\dfrac{z_1}{z_2 z_3}\right)} = \dfrac{\bar{z}_1}{\bar{z}_2 \bar{z}_3}$;   (b) $\left|\dfrac{z_1}{z_2 z_3}\right| = \dfrac{|z_1|}{|z_2||z_3|}$.

10. With the aid of inequalities in Sec. 4, show that when $|z_3| \neq |z_4|$,

$$\left|\frac{z_1 + z_2}{z_3 + z_4}\right| \leq \frac{|z_1| + |z_2|}{\big||z_3| - |z_4|\big|}.$$

11. In each case sketch the set of points determined by the given condition:

(a) $|z - 1 + i| = 1$;   (b) $|z + i| \leq 3$;   (c) $\mathrm{Re}(\bar{z} - i) = 2$;   (d) $|2z - i| = 4$.

12. Apply inequalities in Secs. 3 and 4 to show that

$$|\mathrm{Im}(1 - \bar{z} + z^2)| < 3 \qquad \text{when} \qquad |z| < 1.$$

13. Verify the generalization (4), Sec. 4, of the triangle inequality.

14. By factoring $z^4 - 4z^2 + 3$ into two quadratic factors and then using the first of inequalities (8), Sec. 4, show that if z lies on the circle $|z| = 2$, then

$$\left|\frac{1}{z^4 - 4z^2 + 3}\right| \leq \frac{1}{3}.$$

15. It is shown in Sec. 2 that if $z_1 z_2 = 0$, at least one of the numbers z_1 and z_2 must be zero. Give an alternative proof by appealing to the corresponding result for real numbers and using identity (1), Sec. 4.

16. Use mathematical induction to show that when $n = 2, 3, \ldots$,

(a) $\overline{z_1 + z_2 + \cdots + z_n} = \bar{z}_1 + \bar{z}_2 + \cdots + \bar{z}_n$;   (b) $\overline{z_1 z_2 \cdots z_n} = \bar{z}_1 \bar{z}_2 \cdots \bar{z}_n$.

17. Let $a_0, a_1, a_2, \ldots, a_n$ ($n \geq 1$) denote *real* numbers, and let z be any complex number. With the aid of the results in Exercise 16, show that

$$\overline{a_0 + a_1 z + a_2 z^2 + \cdots + a_n z^n} = a_0 + a_1 \bar{z} + a_2 \bar{z}^2 + \cdots + a_n \bar{z}^n.$$

18. Show that the equation $|z - z_0| = R$ of the circle centered at z_0 with radius R can be written

$$|z|^2 - 2\,\mathrm{Re}(z\bar{z}_0) + |z_0|^2 = R^2.$$

19. Using equations (9), Sec. 3, show that the hyperbola $x^2 - y^2 = 1$ can be written

$$z^2 + \bar{z}^2 = 2.$$

20. Using the fact that $|z_1 - z_2|$ is the distance between two points z_1 and z_2, give a geometric argument that
(a) the equation $|z - 4i| + |z + 4i| = 10$ represents an ellipse whose foci are $(0, \pm 4)$;
(b) the equation $|z - 1| = |z + i|$ represents the line through the origin whose slope is -1.

5. Polar Form

Let r and θ be polar coordinates of the point (x, y) which corresponds to a *nonzero* complex number $z = x + iy$. Since

(1) $x = r \cos \theta$ and $y = r \sin \theta,$

z can be expressed in *polar form* as

(2) $z = r(\cos \theta + i \sin \theta).$

Example 1. The complex number $1 - i$, which lies in the fourth quadrant, becomes

(3) $1 - i = \sqrt{2}\left[\cos\left(-\frac{\pi}{4}\right) + i \sin\left(-\frac{\pi}{4}\right)\right]$

in polar form. Observe that any one of the values

$$\theta = -\frac{\pi}{4} + 2n\pi \qquad\qquad (n = 0, \pm 1, \pm 2, \ldots)$$

can be used here. For instance,

$$1 - i = \sqrt{2}\left(\cos \frac{7\pi}{4} + i \sin \frac{7\pi}{4}\right).$$

The positive number r is the length of the vector representing z; that is, $r = |z|$. The number θ is called an *argument* of z, and we write $\theta = \arg z$. Geometrically, $\arg z$ denotes the angle, measured in radians, that z makes with the positive real axis when z is interpreted as a directed line segment from the origin (Fig. 5). Hence it has any one of an infinite number of real values differing by integral multiples of 2π. These values can be determined from the equation

(4) $\tan \theta = \frac{y}{x},$

where the quadrant containing the point corresponding to z must be specified. In general, then, if $z \neq 0$,

(5) $z = r[\cos(\theta + 2n\pi) + i \sin(\theta + 2n\pi)]$ $(n = 0, \pm 1, \pm 2, \ldots),$

where r is the modulus of z and θ is any particular value of $\arg z$.

If $z = 0$, then θ is undefined. Thus any complex number which is to be written in terms of polar coordinates is understood to be nonzero even when that condition is not explicitly stated.

The *principal value* of $\arg z$, denoted by $\operatorname{Arg} z$, is defined as that unique value of

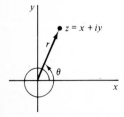

Figure 5

arg z such that $-\pi < \arg z \leqq \pi$. The principal value is used in expression (3) for the number $1 - i$. Note that $\arg z = \text{Arg } z + 2n\pi$ $(n = 0, \pm 1, \pm 2, \ldots)$. Also, when z is a negative real number, $\text{Arg } z = \pi$.

We turn now to an important identity involving arguments; namely,

$$(6) \qquad\qquad \arg(z_1 z_2) = \arg z_1 + \arg z_2.$$

It is to be interpreted as saying that any argument of z_1 plus any argument of z_2 is an argument of the product $z_1 z_2$; and, conversely, any argument of $z_1 z_2$ can be expressed as any argument of z_1 plus some argument of z_2.

To obtain identity (6), we start with z_1 and z_2 in polar form:

$$z_1 = r_1(\cos \theta_1 + i \sin \theta_1), \qquad z_2 = r_2(\cos \theta_2 + i \sin \theta_2).$$

Straightforward multiplication shows that

$$z_1 z_2 = r_1 r_2 [(\cos \theta_1 \cos \theta_2 - \sin \theta_1 \sin \theta_2) + i(\sin \theta_1 \cos \theta_2 + \cos \theta_1 \sin \theta_2)],$$

and this reduces to the polar form of the product:

$$(7) \qquad\qquad z_1 z_2 = r_1 r_2 [\cos(\theta_1 + \theta_2) + i \sin(\theta_1 + \theta_2)].$$

Now θ_1 and θ_2 can be any values of $\arg z_1$ and $\arg z_2$, respectively, and it is clear from equation (7) that $\theta_1 + \theta_2$ is an argument of the product $z_1 z_2$ (Fig. 6). Moreover, since one value of $\arg(z_1 z_2)$ is $\theta_1 + \theta_2$, the totality of such values is

$$\arg(z_1 z_2) = (\theta_1 + \theta_2) + 2n\pi \qquad (n = 0, \pm 1, \pm 2, \ldots).$$

Similarly,

$$\arg z_1 = \theta_1 + 2n_1\pi \qquad (n_1 = 0, \pm 1, \pm 2, \ldots).$$

So when values of $\arg(z_1 z_2)$ and $\arg z_1$ are specified, they correspond to particular choices of n and n_1, respectively; and if we write $\arg z_2 = \theta_2 + 2(n - n_1)\pi$, equation (6) is satisfied. For then

$$(\theta_1 + \theta_2) + 2n\pi = (\theta_1 + 2n_1\pi) + [\theta_2 + 2(n - n_1)\pi].$$

Statement (6) is not always valid when *arg* is replaced everywhere by *Arg*, as illustrated in Example 2.

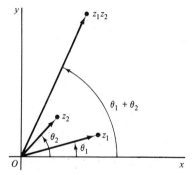

Figure 6

Example 2. When $z_1 = -1$ and $z_2 = i$,

$$\text{Arg}(z_1 z_2) = \text{Arg}(-i) = -\frac{\pi}{2};$$

but

$$\text{Arg } z_1 + \text{Arg } z_2 = \pi + \frac{\pi}{2} = \frac{3\pi}{2}.$$

On the other hand, if we take the values of arg z_1 and arg z_2 just used and select the value $\text{arg}(z_1 z_2) = 3\pi/2$, we find that equation (6) *is* satisfied.

When a nonzero complex number $z = r(\cos\theta + i\sin\theta)$ is multiplied by i, the radius vector corresponding to iz is obtained by rotating the one for z through a right angle in the positive, or counterclockwise, direction without changing its length. This is true because

$$i = 1\left(\cos\frac{\pi}{2} + i\sin\frac{\pi}{2}\right);$$

and, according to equation (7),

$$iz = r\left[\cos\left(\theta + \frac{\pi}{2}\right) + i\sin\left(\theta + \frac{\pi}{2}\right)\right].$$

Equation (7) also shows that the polar form for the unique multiplicative inverse of a nonzero complex number

$$z = r(\cos\theta + i\sin\theta)$$

is

(8)
$$z^{-1} = \frac{1}{r}[\cos(-\theta) + i\sin(-\theta)],$$

the product of these polar forms being unity. Since $z_1/z_2 = z_1 z_2^{-1}$, we thus have the following expression for the quotient of two nonzero complex numbers:

(9)
$$\frac{z_1}{z_2} = \frac{r_1}{r_2}[\cos(\theta_1 - \theta_2) + i\sin(\theta_1 - \theta_2)].$$

This can be used to verify the statement

(10)
$$\text{arg}\left(\frac{z_1}{z_2}\right) = \text{arg } z_1 - \text{arg } z_2,$$

which is analogous to statement (6).

6. Exponential Form

It is often convenient to let $e^{i\theta}$, or $\exp(i\theta)$, denote the quantity $\cos\theta + i\sin\theta$ appearing in the polar form

(1) $$z = r(\cos \theta + i \sin \theta)$$

of a nonzero complex number z. The number z can then be expressed more compactly in *exponential form*:

(2) $$z = re^{i\theta}.$$

The equation

(3) $$e^{i\theta} = \cos \theta + i \sin \theta,$$

which defines the symbol $e^{i\theta}$ for any real value of θ, is known as *Euler's formula*. The choice of the symbol $e^{i\theta}$ is motivated in Sec. 21. We simply note here a few of its properties which suggest that it is a natural choice.

The additive property

(4) $$e^{i\theta_1}e^{i\theta_2} = e^{i(\theta_1 + \theta_2)}$$

is a restatement of expression (7), Sec. 5, for the product of two complex numbers z_1 and z_2 when their moduli r_1 and r_2 are unity:

$$(\cos \theta_1 + i \sin \theta_1)(\cos \theta_2 + i \sin \theta_2) = \cos(\theta_1 + \theta_2) + i \sin(\theta_1 + \theta_2).$$

Writing $e^{-i\theta}$ for $e^{i(-\theta)}$, we find from equation (4) that $e^{i\theta}e^{-i\theta} = 1$; hence $1/e^{i\theta} = e^{-i\theta}$.

Observe that the polar form (8), Sec. 5, of the multiplicative inverse of a nonzero complex number $z = re^{i\theta}$ becomes

(5) $$z^{-1} = \frac{1}{r}e^{i(-\theta)} = \frac{1}{r}e^{-i\theta}$$

in exponential notation. Likewise, when $z_1 = r_1 \exp(i\theta_1)$ and $z_2 = r_2 \exp(i\theta_2)$, expressions (7) and (9) in Sec. 5 can be written

(6) $$z_1 z_2 = r_1 r_2 \exp[i(\theta_1 + \theta_2)]$$

and

(7) $$\frac{z_1}{z_2} = \frac{r_1}{r_2} \exp[i(\theta_1 - \theta_2)],$$

respectively. An important advantage of expressions (5), (6), and (7) over their counterparts in Sec. 5 is the ease with which they can be remembered and used. They, as well as property (4), are obtained formally by applying the usual algebraic rules for real numbers and e^x.

In view of the polar representation (Sec. 5)

$$z = r[\cos(\theta + 2n\pi) + i \sin(\theta + 2n\pi)] \qquad (n = 0, \pm1, \pm2, \ldots),$$

expression (2) is only one of an infinite number of possibilities for the exponential form of z:

(8) $$z = re^{i(\theta + 2n\pi)} \qquad (n = 0, \pm1, \pm2, \ldots).$$

This is also seen geometrically when θ is interpreted as the angle of inclination of the radius vector of length r representing the number $z = re^{i\theta}$. To be specific, z lies on the

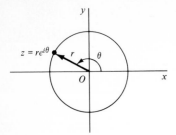

Figure 7

circle of radius r centered at the origin (Fig. 7); and if θ is increased, the point z traverses the circle in the counterclockwise direction. In particular, when θ is increased by 2π, we arrive at the original point. The same is, of course, true when θ is decreased by 2π.

It is thus evident from Fig. 7 that *two nonzero complex numbers $z_1 = r_1 \exp(i\theta_1)$ and $z_2 = r_2 \exp(i\theta_2)$ are equal if and only if $r_1 = r_2$ and $\theta_1 = \theta_2 + 2k\pi$, where k is an integer* $(k = 0, \pm 1, \pm 2, \ldots)$.

Figure 7, with $r = R$, also shows that the equation

$$z = Re^{i\theta} \qquad (0 \leqq \theta \leqq 2\pi) \tag{9}$$

is a parametric representation of the circle $|z| = R$, centered at the origin with radius R. As the parameter θ in Fig. 7 increases from $\theta = 0$ over the interval $0 \leqq \theta \leqq 2\pi$, the point z starts from the positive real axis and traverses the circle once in the counterclockwise direction. More generally, the circle $|z - z_0| = R$, whose center is z_0 and whose radius is R, has the parametric representation

$$z = z_0 + Re^{i\theta} \qquad (0 \leqq \theta \leqq 2\pi). \tag{10}$$

This can be seen vectorially (Fig. 8) by noting that a point z traversing the circle $|z - z_0| = R$ once in the counterclockwise direction corresponds to the sum of the fixed vector z_0 and a vector of length R whose angle of inclination θ varies from $\theta = 0$ to $\theta = 2\pi$.

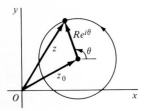

Figure 8

7. Powers and Roots

Integral powers of a nonzero complex number $z = re^{i\theta}$ are given by the formula

$$z^n = r^n e^{in\theta} \qquad (n = 0, \pm 1, \pm 2, \ldots). \tag{1}$$

Since $z^{n+1} = z^n z$ when $n = 1, 2, \ldots$ (Sec. 1), formula (1) is easily verified for positive values of n by means of mathematical induction with the aid of expression (6), Sec. 6, for the product of two complex numbers in exponential form. The formula also holds when $n = 0$ with the convention that $z^0 = 1$. If $n = -1, -2, \ldots$, on the other hand, we define z^n in terms of the inverse of z by means of the equation $z^n = (z^{-1})^{-n}$, or $z^n = (z^{-1})^m$ where $m = -n = 1, 2, \ldots$. Then, since formula (1) is valid for *positive* integral powers, it follows from the exponential form of z^{-1} in Sec. 6 that

$$z^n = \left[\frac{1}{r} e^{i(-\theta)}\right]^m = \left(\frac{1}{r}\right)^m e^{im(-\theta)} = r^n e^{in\theta} \qquad (n = -1, -2, \ldots).$$

Hence formula (1) is valid for all integral powers.

Observe that if $r = 1$, formula (1) becomes

(2) $$(e^{i\theta})^n = e^{in\theta} \qquad (n = 0, \pm 1, \pm 2, \ldots).$$

When written in the form

(3) $$(\cos\theta + i\sin\theta)^n = \cos n\theta + i\sin n\theta \qquad (n = 0, \pm 1, \pm 2, \ldots),$$

this is known as *de Moivre's formula*.

Formula (1) is useful in computing roots of nonzero complex numbers.

Example 1. Let us solve the equation

(4) $$z^n = 1,$$

where n has any one of the values $n = 2, 3, \ldots$, and thus find the nth roots of unity. Inasmuch as $z \neq 0$, we may write $z = re^{i\theta}$ and look for values of r and θ such that

$$(re^{i\theta})^n = 1,$$

or

$$r^n e^{in\theta} = 1e^{i0}.$$

Now, according to the statement in italics in Sec. 6,

$$r^n = 1 \qquad \text{and} \qquad n\theta = 0 + 2k\pi$$

where k is any integer ($k = 0, \pm 1, \pm 2, \ldots$). Consequently, $r = 1$ and $\theta = 2k\pi/n$; and it follows that the complex numbers

$$z = \exp\left(i\frac{2k\pi}{n}\right) \qquad (k = 0, \pm 1, \pm 2, \ldots)$$

are nth roots of unity. Since they appear here in exponential form, we are able to see immediately that they lie on the unit circle centered at the origin and are equally spaced around that circle every $2\pi/n$ radians. Evidently, then, all of the *distinct* nth roots of unity are obtained by writing

(5) $$z = \exp\left(i\frac{2k\pi}{n}\right) = \cos\frac{2k\pi}{n} + i\sin\frac{2k\pi}{n} \qquad (k = 0, 1, 2, \ldots, n-1),$$

and no further roots are obtained with other values of k.

So the number of nth roots of unity is n. When $n = 2$, these roots are, of course, ± 1. When $n \geq 3$, they correspond to points lying at the vertices of a regular polygon

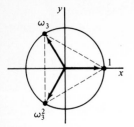

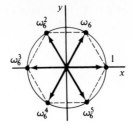

Figure 9   Figure 10

of n sides. This polygon is inscribed in the unit circle centered at the origin, and it has one vertex at the point corresponding to the root $z = 1$ $(k = 0)$. If we write

$$(6) \qquad \qquad \omega_n = \exp\left(i\frac{2\pi}{n}\right),$$

then by property (2), the nth roots of unity are simply

$$1, \omega_n, \omega_n^2, \ldots, \omega_n^{n-1}.$$

Note that $\omega_n^n = 1$. See Fig. 9 for the interpretation of the three cube roots of unity as the vertices of an equilateral triangle. Figure 10 illustrates the case $n = 6$.

The above method can be used to find the nth roots of any nonzero complex number $z_0 = r_0 \exp(i\theta_0)$. Those roots, which are obtained by solving the equation

$$(7) \qquad \qquad z^n = z_0$$

for z, are the numbers

$$(8) \qquad \qquad c_k = \sqrt[n]{r_0} \exp\left[i\left(\frac{\theta_0}{n} + \frac{2k\pi}{n}\right)\right] \qquad (k = 0, 1, \ldots, n - 1),$$

where $\sqrt[n]{r_0}$ denotes the *positive* nth root of r_0. The number $\sqrt[n]{r_0}$ is the length of each of the radius vectors representing the n roots. An argument of the first root c_0 is θ_0/n, and arguments of the other roots are obtained by adding integral multiples of $2\pi/n$. Consequently, as was the case with the nth roots of unity, the roots when $n = 2$ always lie at the opposite ends of a diameter of a circle, one root being the negative of the other; and when $n \geq 3$, they lie at the vertices of a regular polygon of n sides.

If c is any particular nth root of z_0, the set of all nth roots can be written

$$c, c\omega_n, c\omega_n^2, \ldots, c\omega_n^{n-1},$$

where $\omega_n = \exp(i2\pi/n)$, as defined in equation (6). This is because multiplication of any nonzero complex number by ω_n corresponds to increasing the argument of that number by $2\pi/n$.

We shall let $z_0^{1/n}$ denote the set of nth roots of a nonzero complex number z_0. If, in particular, z_0 is a positive real number r_0, the symbol $r_0^{1/n}$ denotes a set of roots and the symbol $\sqrt[n]{r_0}$ in expression (8) is reserved for the one positive root. When the value of θ_0 that is used in expression (8) is the principal value of arg z_0 $(-\pi < \theta_0 \leq \pi)$, the number c_0 is often referred to as the *principal nth root* of z_0. Thus when z_0 is a positive real number r_0, its principal root is $\sqrt[n]{r_0}$.

Note that if $z_0 = 0$, equation (7) has only the solution $z = 0$. Hence the only nth root of zero is zero.

Finally, a convenient way to remember expression (8) is to write z_0 in its most general exponential form (see Sec. 6),

$$(9) \qquad\qquad z_0 = r_0 \exp[i(\theta_0 + 2k\pi)] \qquad\qquad (k = 0, \pm 1, \pm 2, \ldots),$$

and to formally apply laws of fractional exponents for real numbers, keeping in mind that there are precisely n distinct roots:

$$(10) \qquad\qquad z_0^{1/n} = \sqrt[n]{r_0} \exp \frac{i(\theta_0 + 2k\pi)}{n} \qquad\qquad (k = 0, 1, 2, \ldots, n - 1).$$

This formula is, of course, valid when $n = 1$ as well as when $n = 2, 3, \ldots$. The case $n = 1$ was excluded from our discussion only because of its trivial nature.

Example 2. To illustrate formula (10), let us find all values of $(-8i)^{1/3}$, or the three cube roots of $-8i$. We need only write

$$-8i = 8 \exp\left[i\left(-\frac{\pi}{2} + 2k\pi \right) \right] \qquad\qquad (k = 0, \pm 1, \pm 2, \ldots)$$

to see that the desired roots are

$$c_k = 2 \exp\left[i\left(-\frac{\pi}{6} + \frac{2k\pi}{3} \right) \right] \qquad\qquad (k = 0, 1, 2).$$

In cartesian coordinates, then,

$$c_0 = 2 \exp\left[i\left(-\frac{\pi}{6} \right) \right] = 2\left(\cos \frac{\pi}{6} - i \sin \frac{\pi}{6} \right) = \sqrt{3} - i;$$

and, in like manner, we find that $c_1 = 2i$ and $c_2 = -\sqrt{3} - i$. These roots lie at the vertices of an equilateral triangle that is inscribed in a circle of radius 2 centered at the origin. The principal root is $c_0 = \sqrt{3} - i$.

EXERCISES

1. Find one value of $\arg z$ when

   (a) $z = \dfrac{-2}{1 + \sqrt{3}i}$;   (b) $\dfrac{i}{-2 - 2i}$;   (c) $z = (\sqrt{3} - i)^6$.

   Ans. (a) $2\pi/3$;   (c) π.

2. By writing the individual factors on the left in exponential form, performing the needed operations, and finally changing back to cartesian coordinates, show that

   (a) $i(1 - \sqrt{3}i)(\sqrt{3} + i) = 2(1 + \sqrt{3}i)$;   (b) $5i/(2 + i) = 1 + 2i$;

   (c) $(-1 + i)^7 = -8(1 + i)$;   (d) $(1 + \sqrt{3}i)^{-10} = 2^{-11}(-1 + \sqrt{3}i)$.

3. In each case find all the roots in cartesian form, exhibit them geometrically, and point out which is the principal root:

(a) $(2i)^{1/2}$;   (b) $(1 - \sqrt{3}i)^{1/2}$;   (c) $(-1)^{1/3}$;   (d) $(-16)^{1/4}$;   (e) $8^{1/6}$;

(f) $(-4\sqrt{2} + 4\sqrt{2}i)^{1/3}$.

    Suggestion: In part (f) show that $c_0 = \sqrt{2}(1 + i)$, $\omega_3 = (-1 + \sqrt{3}i)/2$, and $\omega_3^2 = (-1 - \sqrt{3}i)/2$. Then use the fact that the desired roots are c_0, $c_0\omega_3$, and $c_0\omega_3^2$.

    Ans.   (a) $\pm(1 + i)$;   (b) $\pm(\sqrt{3} - i)/\sqrt{2}$;   (d) $\pm\sqrt{2}(1 + i), \pm\sqrt{2}(1 - i)$;

    (e) $\pm\sqrt{2}, \pm(1 + \sqrt{3}i)/\sqrt{2}, \pm(-1 + \sqrt{3}i)/\sqrt{2}$;

    (f) $\sqrt{2}(1 + i), [-(\sqrt{3} + 1) + (\sqrt{3} - 1)i]/\sqrt{2}, [(\sqrt{3} - 1) - (\sqrt{3} + 1)i]/\sqrt{2}$.

4. Show that

(a) $|e^{i\theta}| = 1$;   (b) $\overline{e^{i\theta}} = e^{-i\theta}$;   (c) $(e^{i\theta})^2 = e^{i2\theta}$;

(d) $e^{i\theta_1}e^{i\theta_2} \cdots e^{i\theta_n} = e^{i(\theta_1 + \theta_2 + \cdots + \theta_n)}$ $(n = 2, 3, \ldots)$.

5. Verify formula (1), Sec. 7, for z^n when $n = 1, 2, \ldots$.

6. Find a value of arg z when $z_1 \neq 0$:

(a) $z = z_1^n$ $(n = 1, 2, \ldots)$;   (b) $z = z_1^{-1}$.

    Ans.   (a) n arg z_1;   (b) $-$arg z_1.

7. Show that if Re $z_1 > 0$ and Re $z_2 > 0$, then

$$\text{Arg}(z_1z_2) = \text{Arg } z_1 + \text{Arg } z_2,$$

where $\text{Arg}(z_1z_2)$ denotes the principal value of $\text{arg}(z_1z_2)$, etc.

8. Verify the statement (Sec. 5)

$$\arg\left(\frac{z_1}{z_2}\right) = \arg z_1 - \arg z_2.$$

9. According to Sec. 3, the difference $z - z_0$ of two distinct complex numbers can be interpreted vectorially (Fig. 11), where θ denotes the angle of inclination of the vector representing $z - z_0$. By writing the number $z - z_0$ in polar form,

$$z - z_0 = R(\cos\theta + i\sin\theta),$$

show algebraically that the values of $\arg\overline{(z - z_0)}$ are the same as the values of $-\arg(z - z_0)$, as indicated in Fig. 11. Show that $\text{Arg}\overline{(z - z_0)} = -\text{Arg}(z - z_0)$ if and only if $z - z_0$ is not a negative real number.

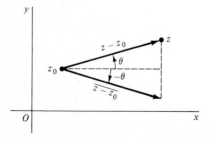

Figure 11

10. (a) Let a denote any fixed real number, and show that the two square roots of $a + i$ are $\pm\sqrt{A}\,\exp(i\alpha/2)$ where $A = \sqrt{a^2 + 1}$ and $\alpha = \text{Arg}(a + i)$.

(b) With the aid of the trigonometric identities

$$\cos^2\left(\frac{\alpha}{2}\right) = \frac{1 + \cos\alpha}{2}, \qquad \sin^2\left(\frac{\alpha}{2}\right) = \frac{1 - \cos\alpha}{2},$$

show that the square roots obtained in part (a) can be written

$$\pm\frac{1}{\sqrt{2}}(\sqrt{A + a} + i\sqrt{A - a}).$$

11. Let z be a nonzero complex number and n a negative integer ($n = -1, -2, \ldots$). Also, write $z = re^{i\theta}$ and $m = -n = 1, 2, \ldots$.

(a) Using the expressions $z^m = r^m e^{im\theta}$ and $z^{-1} = (1/r)e^{i(-\theta)}$, verify that $(z^m)^{-1} = (z^{-1})^m$ and hence that the definition $z^n = (z^{-1})^m$ in Sec. 7 could have been written alternatively as $z^n = (z^m)^{-1}$.

(b) Define $z^{1/n}$ by means of the equation $z^{1/n} = (z^{-1})^{1/m}$ and, by showing that the m values of $(z^{1/m})^{-1}$ and $(z^{-1})^{1/m}$ are the same, verify that $z^{1/n} = (z^{1/m})^{-1}$.

12. Find the four roots of the equation $z^4 + 4 = 0$ and use them to factor $z^4 + 4$ into quadratic factors with real coefficients.

$$\textit{Ans.} \quad (z^2 + 2z + 2)(z^2 - 2z + 2).$$

13. Use de Moivre's formula (Sec. 7) to derive the following trigonometric identities:

(a) $\cos 3\theta = \cos^3\theta - 3\cos\theta\sin^2\theta$; (b) $\sin 3\theta = 3\cos^2\theta\sin\theta - \sin^3\theta$.

14. Obtain expression (8), Sec. 7, for the nth roots of z_0.

15. Given that $z_1 z_2 \neq 0$, use the exponential forms of z_1 and z_2 to prove that

$$\text{Re}(z_1\bar{z}_2) = |z_1|\,|z_2|$$

if and only if $\theta_1 - \theta_2 = 2n\pi$ ($n = 0, \pm1, \pm2, \ldots$), where $\theta_1 = \arg z_1$ and $\theta_2 = \arg z_2$.

16. Given that $z_1 z_2 \neq 0$ and using the result in Exercise 15, modify the derivation of the triangle inequality (3), Sec. 4, to show that

(a) $|z_1 + z_2| = |z_1| + |z_2|$; (b) $|z_1 - z_2| = \big||z_1| - |z_2|\big|$

if and only if $\theta_1 - \theta_2 = 2n\pi$ ($n = 0, \pm1, \pm2, \ldots$), where $\theta_1 = \arg z_1$ and $\theta_2 = \arg z_2$. Interpret these statements geometrically.

17. Solve the equation $|e^{i\theta} - 1| = 2$ for θ ($0 \leq \theta < 2\pi$) and verify the solution geometrically.

$$\textit{Ans.} \quad \pi.$$

18. Establish the identity

$$1 + z + z^2 + \cdots + z^n = \frac{1 - z^{n+1}}{1 - z} \qquad (z \neq 1)$$

and then use it to derive *Lagrange's trigonometric identity*:

$$1 + \cos\theta + \cos 2\theta + \cdots + \cos n\theta = \frac{1}{2} + \frac{\sin[(n + \frac{1}{2})\theta]}{2\sin(\theta/2)} \qquad (0 < \theta < 2\pi).$$

Suggestion: As for the first identity, write $S = 1 + z + z^2 + \cdots + z^n$ and consider the difference $S - zS$. To derive the second identity, write $z = e^{i\theta}$ in the first one.

19. Show that if c is any nth root of unity other than unity itself, then

$$1 + c + c^2 + \cdots + c^{n-1} = 0.$$

Suggestion: Use the first identity in Exercise 18.

20. (*a*) Prove that the usual quadratic formula solves the quadratic equation

$$az^2 + bz + c = 0 \qquad\qquad (a \neq 0)$$

when the coefficients a, b, and c are complex numbers. Specifically, by completing the square on the left-hand side, prove that the roots of the equation are

$$z = \frac{-b + (b^2 - 4ac)^{1/2}}{2a},$$

where the two square roots are to be considered when $b^2 - 4ac \neq 0$.

(*b*) Use the result in part (*a*) to find the roots of the equation

$$z^2 + 2z + (1 - i) = 0.$$

$$Ans. \quad (b)\ \left(-1 + \frac{1}{\sqrt{2}}\right) + \frac{i}{\sqrt{2}},\ \left(-1 - \frac{1}{\sqrt{2}}\right) - \frac{i}{\sqrt{2}}.$$

8. Regions in the Complex Plane

In this section we are concerned with sets of complex numbers, or points in the z plane, and their closeness to one another. Our basic tool is the concept of an ε *neighborhood*

$$(1) \qquad\qquad |z - z_0| < \varepsilon$$

of a given point z_0. It consists of all points z lying inside but not on a circle centered at z_0 and with a specified positive radius ε (Fig. 12). When the value of ε is understood or is immaterial in the discussion, the set (1) is often referred to as just a neighborhood.

A point z_0 is said to be an *interior point* of a set S whenever there is some neighborhood of z_0 that contains only points of S; it is called an *exterior point* of S when there exists a neighborhood of it containing no points of S. If z_0 is neither of these, it is a *boundary point* of S. A boundary point is therefore a point all of whose neighborhoods contain points in S and points not in S. The totality of all boundary points is called the *boundary* of S. The circle $|z| = 1$, for instance, is the boundary of each of the sets

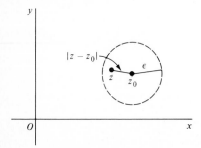

Figure 12

(2) $$|z| < 1 \quad \text{and} \quad |z| \leq 1.$$

A set is *open* if it contains none of its boundary points. It is left as an exercise to show that a set is open if and only if each of its points is an interior point. A set is *closed* if it contains all its boundary points; and the *closure* $\overline{S}$ of a set S is the closed set consisting of all points in S together with the boundary of S. Note that the first of sets (2) is open and that the second is the closure of both those sets.

Some sets are, of course, neither open nor closed. For a set to be not open, there must be a boundary point that is contained in the set; and if a set is not closed, there exists a boundary point not contained in the set. Observe that the punctured disk $0 < |z| \leq 1$ is neither open nor closed. The set of all complex numbers is, on the other hand, both open and closed since it has no boundary points.

An open set S is *connected* if each pair of points z_1 and z_2 in it can be joined by a polygonal path, consisting of a finite number of line segments joined end to end, which lies entirely in S. The open set $|z| < 1$ is connected. The annulus $1 < |z| < 2$ is, of course, open and it is also connected (see Fig. 13). An open set that is connected is called a *domain*. Note that any neighborhood is a domain. A domain together with some, none, or all of its boundary points is referred to as a *region*.

A set S is *bounded* if every point of S lies inside some circle $|z| = R$; otherwise, it is *unbounded*. Both of the sets (2) are bounded regions.

A point z_0 is said to be an *accumulation point* of a set S if each neighborhood of z_0 contains at least one point of S distinct from z_0. It follows that if a set S is closed, then it contains each of its accumulation points. For if an accumulation point z_0 were not in S, it would be a boundary point of S; but this contradicts the fact that a closed set contains all its boundary points. It is left as an exercise to show that the converse is, in fact, true. Thus, a set is closed if and only if it contains all its accumulation points.

Evidently a point z_0 is *not* an accumulation point of set S whenever there exists some neighborhood of z_0 which does not contain points of S distinct from z_0. Note that the origin is the only accumulation point of the set $z_n = i/n$ $(n = 1, 2, \ldots)$.

It is sometimes convenient to include with the complex plane the *point at infinity*, denoted by ∞. The complex plane together with this point is called the *extended complex plane*. To visualize the point at infinity, one can think of the complex plane

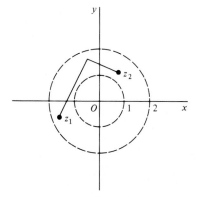

Figure 13

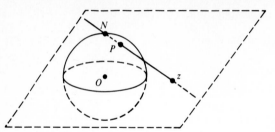

Figure 14

as passing through the equator of a unit sphere centered at the point $z = 0$ (Fig. 14). To each point z in the plane there corresponds exactly one point P on the surface of the sphere. The point P is determined by the intersection of the line through the point z and the north pole N of the sphere with that surface. In like manner, to each point P on the surface of the sphere, other than the north pole N, there corresponds exactly one point z in the plane. By letting the point N of the sphere correspond to the point at infinity, we obtain a one to one correspondence between the points of the sphere and the points of the extended complex plane. The sphere is known as the *Riemann sphere*, and the correspondence is called a *stereographic projection*.

Observe that the exterior of the unit circle centered at the origin in the complex plane corresponds to the upper hemisphere with the equator and the point N deleted. Moreover, for each small positive number ε, those points in the complex plane exterior to the circle $|z| = 1/\varepsilon$ correspond to points on the sphere close to N. We thus call the set $|z| > 1/\varepsilon$ an ε *neighborhood*, or neighborhood, of ∞.

Let us agree that in referring to a point z we mean a point in the *finite* plane. Hereafter, when the point at infinity is to be considered, it will be specifically mentioned.

EXERCISES

1. Sketch the following sets and determine which are both open and connected (and therefore domains):

   (a) $|z - 2 + i| \leq 1$; (b) $|2z + 3| > 4$; (c) Im $z > 1$;
   (d) Im $z = 1$; (e) $0 \leq \arg z \leq \pi/4$ $(z \neq 0)$; (f) $|z - 4| \geq |z|$;
   (g) $0 < |z - z_0| < \delta$ where z_0 is a fixed point and δ is a positive number.

   Ans. $(b), (c), (g)$ are domains.

2. Which set in Exercise 1 is neither open nor closed?

   Ans. (e).

3. Which sets in Exercise 1 are bounded?

   Ans. $(a), (g)$.

4. In each case sketch the closure of the set:

   (a) $-\pi < \arg z < \pi$ $(z \neq 0)$; (b) $|\text{Re } z| < |z|$; (c) $\text{Re}(1/z) \leq \dfrac{1}{2}$;
   (d) $\text{Re}(z^2) > 0$.

5. Let S be the open set consisting of all points z such that $|z| < 1$ or $|z - 2| < 1$. Show why S is not connected.

6. Show that a set S is open if and only if each point in S is an interior point.

7. Determine the accumulation points of each of the following sets:

(a) $z_n = i^n$ $(n = 1, 2, \ldots)$;   (b) $z_n = i^n/n$ $(n = 1, 2, \ldots)$;

(c) $0 \leqq \arg z < \pi/2$ $(z \neq 0)$;   (d) $z_n = (-1)^n(1 + i)(n - 1)/n$ $(n = 1, 2, \ldots)$.

Ans. (a) None;   (b) 0;   (d) $\pm(1 + i)$.

8. Prove that if a set contains each of its accumulation points, it must be a closed set.

9. Show that any point z_0 of a domain is an accumulation point of that domain.

10. Prove that a finite set of points $z_1, z_2, \ldots, z_n$ cannot have any accumulation points.

11. Show that a set S is unbounded if and only if every neighborhood of the point at infinity contains at least one point in S.

TWO

ANALYTIC FUNCTIONS

We now consider functions of a complex variable and develop a theory of differentiation for them. The main goal of the chapter is to introduce analytic functions, which play a central role in complex analysis.

9. Functions of a Complex Variable

Let S be a set of complex numbers. A *function* f defined on S is a rule that assigns to each z in S a complex number w. The number w is called the *value* of f at z and is denoted by $f(z)$; that is,

$$w = f(z).$$

The set S is called the *domain of definition* of f.*

It is not always convenient to use different notation to distinguish between a given function and its values. For example, if f is defined on the half plane Re $z > 0$ by means of the equation $w = 1/z$, it may also be referred to as the function $w = 1/z$, or simply the function $1/z$, where Re $z > 0$.

It must be emphasized that both a domain of definition and a rule are needed in order for a function to be well defined. When the domain of definition is not mentioned, we agree that the largest possible set is to be taken. Thus, if we speak only of the function $1/z$, the domain of definition is understood to be the set of all nonzero points in the plane.

A generalization of the concept of function is a rule that assigns more than one value to a point z in the domain of definition. These *multiple-valued functions* occur in the theory of functions of a complex variable just as they do in the case of real variables. An example is $z^{1/2}$, which has two values at each nonzero point in the complex plane. Our study of multiple-valued functions usually involves certain single-valued functions where just one of the possible values assigned to each point is taken.

---

*Although the domain of definition is often a domain as defined in Sec. 8, it need not be.

Let us agree that *the term* function *signifies a single-valued function* unless the contrary is clearly indicated.

Suppose that $w = u + iv$ is the value of a function f at $z = x + iy$; that is,

$$u + iv = f(x + iy).$$

Each of the real numbers u and v depends on the real variables x and y.

Example 1. If $f(z) = z^2$, then

$$f(z) = (x + iy)^2 = x^2 - y^2 + i2xy;$$

hence

$$u = x^2 - y^2 \quad \text{and} \quad v = 2xy.$$

This example illustrates how a given function of the complex variable z can be expressed in terms of a pair of real-valued functions of the real variables x and y:

(1) $$f(z) = u(x, y) + iv(x, y).$$

If, on the other hand, $u(x, y)$ and $v(x, y)$ are two given real-valued functions of the real variables x and y, equation (1) can be used to define a function of the complex variable $z = x + iy$.

Example 2. If

$$u(x, y) = y \int_0^\infty e^{-xt}\, dt \quad \text{and} \quad v(x, y) = \sum_{n=0}^\infty y^n,$$

one can write

(2) $$f(z) = y \int_0^\infty e^{-xt}\, dt + i \sum_{n=0}^\infty y^n.$$

The domain of definition of the function f here is understood to be the semi-infinite strip $x > 0$, $-1 < y < 1$; for the improper integral and the infinite series *both* converge only when x and y are so restricted.

If in equation (1) the number $v(x, y)$ is always zero, then the number $f(z)$ is always real. An example of such a *real-valued function of a complex variable* is $f(z) = |z|^2$.

If n is zero or a positive integer and if $a_0, a_1, a_2, \ldots, a_n$ are complex constants, where $a_n \neq 0$, the function

$$P(z) = a_0 + a_1 z + a_2 z^2 + \cdots + a_n z^n \qquad (a_n \neq 0)$$

is a *polynomial* of degree n. Note that the sum here has a finite number of terms and that the domain of definition is the entire z plane. Quotients of polynomials, $P(z)/Q(z)$, are called *rational functions* and are defined at each point z except where $Q(z) = 0$. Polynomials and rational functions constitute elementary, but important, classes of functions of a complex variable.

10. Mappings

Properties of a real-valued function of a real variable are often exhibited by the graph of the function. But when $w = f(z)$ where z and w are complex, no such convenient graphical representation of the function f is available because each of the numbers z and w is located in a plane rather than on a line. One can, however, display some information about the function by indicating pairs of corresponding points $z = (x, y)$ and $w = (u, v)$. To do this, it is generally simpler to draw the z and w planes separately.

When a function f is thought of in this way, it is often referred to as a *mapping* or *transformation*. The *image* of a point z in the domain of definition S is the point $w = f(z)$, and the set of images of all points in a set T that is contained in S is called the image of T. The image of the entire domain of definition S is called the *range* of f. The *inverse image* of a point w is the set of all points z in the domain of definition of f which have w as their image. The inverse image of a point may contain just one point, many points, or none at all. The last case occurs, of course, when w is not in the range of f.

Terms such as *translation, rotation,* and *reflection* are used to convey dominant geometric characteristics of certain mappings. In such cases it is sometimes convenient to consider the z and w planes to be the same. Since

$$z + 1 = (x + 1) + iy,$$

the mapping $w = z + 1$, for instance, can be thought of as a translation of each point z to a position one unit to the right. The mapping $w = iz$ rotates each nonzero point z counterclockwise through a right angle about the origin (Sec. 5), and the mapping $w = \bar{z}$ transforms each point z into its reflection in the real axis (Sec. 3).

More information is usually exhibited by sketching images of curves and regions than by simply indicating images of individual points.

Example. Let us show that under the transformation

(1)
$$w = z + \frac{1}{z},$$

the image of the domain consisting of the points in the upper half plane $y > 0$ which are exterior to the circle $|z| = 1$ is the entire upper half plane $v > 0$ and that the image of the boundary of the domain in the z plane is the u axis.

We begin by verifying that the boundary in the z plane is mapped as stated, with the corresponding points shown in Fig. 15. The image of a point $z = x$ $(x > 1)$ on the

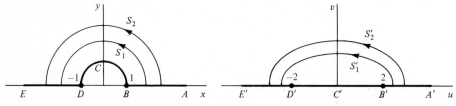

Figure 15. $w = z + \dfrac{1}{z}$.

x axis and to the right of the point $z = 1$ is, according to equation (1), the point $w = x + (1/x)$. This image point represents a real number in the w plane; and since the derivative of the function $x + (1/x)$ is always positive when $x > 1$, the value of w increases as x increases from $x = 1$. Hence, as the point $z = x$ $(x \geq 1)$ moves to the right along the x axis, its image moves to the right along the u axis, starting at the point $w = 2$. Similarly, if $z = -x$ $(x \geq 1)$ and z moves to the left from the point $z = -1$ along the x axis, the image point $w = -[x + (1/x)]$ moves to the left along the u axis, this time from the point $w = -2$.

To see that the image of the upper half of the circle $|z| = 1$ is the segment $-2 \leq u \leq 2$ of the u axis, we express z in the exponential form $z = re^{i\theta}$ and write transformation (1) as

$$(2) \qquad w = re^{i\theta} + \frac{1}{re^{i\theta}}.$$

The image of a typical point $z = e^{i\theta}$ $(0 \leq \theta \leq \pi)$ on the upper half of the circle $|z| = 1$ is then found by writing $r = 1$ in equation (2):

$$w = e^{i\theta} + \frac{1}{e^{i\theta}} = e^{i\theta} + e^{-i\theta} = 2 \cos \theta.$$

As the value of θ increases from $\theta = 0$ to $\theta = \pi$, the point $z = e^{i\theta}$ traverses the upper half of the circle in the counterclockwise direction and its image $w = 2 \cos \theta$ moves to the left along the u axis from $w = 2$ to $w = -2$.

We now show how the points above the boundary in the z plane correspond to the points above the one in the w plane. According to equation (2), the image of a point $z = r_1 e^{i\theta}$ $(0 \leq \theta \leq \pi)$ on the upper half S_1 (Fig. 15) of a circle $|z| = r_1$, where $r_1 > 1$, can be written

$$w = r_1 e^{i\theta} + \frac{1}{r_1 e^{i\theta}} = r_1(\cos \theta + i \sin \theta) + \frac{1}{r_1}(\cos \theta - i \sin \theta).$$

That is,

$$w = a \cos \theta + ib \sin \theta$$

where a and b are the positive numbers

$$a = r_1 + \frac{1}{r_1}, \qquad b = r_1 - \frac{1}{r_1}.$$

The real and imaginary parts

$$(3) \qquad u = a \cos \theta, \qquad v = b \sin \theta$$

of the number $w = u + iv$ are evidently related by the equation

$$\frac{u^2}{a^2} + \frac{v^2}{b^2} = 1,$$

which represents an ellipse with foci at the points $w = \pm\sqrt{a^2 - b^2} = \pm 2$. Since the values of $\cos \theta$ decrease steadily from 1 to -1 as θ increases from 0 to π, and since

$\sin \theta \geqq 0$ for those values of θ, it follows from equations (3) that the image point w moves to the left on the upper half S_1' of the ellipse as the point z traverses the semicircle S_1 from right to left.

When a radius r_2 larger than r_1 is taken, the upper half S_2 of the circle $|z| = r_2$ and its image S_2' are both larger, as indicated in Fig. 15. In fact, as the semicircles increase in size, their images, which are the upper halves of ellipses with foci at $w = \pm 2$, fill out the entire upper half plane $v > 0$. If the semicircles are thought of as shrinking in size toward the circular part of the boundary, the semiellipses tend to become the segment $-2 \leqq u \leqq 2$ of the u axis. Since any given point w above the u axis lies on one and only one of the semiellipses, it must then be the image of exactly one point in the domain in the z plane. This completes the verification of the mapping, which is also shown in Fig. 17 of Appendix 2.

Such a geometric interpretation of a function as a mapping is developed more extensively in Chap. 7. A reader who wishes to see further examples of mappings at this time has all the background needed to read Secs. 63 and 67, where the mappings $w = Bz + C$ $(B \neq 0)$ and $w = z^2$ are discussed.

11. Limits

Let a function f be defined at all points z in some neighborhood (Sec. 8) of z_0, except possibly for the point z_0 itself. The statement that the *limit* of $f(z)$ as z approaches z_0 is a number w_0,

$$(1) \qquad \qquad \lim_{z \to z_0} f(z) = w_0 ,$$

means that the point $w = f(z)$ can be made arbitrarily close to w_0 if we choose the point z close enough to z_0 but distinct from it. We now express the definition of limit in a precise and usable form.

Statement (1) means that for each positive number ε there is a positive number δ such that

$$(2) \qquad \qquad |f(z) - w_0| < \varepsilon \qquad \text{whenever} \qquad 0 < |z - z_0| < \delta.$$

Geometrically, this definition says that for each ε neighborhood $|w - w_0| < \varepsilon$ of w_0 there is a δ neighborhood $|z - z_0| < \delta$ of z_0 such that the images w of all points z in the δ neighborhood, with the possible exception of z_0, lie in the ε neighborhood (Fig. 16). Since condition (2) applies to *all* points in the domain $0 < |z - z_0| < \delta$, the symbol $z \to z_0$ in statement (1) implies that z is allowed to approach z_0 in an arbitrary

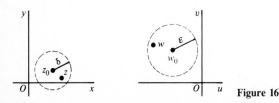

Figure 16

manner, not just from some particular direction. Note, however, that even though all points in the domain $0 < |z - z_0| < \delta$ are to be considered, their images need not constitute the entire neighborhood $|w - w_0| < \varepsilon$. If f has the constant value w_0, for instance, the image of z is always the center of that neighborhood. Note, too, that once a δ has been found, it can be replaced by any smaller positive number, such as $\delta/2$.

Definition (2) requires that f be defined at all points other that z_0 in some neighborhood of z_0. Such a neighborhood, of course, always exists when z_0 is an interior point of a region on which f is defined. We can extend the definition of limit to the case in which z_0 is a boundary point of that region by agreeing that the first of inequalities (2) need be satisfied by only those points z which lie in *both* the region and the domain $0 < |z - z_0| < \delta$.

While definition (2) provides a means of testing whether a given point w_0 is a limit, it does not directly provide a method for determining that limit. Theorems on limits, presented in the next section, will enable us to actually find many limits.

Example 1. Let us show that if $f(z) = iz/2$ in the open disk $|z| < 1$, then

$$\lim_{z \to 1} f(z) = \frac{i}{2},$$

the point $z = 1$ being on the boundary of the domain of definition. Observe that when z is in the region $|z| < 1$,

$$\left| f(z) - \frac{i}{2} \right| = \left| \frac{iz}{2} - \frac{i}{2} \right| = \frac{|z - 1|}{2}.$$

Hence, for any such z and any positive number ε,

$$\left| f(z) - \frac{i}{2} \right| < \varepsilon \qquad \text{whenever} \qquad 0 < |z - 1| < 2\varepsilon.$$

Thus condition (2) is satisfied by points in the region $|z| < 1$ when δ is equal to 2ε (Fig. 17) or any smaller positive number.

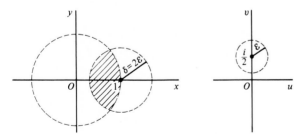

Figure 17

Example 2. To further illustrate definition (2), we now show that

(3) $$\lim_{z \to 2i} (2x + iy^2) = 4i \qquad\qquad (z = x + iy).$$

For each positive number ε we must find a positive number δ such that

(4) $|2x + iy^2 - 4i| < \varepsilon$ whenever $0 < |z - 2i| < \delta$.

To do this, we write

$$|2x + iy^2 - 4i| \leq 2|x| + |y^2 - 4| = 2|x| + |y - 2||y + 2|$$

and thus note that the first of inequalities (4) will be satisfied if

$$2|x| < \frac{\varepsilon}{2}   \text{and} |y - 2||y + 2| < \frac{\varepsilon}{2}.$$

The first of these inequalities is, of course, satisfied if $|x| < \varepsilon/4$. To establish conditions on y such that the second one holds, we restrict y so that $|y - 2| < 1$ and observe then that

$$|y + 2| = |(y - 2) + 4| \leq |y - 2| + 4 < 5.$$

Hence if $|y - 2| < \min\{\varepsilon/10, 1\}$, where $\min\{\varepsilon/10, 1\}$ denotes the minimum, or smaller, of the two numbers $\varepsilon/10$ and 1, it follows that

$$|y - 2||y + 2| < \left(\frac{\varepsilon}{10}\right)5 = \frac{\varepsilon}{2}.$$

An appropriate value of δ is now easily seen (Fig. 18) from the conditions that $|x|$ be less than $\varepsilon/4$ and that $|y - 2|$ be less than $\min\{\varepsilon/10, 1\}$:

$$\delta = \min\left\{\frac{\varepsilon}{10}, 1\right\}.$$

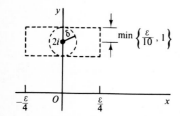

Figure 18

In introducing the concept of limit in the first paragraph of this section, we tacitly assumed that when a limit of a function $f(z)$ exists at a point z_0, it is unique. This is indeed the case. For suppose that

$$\lim_{z \to z_0} f(z) = w_0   \text{and}   \lim_{z \to z_0} f(z) = w_1.$$

Then for an arbitrary positive number ε there are positive numbers δ_0 and δ_1 such that

and
$$|f(z) - w_0| < \varepsilon   \text{whenever}   0 < |z - z_0| < \delta_0$$
$$|f(z) - w_1| < \varepsilon   \text{whenever}   0 < |z - z_0| < \delta_1.$$

So if $0 < |z - z_0| < \delta$ where δ denotes the smaller of the two numbers δ_0 and δ_1, then

$$|[f(z) - w_0] - [f(z) - w_1]| \leq |f(z) - w_0| + |f(z) - w_1| < 2\varepsilon;$$

that is, $|w_1 - w_0| < 2\varepsilon$. But $w_1 - w_0$ is a constant, and ε can be chosen arbitrarily small. Hence $w_1 - w_0 = 0$, or $w_1 = w_0$.

A meaning is readily given to the statement

(5) $$\lim_{z \to z_0} f(z) = w_0$$

when either z_0 or w_0, or possibly each of these numbers, is the point at infinity. In definition (2) we simply replace the appropriate neighborhoods of z_0 and w_0 by neighborhoods of ∞ (Sec. 8). For example, the statement

(6) $$\lim_{z \to \infty} f(z) = w_0$$

means that for each positive number ε there is a positive number δ such that

$$|f(z) - w_0| < \varepsilon \qquad \text{whenever} \qquad |z| > \frac{1}{\delta}.$$

That is, the point $w = f(z)$ lies in the ε neighborhood $|w - w_0| < \varepsilon$ of w_0 whenever the point z lies in the δ neighborhood $|z| > 1/\delta$ of the point at infinity.

Example 3. Observe that

$$\lim_{z \to \infty} \frac{1}{z^2} = 0$$

since

$$\left| \frac{1}{z^2} - 0 \right| < \varepsilon \qquad \text{whenever} \qquad |z| > \frac{1}{\sqrt{\varepsilon}}.$$

Here $\delta = \sqrt{\varepsilon}$.

When w_0 is the point at infinity and z_0 lies in the finite plane, we write

(7) $$\lim_{z \to z_0} f(z) = \infty$$

if for each ε there is a corresponding δ such that

$$|f(z)| > \frac{1}{\varepsilon} \qquad \text{whenever} \qquad 0 < |z - z_0| < \delta.$$

Example 4. As expected,

$$\lim_{z \to 0} \frac{1}{z^2} = \infty.$$

For

$$\left| \frac{1}{z^2} \right| > \frac{1}{\varepsilon} \qquad \text{whenever} \qquad 0 < |z - 0| < \sqrt{\varepsilon}.$$

The definition of limit when both z_0 and w_0 are the point at infinity is similar.

12. Theorems on Limits

We can expedite our treatment of limits by establishing a connection between the limit of a function of a complex variable and limits of real-valued functions of two real variables. Since limits of the latter type are studied in calculus, we use their definition and properties freely.

Theorem 1. *Suppose that*

$$f(z) = u(x, y) + iv(x, y), \qquad z_0 = x_0 + iy_0, \qquad and \qquad w_0 = u_0 + iv_0.$$

Then

(1)
$$\lim_{z \to z_0} f(z) = w_0$$

if and only if

(2)
$$\lim_{(x, y) \to (x_0, y_0)} u(x, y) = u_0 \qquad and \qquad \lim_{(x, y) \to (x_0, y_0)} v(x, y) = v_0.$$

To prove the theorem, we first assume that statement (1) is true and show that conditions (2) follow. According to the definition of limit in Sec. 11, there is for each positive number ε a positive number δ such that

$$|(u - u_0) + i(v - v_0)| < \varepsilon$$

whenever
$$0 < |(x - x_0) + i(y - y_0)| < \delta.$$

Since

$$|u - u_0| \leqq |(u - u_0) + i(v - v_0)|$$

and

$$|v - v_0| \leqq |(u - u_0) + i(v - v_0)|,$$

it follows that

$$|u - u_0| < \varepsilon \qquad and \qquad |v - v_0| < \varepsilon$$

whenever
$$0 < \sqrt{(x - x_0)^2 + (y - y_0)^2} < \delta.$$

So conditions (2) hold.

Let us now start with conditions (2). For each positive number ε there exist positive numbers δ_1 and δ_2 such that

$$|u - u_0| < \frac{\varepsilon}{2} \qquad \text{whenever} \qquad 0 < \sqrt{(x - x_0)^2 + (y - y_0)^2} < \delta_1$$

and

$$|v - v_0| < \frac{\varepsilon}{2} \qquad \text{whenever} \qquad 0 < \sqrt{(x - x_0)^2 + (y - y_0)^2} < \delta_2.$$

Let δ denote the smaller of the two numbers δ_1 and δ_2. Since

$$\left|(u - u_0) + i(v - v_0)\right| \leq |u - u_0| + |v - v_0|,$$

it follows that

$$\left|(u + iv) - (u_0 + iv_0)\right| < \varepsilon$$

whenever $$0 < \left|(x + iy) - (x_0 + iy_0)\right| < \delta.$$

This is statement (1), and the proof of the theorem is complete.

Theorem 2. *Suppose that*

(3) $$\lim_{z \to z_0} f(z) = w_0 \qquad and \qquad \lim_{z \to z_0} F(z) = W_0.$$

Then

(4) $$\lim_{z \to z_0} \left[f(z) + F(z)\right] = w_0 + W_0,$$

(5) $$\lim_{z \to z_0} \left[f(z)F(z)\right] = w_0 W_0;$$

and if $W_0 \neq 0$,

(6) $$\lim_{z \to z_0} \frac{f(z)}{F(z)} = \frac{w_0}{W_0}.$$

This fundamental theorem can be proved directly by using the definition of the limit of a function of a complex variable. But, with the aid of Theorem 1, it follows almost immediately from theorems on limits of real-valued functions of two real variables.

Consider the verification of property (5), for example, and write

$$f(z) = u(x, y) + iv(x, y), \qquad F(z) = U(x, y) + iV(x, y),$$

$$z_0 = x_0 + iy_0, \qquad w_0 = u_0 + iv_0, \qquad W_0 = U_0 + iV_0.$$

Then, according to hypotheses (3) and Theorem 1, the limits as (x, y) approaches (x_0, y_0) of the functions u, v, U, and V exist and have the values u_0, v_0, U_0, and V_0, respectively. So the real and imaginary components of the function

$$f(z)F(z) = (uU - vV) + i(vU + uV)$$

have the limits $u_0 U_0 - v_0 V_0$ and $v_0 U_0 + u_0 V_0$, respectively, as (x, y) approaches (x_0, y_0). Hence, by Theorem 1 again, $f(z)F(z)$ has the limit

$$(u_0 U_0 - v_0 V_0) + i(v_0 U_0 + u_0 V_0)$$

as z approaches z_0; and this is equal to $w_0 W_0$. Property (5) is thus established. Corresponding verifications of properties (4) and (6) can be given.

From the definition of limit we see that for any z_0,

$$\lim_{z \to z_0} z = z_0$$

since one can write $\delta = \varepsilon$ when $f(z) = z$. Then, by property (5) and mathematical induction,

$$\lim_{z \to z_0} z^n = z_0^n \qquad\qquad (n = 1, 2, \ldots).$$

Also,

$$\lim_{z \to z_0} c = c$$

where c is a complex constant. So, in view of properties (4) and (5), the limit of a polynomial

$$P(z) = a_0 + a_1 z + a_2 z^2 + \cdots + a_n z^n$$

as z approaches a point z_0 is the value of the polynomial at that point:

$$\lim_{z \to z_0} P(z) = P(z_0). \tag{7}$$

Another useful property of limits is that

$$\text{(8)} \qquad \text{if} \quad \lim_{z \to z_0} f(z) = w_0, \quad \text{then} \quad \lim_{z \to z_0} |f(z)| = |w_0|.$$

This is easily proved by using the definition of limit and the fact that (see Sec. 4)

$$\big| |f(z)| - |w_0| \big| \leq |f(z) - w_0|.$$

Finally, we remind the reader that the results of this section involve points in the *finite* plane only. As indicated in Sec. 8, the point at infinity is not to be considered unless it is specifically mentioned.

EXERCISES

1. For each of the functions defined below, describe the domain of definition that is understood:

$$\text{(a) } f(z) = \frac{1}{z^2 + 1}; \quad \text{(b) } f(z) = \text{Arg}\left(\frac{1}{z}\right); \quad \text{(c) } f(z) = \frac{z}{z + \bar{z}}; \quad \text{(d) } f(z) = \frac{1}{1 - |z|^2}.$$

Ans. (a) $z \neq \pm i$; (c) Re $z \neq 0$.

2. Describe the largest possible domain of definition of the function

$$g(z) = \frac{y}{x} + \frac{i}{1 - y} \qquad\qquad (z = x + iy).$$

Show that if z is a point in the semi-infinite strip $x > 0, -1 < y < 1$, then $g(z) = f(z)$ where f is the function

$$f(z) = y \int_0^\infty e^{-xt}\, dt + i \sum_{n=0}^\infty y^n$$

in Example 2 of Sec. 9.

3. Write the function $f(z) = z^3 + z + 1$ in the form $f(z) = u(x, y) + iv(x, y)$.

4. Suppose that $f(z) = x^2 - y^2 - 2y + i(2x - 2xy)$. Express the right-hand side here in terms of z and simplify.

Suggestion: Identities (9), Sec. 3, can be used.

$$Ans. \quad \bar{z}^2 + 2iz.$$

5. By modifying the discussion in the example in Sec. 10, verify that the transformation $w = z + (1/z)$ maps (*a*) the entire circle $|z| = 1$ onto the segment $-2 \le u \le 2$ of the u axis and the domain outside that circle onto the rest of the w plane; (*b*) the region of the z plane shown in Fig. 18, Appendix 2, onto the region of the w plane indicated there.

6. Let a, b, c, and z_0 denote complex constants. Use definition (2), Sec. 11, of limit to prove that

   (*a*) $\lim\limits_{z \to z_0} c = c$;   (*b*) $\lim\limits_{z \to z_0} (az + b) = az_0 + b$ $(a \neq 0)$;

   (*c*) $\lim\limits_{z \to z_0} (z^2 + c) = z_0^2 + c$;   (*d*) $\lim\limits_{z \to z_0} \operatorname{Re} z = \operatorname{Re} z_0$;   (*e*) $\lim\limits_{z \to z_0} \bar{z} = \bar{z}_0$;

   (*f*) $\lim\limits_{z \to 1-i} [x + i(2x + y)] = 1 + i$ $(z = x + iy)$;   (*g*) $\lim\limits_{z \to 0} (\bar{z}^2/z) = 0$.

7. Prove statement (4) in Theorem 2, Sec. 12, (*a*) by using Theorem 1, Sec. 12, and properties of limits of real-valued functions of two real variables; (*b*) directly from definition (2), Sec. 11, of limit.

8. Let n be a positive integer and let $P(z)$ and $Q(z)$ be polynomials, where $Q(z_0) \neq 0$. Use Theorem 2, Sec. 12, and established limits to find

   (*a*) $\lim\limits_{z \to z_0} \dfrac{1}{z^n}$ $(z_0 \neq 0)$;   (*b*) $\lim\limits_{z \to i} \dfrac{iz^3 - 1}{z + i}$;   (*c*) $\lim\limits_{z \to z_0} \dfrac{P(z)}{Q(z)}$.

$$Ans. \qquad (a)\ 1/z_0^n; \quad (b)\ 0; \quad (c)\ P(z_0)/Q(z_0).$$

9. Write $\Delta z = z - z_0$ and show that

$$\lim_{z \to z_0} f(z) = w_0 \qquad \text{if and only if} \qquad \lim_{\Delta z \to 0} f(z_0 + \Delta z) = w_0.$$

10. Show that

$$\lim_{z \to z_0} f(z)g(z) = 0 \qquad \text{if} \qquad \lim_{z \to z_0} f(z) = 0$$

and if there exists a positive number M such that $|g(z)| \le M$ for all z in some neighborhood of z_0.

11. Let z_0 and w_0 denote points in the finite complex plane. By appealing to definitions of limits in Sec. 11, show that

   (*a*) $\lim\limits_{z \to \infty} f(z) = w_0$ if and only if $\lim\limits_{z \to 0} f\!\left(\dfrac{1}{z}\right) = w_0$;

   (*b*) $\lim\limits_{z \to z_0} f(z) = \infty$ if and only if $\lim\limits_{z \to z_0} \dfrac{1}{f(z)} = 0$.

12. Interpret statement (5), Sec. 11, when both z_0 and w_0 are the point at infinity. Then show that

$$\lim_{z \to \infty} f(z) = \infty \qquad \text{if and only if} \qquad \lim_{z \to 0} \frac{1}{f(1/z)} = 0.$$

13. With the aid of results in Exercises 11 and 12, show that

   (*a*) $\lim\limits_{z \to \infty} \dfrac{4z^2}{(z-1)^2} = 4$;   (*b*) $\lim\limits_{z \to 1} \dfrac{1}{(z-1)^3} = \infty$;   (*c*) $\lim\limits_{z \to \infty} \dfrac{z^2 + 1}{z - 1} = \infty$.

14. Show that a limit of type (6), Sec. 11, is unique.

15. Prove property (8), Sec. 12, of limits.

16. Another interpretation of a function $w = f(z) = u(x, y) + iv(x, y)$ is that of a *vector field* in the domain of definition of f. The equation assigns a vector w with components $u(x, y)$ and $v(x, y)$ to each point (x, y) in the domain of definition. Indicate graphically the vector fields represented by the equations (a) $w = iz$; (b) $w = z/|z|$.

13. Continuity

A function f is *continuous* at a point z_0 if all three of the following conditions are satisfied:

$$(1) \qquad\qquad \lim_{z \to z_0} f(z) \text{ exists,}$$

$$(2) \qquad\qquad f(z_0) \text{ exists,}$$

$$(3) \qquad\qquad \lim_{z \to z_0} f(z) = f(z_0).$$

Observe that statement (3) actually contains statements (1) and (2); for the existence of the quantity on each side of the equation there is implicit. Statement (3) says that for each positive number ε there is a positive number δ such that

$$(4) \qquad |f(z) - f(z_0)| < \varepsilon \qquad \text{whenever} \qquad |z - z_0| < \delta.$$

A function of a complex variable is said to be continuous in a region R if it is continuous at each point in R.

If two functions are continuous at a point, their sum and product are also continuous at that point; their quotient is continuous at any such point where the denominator is not zero. These observations are immediate consequences of Theorem 2, Sec. 12. Note, too, that a polynomial is continuous in the entire plane because of equation (7), Sec. 12.

It follows directly from definition (4) that *a composition of continuous functions is continuous*. To show this, we let $w = f(z)$ be a function defined for all z in a neighborhood of a point z_0; and we let $g(w)$ be a function whose domain of definition contains the image (Sec. 10) of that neighborhood. The composition $g[f(z)]$ is then defined for all z in the neighborhood of z_0. Suppose now that f is continuous at z_0 and that g is continuous at the point $w_0 = f(z_0)$. In view of the continuity of g at w_0, we know that for each positive number ε there is a positive number γ such that

$$|g[f(z)] - g[f(z_0)]| < \varepsilon \qquad \text{whenever} \qquad |f(z) - f(z_0)| < \gamma.$$

But corresponding to γ there exists a positive number δ such that the second of these inequalities is satisfied whenever $|z - z_0| < \delta$. The continuity of the composition $g[f(z)]$ at z_0 is thus established.

From Theorem 1, Sec. 12, it follows that a function f of a complex variable is continuous at a point $z_0 = (x_0, y_0)$ if and only if its component functions u and v are continuous there.

Example 1. The function

$$f(z) = xy^2 + i(2x - y)$$

is continuous everywhere in the complex plane because the component functions are polynomials in x and y and are therefore continuous at each point (x, y).

Example 2. Likewise, the function

$$f(z) = e^{xy} + i \sin(x^2 - 2xy^3)$$

is continuous for all z because of the continuity of polynomials in x and y as well as the continuity of the exponential and sine functions.

Various properties of continuous functions of a complex variable can be deduced from corresponding properties of continuous real-valued functions of two real variables.*

Suppose, for example, that a function $f(z) = u(x, y) + iv(x, y)$ is continuous in a region R which is both closed and bounded. The function

$$\sqrt{[u(x, y)]^2 + [v(x, y)]^2}$$

is then continuous in R and thus reaches a maximum value somewhere in that region. That is, f is *bounded* in R and $|f(z)|$ reaches a maximum value somewhere in R. More precisely, there exists a nonnegative real number M such that

(5) $$|f(z)| \leq M \qquad \text{for all } z \text{ in } R,$$

where equality holds for at least one such z.

Another result which follows from the corresponding one for real-valued functions of two real variables is that a function f which is continuous in a closed bounded region R is *uniformly continuous* there. That is, a single value of δ, independent of z_0, may be chosen such that condition (4) is satisfied at each point z_0 in R.

14. Derivatives

Let f be a function whose domain of definition contains a neighborhood of a point z_0. The *derivative* of f at z_0, written $f'(z_0)$, is defined by the equation

(1) $$f'(z_0) = \lim_{z \to z_0} \frac{f(z) - f(z_0)}{z - z_0},$$

provided this limit exists. The function f is said to be *differentiable* at z_0 when its derivative at z_0 exists.

By expressing the variable z in definition (1) in terms of the new complex variable

$$\Delta z = z - z_0,$$

---

*For such properties quoted here, see, for instance, A. E. Taylor and W. R. Mann, "Advanced Calculus," 3d ed., pp. 125–126 and 529–532, 1983.

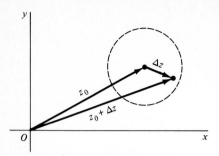

Figure 19

we can write that definition as

$$(2) \qquad f'(z_0) = \lim_{\Delta z \to 0} \frac{f(z_0 + \Delta z) - f(z_0)}{\Delta z}.$$

Note that because f is defined throughout a neighborhood of z_0, the number $f(z_0 + \Delta z)$ is always defined for $|\Delta z|$ sufficiently small (Fig. 19).

When taking form (2) of the definition of derivative, we often drop the subscript on z_0 and introduce the number

$$\Delta w = f(z + \Delta z) - f(z),$$

which denotes the change in the value of f corresponding to a change Δz in the variable z. Then, if we write dw/dz for $f'(z)$, equation (2) becomes

$$(3) \qquad \frac{dw}{dz} = \lim_{\Delta z \to 0} \frac{\Delta w}{\Delta z}.$$

Example 1. Suppose that $f(z) = z^2$. At any point z,

$$\lim_{\Delta z \to 0} \frac{\Delta w}{\Delta z} = \lim_{\Delta z \to 0} \frac{(z + \Delta z)^2 - z^2}{\Delta z} = \lim_{\Delta z \to 0} (2z + \Delta z) = 2z$$

since $2z + \Delta z$ is a polynomial in Δz. Hence $dw/dz = 2z$, or $f'(z) = 2z$.

Example 2. Let us now examine the function $f(z) = |z|^2$. Here

$$\frac{\Delta w}{\Delta z} = \frac{|z + \Delta z|^2 - |z|^2}{\Delta z} = \frac{(z + \Delta z)(\bar{z} + \overline{\Delta z}) - z\bar{z}}{\Delta z} = \bar{z} + \overline{\Delta z} + z\frac{\overline{\Delta z}}{\Delta z}.$$

When $z = 0$, this reduces to $\Delta w/\Delta z = \overline{\Delta z}$; hence $dw/dz = 0$ at the origin. If the limit of $\Delta w/\Delta z$ exists when $z \neq 0$, that limit may be found by letting the variable $\Delta z = \Delta x + i\,\Delta y$ approach 0 in any manner. In particular, when Δz approaches 0 through the real values $\Delta z = \Delta x + i0$, we may write $\overline{\Delta z} = \Delta z$. Hence if the limit of $\Delta w/\Delta z$ exists, its value must be $\bar{z} + z$. However, when Δz approaches 0 through the pure imaginary values $\Delta z = 0 + i\,\Delta y$, so that $\overline{\Delta z} = -\Delta z$, the limit is found to be $\bar{z} - z$. Since a limit is unique, it follows that $\bar{z} + z = \bar{z} - z$, or $z = 0$, if dw/dz exists. But $z \neq 0$, and we may conclude from this contradiction that dw/dz exists *only* at the origin.

This example shows that a function can be differentiable at a certain point but nowhere else in any neighborhood of that point. Since the real and imaginary parts of $f(z) = |z|^2$ are

(4) $$u(x, y) = x^2 + y^2 \quad \text{and} \quad v(x, y) = 0,$$

respectively, it also shows that the real and imaginary components of a function of a complex variable can have continuous partial derivatives of all orders at a point and yet the function may not even be differentiable there.

The function $f(z) = |z|^2$ is continuous at each point in the plane since its components (4) are continuous at each point. So the continuity of a function at a point does not imply the existence of a derivative there. It is, however, true that *the existence of the derivative of a function at a point implies the continuity of the function at that point*. To see this, we assume that $f'(z_0)$ exists and write

$$\lim_{z \to z_0}[f(z) - f(z_0)] = \lim_{z \to z_0}\frac{f(z) - f(z_0)}{z - z_0} \lim_{z \to z_0}(z - z_0) = f'(z_0) \cdot 0 = 0,$$

from which it follows that

$$\lim_{z \to z_0} f(z) = f(z_0).$$

This is the statement of continuity of f at z_0 (Sec. 13).

15. Differentiation Formulas

The definition of derivative in Sec. 14 is identical in form to that of the derivative of a real-valued function of a real variable. In fact, the basic differentiation formulas given below can be derived from that definition, together with various theorems on limits, by essentially the same steps used in calculus. In these formulas the derivative of a function f at a point z is denoted by either $f'(z)$ or $d[f(z)]/dz$, depending on which notation is more convenient.

Let c be a complex constant, and let f be a function whose derivative exists at a point z. It is easy to show that

(1) $$\frac{d}{dz}c = 0, \qquad \frac{d}{dz}z = 1, \qquad \frac{d}{dz}[cf(z)] = cf'(z).$$

Also, if n is a positive integer,

(2) $$\frac{d}{dz}z^n = nz^{n-1}.$$

This formula remains valid when n is a negative integer, provided $z \neq 0$.

If the derivatives of two functions f and F exist at a point z, then

(3) $$\frac{d}{dz}[f(z) + F(z)] = f'(z) + F'(z),$$

(4) $$\frac{d}{dz}[f(z)F(z)] = f(z)F'(z) + f'(z)F(z);$$

and when $F(z) \neq 0$,

(5)
$$\frac{d}{dz}\left[\frac{f(z)}{F(z)}\right] = \frac{F(z)f'(z) - f(z)F'(z)}{[F(z)]^2}.$$

Let us derive formula (4). To do this, we write the following expression for the change in the product $f(z)F(z)$:

$$f(z + \Delta z)F(z + \Delta z) - f(z)F(z)$$
$$= f(z)[F(z + \Delta z) - F(z)] + [f(z + \Delta z) - f(z)]F(z + \Delta z).$$

If we divide both sides of this equation by Δz and then let Δz tend to zero, we arrive at the desired formula for $d[f(z)F(z)]/dz$. Here we have used the fact that F is continuous at the point z, since $F'(z)$ exists; thus $F(z + \Delta z) \to F(z)$ as $\Delta z \to 0$ (see Exercise 9, Sec. 12).

There is also a chain rule for differentiating composite functions. Suppose that f has a derivative at z_0 and that g has a derivative at the point $f(z_0)$. Then the function $F(z) = g[f(z)]$ has a derivative at z_0, and

(6)
$$F'(z_0) = g'[f(z_0)]f'(z_0).$$

If we write $w = f(z)$ and $W = g(w)$, so that $W = F(z)$, the chain rule becomes

$$\frac{dW}{dz} = \frac{dW}{dw}\frac{dw}{dz}.$$

Example. To find the derivative of $(2z^2 + i)^5$, write $w = 2z^2 + i$ and $W = w^5$. Then

$$\frac{d}{dz}(2z^2 + i)^5 = 5w^4 4z = 20z(2z^2 + i)^4.$$

To start the proof of formula (6), choose a specific point z_0 where $f'(z_0)$ exists. Write $w_0 = f(z_0)$ and also assume that $g'(w_0)$ exists. There is, then, some ε neighborhood $|w - w_0| < \varepsilon$ of w_0 such that for all points w in that neighborhood we can define a function Φ which has the values $\Phi(w_0) = 0$ and

(7)
$$\Phi(w) = \frac{g(w) - g(w_0)}{w - w_0} - g'(w_0) \qquad \text{when } w \neq w_0.$$

Note that in view of the definition of derivative,

(8)
$$\lim_{w \to w_0} \Phi(w) = 0.$$

Hence Φ is continuous at w_0.

Now expression (7) can be put in the form

(9)
$$g(w) - g(w_0) = [g'(w_0) + \Phi(w)](w - w_0) \qquad (|w - w_0| < \varepsilon),$$

which is valid even when $w = w_0$; and since $f'(z_0)$ exists and f is therefore continuous at z_0, we can choose a positive number δ such that the point $f(z)$ lies in the ε neighborhood $|w - w_0| < \varepsilon$ of w_0 if z lies in the δ neighborhood $|z - z_0| < \delta$ of z_0. Thus it is legitimate to replace the variable w in equation (9) by $f(z)$ when z is any point in the neighborhood $|z - z_0| < \delta$. With that substitution, as well as the one $w_0 = f(z_0)$, equation (9) becomes

(10) $\dfrac{g[f(z)] - g[f(z_0)]}{z - z_0} = \{g'[f(z_0)] + \Phi[f(z)]\}\dfrac{f(z) - f(z_0)}{z - z_0}$ $(0 < |z - z_0| < \delta)$,

where we must stipulate that $z \neq z_0$ so that we are not dividing by zero. As already noted, f is continuous at z_0 and Φ is continuous at the point $w_0 = f(z_0)$. Thus the composition $\Phi[f(z)]$ is continuous at z_0; and, since $\Phi(w_0) = 0$,

$$\lim_{z \to z_0} \Phi[f(z)] = 0.$$

So equation (10) becomes

$$F'(z_0) = g'[f(z_0)]f'(z_0)$$

in the limit as z approaches z_0.

EXERCISES

1. Using results in Sec. 15, show that the derivative of a polynomial

$$P(z) = a_0 + a_1z + a_2z^2 + \cdots + a_nz^n \qquad (a_n \neq 0)$$

   of degree n $(n \geq 1)$ exists everywhere and that $P'(z) = a_1 + 2a_2z + \cdots + na_nz^{n-1}$.

2. Use results in Sec. 15 and Exercise 1 to find $f'(z)$ when

   (a) $f(z) = 3z^2 - 2z + 4$;   (b) $f(z) = (1 - 4z^2)^3$;

   (c) $f(z) = (z - 1)/(2z + 1)$ $(z \neq -1/2)$;   (d) $f(z) = (1 + z^2)^4/z^2$ $(z \neq 0)$.

3. Apply the definition of derivative directly to prove that $f'(z) = -1/z^2$ when $f(z) = 1/z$, provided $z \neq 0$.

4. Derive formula (3), Sec. 15, for the derivative of the sum of two functions.

5. (a) With the aid of formula (4), Sec. 15, use mathematical induction to derive formula (2) in that section for the derivative of z^n when n is a positive integer.

   (b) Use the binomial formula (Exercise 17, Sec. 2) to give an alternative derivation of the result in part (a).

6. Extend the result in Exercise 5 to include the case in which n is a negative integer and $z \neq 0$.

7. Apply the definition of derivative to show that if $f(z) = \text{Re } z$, then $f'(z)$ does not exist anywhere.

8. Show that the function $f(z) = \bar{z}$ is nowhere differentiable.

9. Determine whether the function $f(z) = \text{Im } z$ has a derivative at any point.

10. Let f be a function that is defined in a neighborhood of a point z_0 and continuous at that point. Show that if $f(z_0) \neq 0$, there must be a neighborhood of z_0 throughout which $f(z) \neq 0$.

    Suggestion: Write $\varepsilon = |f(z_0)|/2$ in definition (4), Sec. 13, of continuity. Then use the inequality (Sec. 4) $-(|z_1| - |z_2|) \leq |z_1 - z_2|$ to show that there is a neighborhood of z_0 throughout which $|f(z)| > |f(z_0)|/2$.

16. Cauchy-Riemann Equations

In this section we obtain a pair of equations which the first-order partial derivatives of the component functions u and v of a function

(1)
$$f(z) = u(x, y) + iv(x, y)$$

must satisfy at a point $z_0 = (x_0, y_0)$ when the derivative of f exists there. We also give formulas for $f'(z_0)$ in terms of those partial derivatives.

Suppose that the derivative

(2)
$$f'(z_0) = \lim_{\Delta z \to 0} \frac{f(z_0 + \Delta z) - f(z_0)}{\Delta z}$$

exists. Writing $z_0 = x_0 + iy_0$ and $\Delta z = \Delta x + i\Delta y$, we then have, by Theorem 1 of Sec. 12, the expressions

(3)
$$\text{Re}[f'(z_0)] = \lim_{(\Delta x, \Delta y) \to (0, 0)} \text{Re}\left[\frac{f(z_0 + \Delta z) - f(z_0)}{\Delta z}\right],$$

(4)
$$\text{Im}[f'(z_0)] = \lim_{(\Delta x, \Delta y) \to (0, 0)} \text{Im}\left[\frac{f(z_0 + \Delta z) - f(z_0)}{\Delta z}\right]$$

where

(5)
$$\frac{f(z_0 + \Delta z) - f(z_0)}{\Delta z}$$
$$= \frac{u(x_0 + \Delta x, y_0 + \Delta y) - u(x_0, y_0) + i[v(x_0 + \Delta x, y_0 + \Delta y) - v(x_0, y_0)]}{\Delta x + i\Delta y}.$$

It is important now to keep in mind that expressions (3) and (4) are valid as $(\Delta x, \Delta y)$ tends to $(0, 0)$ in any manner that we may choose.

In particular, let $(\Delta x, \Delta y)$ tend to $(0, 0)$ horizontally through the points $(\Delta x, 0)$, as indicated in Fig. 20. This means that $\Delta y = 0$ in equation (5), and we find that

$$\text{Re}[f'(z_0)] = \lim_{\Delta x \to 0} \frac{u(x_0 + \Delta x, y_0) - u(x_0, y_0)}{\Delta x},$$

$$\text{Im}[f'(z_0)] = \lim_{\Delta x \to 0} \frac{v(x_0 + \Delta x, y_0) - v(x_0, y_0)}{\Delta x}.$$

That is,

(6)
$$f'(z_0) = u_x(x_0, y_0) + iv_x(x_0, y_0),$$

where $u_x(x_0, y_0)$ and $v_x(x_0, y_0)$ denote the first-order partial derivatives with respect to x of the functions u and v at (x_0, y_0).

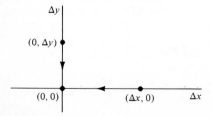

Figure 20

We might have let $(\Delta x, \Delta y)$ tend to zero vertically through the points $(0, \Delta y)$. In that case, $\Delta x = 0$ in equation (5); and we obtain the expression

(7) $$f'(z_0) = v_y(x_0, y_0) - iu_y(x_0, y_0),$$

which can also be written

$$f'(z_0) = -i[u_y(x_0, y_0) + iv_y(x_0, y_0)].$$

Equations (6) and (7) are not only formulas for finding $f'(z_0)$ in terms of partial derivatives of the component functions u and v; they also provide necessary conditions for the existence of $f'(z_0)$. For, equating the real parts and the imaginary parts on the right-hand sides of these equations, we see that the existence of $f'(z_0)$ requires that

(8) $$u_x(x_0, y_0) = v_y(x_0, y_0) \qquad \text{and} \qquad u_y(x_0, y_0) = -v_x(x_0, y_0).$$

Equations (8) are the *Cauchy-Riemann equations,* so named in honor of the French mathematician A. L. Cauchy (1789–1857), who discovered and used them, and in honor of the German mathematician G. F. B. Riemann (1826–1866), who made them fundamental in his development of the theory of functions of a complex variable.

We summarize the above results as follows.

Theorem. *Suppose that*

$$f(z) = u(x, y) + iv(x, y)$$

and that $f'(z_0)$ exists at a point $z_0 = x_0 + iy_0$. Then the first-order partial derivatives of u and v with respect to x and y must exist at (x_0, y_0), and they must satisfy the Cauchy-Riemann equations (8) at that point. Also, $f'(z_0)$ is given in terms of these partial derivatives by either equation (6) or (7).

Example 1. Consider the function

$$f(z) = z^2 = x^2 - y^2 + i2xy.$$

We proved in Example 1 of Sec. 14 that its derivative exists everywhere and that $f'(z) = 2z$. To verify that the Cauchy-Riemann equations are satisfied everywhere, we note that $u(x, y) = x^2 - y^2$ and $v(x, y) = 2xy$. Thus

$$u_x(x, y) = 2x = v_y(x, y), \qquad u_y(x, y) = -2y = -v_x(x, y).$$

Moreover, according to equation (6) or (7),

$$f'(z) = 2x + i2y = 2(x + iy) = 2z.$$

Since the Cauchy-Riemann equations are necessary conditions for the existence of the derivative of a function f at a point z_0, they can often be used to locate points at which f does *not* have a derivative.

Example 2. The function $f(z) = |z|^2$ is discussed in Example 2 of Sec. 14. In this case, $u(x, y) = x^2 + y^2$ and $v(x, y) = 0$; and so $u_x(x, y) = 2x$ and $v_y(x, y) = 0$, while $u_y(x, y) = 2y$ and $v_x(x, y) = 0$. Since the Cauchy-Riemann equations are not satisfied unless $x = y = 0$, the derivative $f'(z)$ cannot exist if $z \neq 0$. Note that the above theorem does not guarantee the existence of $f'(0)$; the theorem of Sec. 17 does, however, do this.

17. Sufficient Conditions

Satisfaction of the Cauchy-Riemann equations at a point $z_0 = (x_0, y_0)$ is not sufficient to ensure the existence of the derivative of a function $f(z)$ at that point. (See Exercise 6, Sec. 18.) But with certain continuity conditions, we have the following useful theorem.

Theorem. *Let the function*

$$f(z) = u(x, y) + iv(x, y)$$

be defined throughout some ε neighborhood of a point $z_0 = x_0 + iy_0$. Suppose that the first-order partial derivatives of the functions u and v with respect to x and y exist everywhere in that neighborhood and that they are continuous at (x_0, y_0). Then if those partial derivatives satisfy the Cauchy-Riemann equations

$$u_x = v_y, \qquad u_y = -v_x$$

at (x_0, y_0), the derivative $f'(z_0)$ exists.

To start the proof, we write $\Delta z = \Delta x + i\Delta y$, where $0 < |\Delta z| < \varepsilon$, and

$$\Delta w = f(z_0 + \Delta z) - f(z_0).$$

Thus

$$\Delta w = \Delta u + i\Delta v$$

where

$$
(1) \qquad
\begin{aligned}
\Delta u &= u(x_0 + \Delta x, y_0 + \Delta y) - u(x_0, y_0), \\
\Delta v &= v(x_0 + \Delta x, y_0 + \Delta y) - v(x_0, y_0).
\end{aligned}
$$

Now in view of the continuity of the first-order partial derivatives of u and v at the point (x_0, y_0),

$$
(2) \qquad
\begin{aligned}
\Delta u &= u_x(x_0, y_0)\Delta x + u_y(x_0, y_0)\Delta y + \varepsilon_1\sqrt{(\Delta x)^2 + (\Delta y)^2}, \\
\Delta v &= v_x(x_0, y_0)\Delta x + v_y(x_0, y_0)\Delta y + \varepsilon_2\sqrt{(\Delta x)^2 + (\Delta y)^2}
\end{aligned}
$$

where ε_1 and ε_2 tend to 0 as $(\Delta x, \Delta y)$ approaches $(0, 0)$. Hence

$$
(3) \qquad
\begin{aligned}
\Delta w &= u_x(x_0, y_0)\Delta x + u_y(x_0, y_0)\Delta y + \varepsilon_1\sqrt{(\Delta x)^2 + (\Delta y)^2} \\
&\quad + i[v_x(x_0, y_0)\Delta x + v_y(x_0, y_0)\Delta y + \varepsilon_2\sqrt{(\Delta x)^2 + (\Delta y)^2}].
\end{aligned}
$$

The existence of expressions of type (2) for functions of two real variables with continuous first-order partial derivatives is established in advanced calculus in connection with differentials.*

*See, for instance, A. E. Taylor and W. R. Mann, "Advanced Calculus," 3d ed., pp. 150–151 and 197–198, 1983.

Since the Cauchy-Riemann equations are satisfied at (x_0, y_0), we can replace $u_y(x_0, y_0)$ by $-v_x(x_0, y_0)$ and $v_y(x_0, y_0)$ by $u_x(x_0, y_0)$ in equation (3) and then divide through by Δz to get

(4) $$\frac{\Delta w}{\Delta z} = u_x(x_0, y_0) + iv_x(x_0, y_0) + (\varepsilon_1 + i\varepsilon_2)\frac{\sqrt{(\Delta x)^2 + (\Delta y)^2}}{\Delta z}.$$

But $\sqrt{(\Delta x)^2 + (\Delta y)^2} = |\Delta z|$, and so

$$\left|\frac{\sqrt{(\Delta x)^2 + (\Delta y)^2}}{\Delta z}\right| = 1.$$

Also, $\varepsilon_1 + i\varepsilon_2$ tends to 0 as $(\Delta x, \Delta y)$ approaches $(0, 0)$. So the last term on the right in equation (4) tends to 0 as the variable $\Delta z = \Delta x + i\,\Delta y$ tends to 0. This means that the limit of the left-hand side of equation (4) exists and that

(5) $$f'(z_0) = u_x(x_0, y_0) + iv_x(x_0, y_0).$$

Example 1. Suppose that

$$f(z) = e^x(\cos y + i \sin y),$$

where y is to be taken in radians when $\cos y$ and $\sin y$ are evaluated. Then

$$u(x, y) = e^x \cos y \qquad \text{and} \qquad v(x, y) = e^x \sin y.$$

Since $u_x = v_y$ and $u_y = -v_x$ everywhere and since those derivatives are everywhere continuous, the conditions in the theorem are satisfied at all points in the complex plane. Thus $f'(z)$ exists everywhere and, according to equation (5),

$$f'(z) = u_x(x, y) + iv_x(x, y) = e^x(\cos y + i \sin y).$$

Note that $f'(z) = f(z)$.

Example 2. It also follows from the theorem of this section that the function

$$f(z) = |z|^2 = (x^2 + y^2) + i0$$

has a derivative at $z = 0$; in fact, $f'(0) = 0 + i0 = 0$. We saw in Example 2, Sec. 16, that this function does not have a derivative at any nonzero point since, even though the component functions u and v exist and are continuous everywhere, the Cauchy-Riemann equations are not satisfied when $z \neq 0$.

18. Polar Coordinates

When $z_0 \neq 0$, the theorem in Sec. 17 is given in polar coordinates by means of the coordinate transformation (Sec. 5)

(1) $$x = r \cos \theta, \qquad y = r \sin \theta.$$

Depending on whether we write

$$z = x + iy \qquad \text{or} \qquad z = r(\cos \theta + i \sin \theta),$$

the real and imaginary parts of $w = u + iv$, where $w = f(z)$, are expressed in terms of either the variables x and y or r and θ. Suppose that the first-order partial derivatives of u and v with respect to x and y exist everywhere in some neighborhood of a given nonzero point z_0 and are continuous at that point. The first-order derivatives with respect to r and θ also have these properties, and the chain rule for differentiating real-valued functions of two real variables can be used to write them in terms of the ones with respect to x and y. Similar statements apply if we start with the derivatives with respect to r and θ. Suppose that in addition to the conditions just stated, the Cauchy-Riemann equations $u_x = v_y$, $u_y = -v_x$ are satisfied at z_0. It is straightforward to show (Exercise 7) that the equations

$$(2) \qquad\qquad u_r = \frac{1}{r} v_\theta, \qquad \frac{1}{r} u_\theta = -v_r$$

are also satisfied there. In fact, equations (2) are an alternative form of the Cauchy-Riemann equations.

We can now restate the theorem in Sec. 17 using polar coordinates.

Theorem. *Let the function*

$$f(z) = u(r, \theta) + iv(r, \theta)$$

be defined throughout some ε neighborhood of a nonzero point

$$z_0 = r_0(\cos \theta_0 + i \sin \theta_0).$$

Suppose that the first-order partial derivatives of the functions u and v with respect to r and θ exist everywhere in that neighborhood and that they are continuous at (r_0, θ_0). Then if those partial derivatives satisfy the polar form (2) of the Cauchy-Riemann equations at (r_0, θ_0), the derivative $f'(z_0)$ exists.

The derivative $f'(z_0)$ here can be expressed in terms of partial derivatives of u and v with respect to r as follows (Exercise 8):

$$(3) \qquad f'(z_0) = (\cos \theta_0 - i \sin \theta_0)[u_r(r_0, \theta_0) + iv_r(r_0, \theta_0)].$$

Note that in view of Euler's formula $e^{i\theta} = \cos \theta + i \sin \theta$ (Sec. 6), this can be written more compactly as

$$(4) \qquad f'(z_0) = e^{-i\theta_0}[u_r(r_0, \theta_0) + iv_r(r_0, \theta_0)].$$

Example. Consider the function

$$f(z) = \frac{1}{z} = \frac{1}{re^{i\theta}}.$$

Observe that $u(r, \theta) = (\cos \theta)/r$ and $v(r, \theta) = -(\sin \theta)/r$ and that the conditions in the theorem are satisfied at any nonzero point $z = re^{i\theta}$ in the plane. Hence the derivative of f exists there; and, according to expression (4),

$$f'(z) = e^{-i\theta}\left(-\frac{\cos \theta}{r^2} + i\frac{\sin \theta}{r^2}\right) = -\frac{1}{(re^{i\theta})^2} = -\frac{1}{z^2}.$$

EXERCISES

1. Use the theorem in Sec. 16 to show that $f'(z)$ does not exist at any point if $f(z)$ is

   (a) $\bar{z}$; (b) $z - \bar{z}$; (c) $2x + ixy^2$; (d) $e^x e^{-iy}$.

2. Using the theorem in Sec. 17 and expression (5) in that section, show that $f'(z)$ and its derivative $f''(z)$ exist everywhere and find $f'(z)$ and $f''(z)$ when

   (a) $f(z) = iz + 2$; (b) $f(z) = e^{-x}e^{-iy}$; (c) $f(z) = z^3$;
   (d) $f(z) = \cos x \cosh y - i \sin x \sinh y$.

   Ans.   (b) $f'(z) = -f(z), f''(z) = f(z)$;   (d) $f''(z) = -f(z)$.

3. From results obtained in Secs. 16 and 17, determine where $f'(z)$ exists and find its value when

   (a) $f(z) = 1/z$; (b) $f(z) = x^2 + iy^2$; (c) $f(z) = z \operatorname{Im} z$.

   Ans.   (a) $f'(z) = -1/z^2$ $(z \neq 0)$;   (b) $f'(x + ix) = 2x$;   (c) $f'(0) = 0$.

4. Using the theorem in Sec. 18, show that the function

   $$g(z) = \sqrt{r}e^{i\theta/2} \qquad\qquad (r > 0, -\pi < \theta < \pi)$$

   has a derivative at each point in its domain of definition. Then show that

   $$g'(z) = \frac{1}{2g(z)}.$$

   (Note that for each point z in the domain of definition, g has one of the two values of the multiple-valued function $z^{1/2}$ at z.)

5. Consider the function $f(z) = x^3 + i(1 - y)^3$. Show why it is legitimate to write

   $$f'(z) = u_x(x, y) + iv_x(x, y) = 3x^2$$

   only when $z = i$.

6. Let f denote the function whose values are $f(0) = 0$ and

   $$f(z) = \frac{(\bar{z})^2}{z} \qquad\qquad \text{when } z \neq 0.$$

   Show that the Cauchy-Riemann equations are satisfied at the point $z = 0$ but that the derivative of f fails to exist there.

   Suggestion: To show that $f'(0)$ does not exist, use definition (1), Sec. 14, of derivative and first let z approach 0 along the real axis $y = 0$. Then let it approach 0 along the line $y = x$.

7. Let $u(x, y)$ and $v(x, y)$ denote real-valued functions whose first-order partial derivatives with respect to x and y exist everywhere in a neighborhood of a nonzero point P in the xy plane and are continuous at that point.

   (a) Use coordinate transformation (1), Sec. 18, and the chain rule to show that

   $$u_r = u_x \cos \theta + u_y \sin \theta, \qquad u_\theta = -u_x r \sin \theta + u_y r \cos \theta$$

   at the point P.

   (b) With the aid of the expressions obtained in part (a) and similar ones for v_r and v_θ, show that equations (2), Sec. 18, are satisfied at P if $u_x = v_y$ and $u_y = -v_x$ there.

   (c) Solve the pair of equations obtained in part (a) for u_x and u_y to show that

$$u_x = u_r \cos\theta - u_\theta \frac{\sin\theta}{r}, \qquad u_y = u_r \sin\theta + u_\theta \frac{\cos\theta}{r}$$

at P. Then use these expressions and similar ones for v_x and v_y to show that $u_x = v_y$ and $u_y = -v_x$ at P if equations (2), Sec. 18, are satisfied there. Thus show that equations (2), Sec. 18, are the Cauchy-Riemann equations in polar coordinates.

8. Show that when $z_0 \neq 0$ and the conditions in the theorem of Sec. 18 are satisfied, the expression $f'(z_0) = u_x + iv_x$ in Sec. 17 can be used to write

$$f'(z_0) = (\cos\theta_0 - i\sin\theta_0)(u_r + iv_r),$$

where all the partial derivatives are to be evaluated at z_0.

  Suggestion: Use the expression for u_x and the corresponding one for v_x obtained in Exercise 7(c) and then refer to the polar form (2), Sec. 18, of the Cauchy-Riemann equations.

9. (a) With the aid of the polar form (2), Sec. 18, of the Cauchy-Riemann equations, show that the expression for $f'(z_0)$ obtained in Exercise 8 can be written

$$f'(z_0) = \frac{-i}{z_0}(u_\theta + iv_\theta).$$

(b) Use the expression for $f'(z_0)$ found in part (a) to show that the derivative of the function $f(z) = 1/z \ (z \neq 0)$ in the example of Sec. 18 is $f'(z) = -1/z^2$.

19. Analytic Functions

We are now ready to introduce the concept of an analytic function. A function f of the complex variable z is *analytic* at a point z_0 if its derivative exists not only at z_0 but also at each point z in some neighborhood of z_0. Note how it follows that if f is analytic at z_0, it is actually analytic at each point in a neighborhood of z_0. A function f is said to be analytic in a region R if it is analytic at each point in R. The term *holomorphic* is also used in the literature to denote analyticity.

  If $f(z) = z^2$, then f is analytic everywhere. But the function $f(z) = |z|^2$ is not analytic at any point since its derivative exists only at $z = 0$ and not throughout any neighborhood. (See Example 2, Sec. 14.)

  If a function f is analytic in a region R, then about each point z of R there must be a neighborhood on which f is defined. This means that z must be an interior point of the domain of definition, and so analytic functions are usually defined on domains. If, however, we should speak of an analytic function f on the closed disk $|z| \leq 1$, for example, it is to be understood that f is analytic throughout some domain containing that disk.

  An *entire function* is a function that is analytic at each point in the entire finite plane. Since the derivative of a polynomial exists everywhere, it follows that *every polynomial is an entire function*.

  If a function f fails to be analytic at a point z_0 but is analytic at some point in every neighborhood of z_0, then z_0 is called a *singular point*, or *singularity*, of f. Consider, for example, the function $f(z) = 1/z \ (z \neq 0)$, whose derivative is $f'(z) = -1/z^2$. It is analytic at every point except for $z = 0$, where it is not even defined. The point $z = 0$

is therefore a singular point. The function $f(z) = |z|^2$, on the other hand, has no singular points since it is nowhere analytic.

A necessary, but by no means sufficient, condition for a function f to be analytic in a domain D is clearly the continuity of f throughout D. Satisfaction of the Cauchy-Riemann equations is also necessary, but not sufficient. Sufficient conditions for analyticity in D are provided by the theorems in Secs. 17 and 18.

Other useful sufficient conditions are obtained from the differentiation formulas in Sec. 15. The derivatives of the sum and product of two functions exist wherever the functions themselves have derivatives. Thus, *if two functions are analytic in a domain D, their sum and their product are both analytic in D*. Similarly, *their quotient is analytic in D provided the function in the denominator does not vanish at any point in D*. In particular, the quotient $P(z)/Q(z)$ of two polynomials is analytic in any domain throughout which $Q(z) \neq 0$.

From the chain rule for the derivative of a composite function we find that *a composition of two analytic functions is analytic*. We now illustrate this with an example involving polar coordinates.

Example. Let f be the entire function $f(z) = z^2$. According to Exercise 4, Sec. 18, the function

$$g(z) = z^{1/2} = \sqrt{r}e^{i\theta/2} \qquad (r > 0, -\pi < \theta < \pi)$$

is analytic everywhere in the indicated domain of definition, consisting of all points in the z plane except for the origin and points lying on the negative real axis. As usual with compositions $g[f(z)]$ (Secs. 13 and 15), we find it convenient here to write $w = f(z)$ and think of g as being defined in the w plane; thus

$$(1) \qquad\qquad g(w) = \sqrt{\rho}e^{i\phi/2} \qquad (\rho > 0, -\pi < \phi < \pi),$$

where $w = \rho e^{i\phi}$.

In order for $g[f(z)]$ to be well defined, we need to restrict f to a domain D such that the range of f is contained in the domain on which g is defined. The largest possible domain D with this property is readily found by writing $w = f(z)$ $(z \neq 0)$ in the form $w = r^2 e^{i2\theta}$ and noting that $-\pi < 2\theta < \pi$ when $-\pi/2 < \theta < \pi/2$. It follows that when $r > 0$ and $-\pi/2 < \theta < \pi/2$,

$$(2) \qquad\qquad \rho = r^2 \quad\text{and}\quad \phi = 2\theta$$

where $\rho > 0$ and $-\pi < \phi < \pi$. The half plane $r > 0$, $-\pi/2 < \theta < \pi/2$, which is the right half of the z plane with the imaginary axis excluded, is therefore the largest possible domain D that can be taken. The analyticity of f and g ensures that $g[f(z)]$ is analytic in the half plane $r > 0$, $-\pi/2 < \theta < \pi/2$, which is the domain to which f is restricted.

Verification of the analyticity of $g[f(z)]$ in this particular example is easily made by noting how it follows from the relation $w = f(z)$ and equations (1) and (2) that

$$g[f(z)] = g(w) = \sqrt{r^2}e^{i2\theta/2} = re^{i\theta} = z$$

for any point z in the domain $r > 0$, $-\pi/2 < \theta < \pi/2$.

Finally, a useful and expected property of analytic functions is that *a function $f(z)$ is constant throughout a domain D when $f'(z) = 0$ everywhere in D*. This is easy to show by writing $f(z) = u(x, y) + iv(x, y)$ and noting that

$$u_x + iv_x = v_y - iu_y = 0$$

when $f'(z) = 0$. Thus, when $f'(z) = 0$ everywhere in D,

$$u_x = u_y = v_x = v_y = 0$$

at each point in D. Now u_x and u_y are the x and y components of the vector grad u, and the component of grad u at a given point in a specific direction is the directional derivative of u there in that direction. So the fact that u_x and u_y are always zero means that grad u is always the zero vector and hence that any directional derivative of u is zero. Consequently, u is constant along any line segment lying entirely in D; and since there is always a finite number of such line segments, joined end to end, connecting any two points in D (Sec. 8), the values of u at those points must be the same. We may conclude, then, that there is a real constant c_1 such that $u(x, y) = c_1$ throughout D. Similarly, $v(x, y) = c_2$; and it follows that $f(z) = c_1 + ic_2$ at each point in D.

20. Harmonic Functions

A real-valued function h of two real variables x and y is said to be *harmonic* in a given domain of the xy plane if throughout that domain it has continuous partial derivatives of the first and second order and satisfies the partial differential equation

(1) $$h_{xx}(x, y) + h_{yy}(x, y) = 0,$$

known as *Laplace's equation*. We now assume that a function

$$f(z) = u(x, y) + iv(x, y)$$

is analytic in a domain D and show that the component functions u and v are harmonic in D.

To do this, we need a result which is proved in Chap. 4 (Sec. 39). Namely, if a function of a complex variable is analytic at a point, then its real and imaginary components have continuous partial derivatives of all orders at that point.

Since f is analytic in D, the first-order partial derivatives of its component functions satisfy the Cauchy-Riemann equations throughout D:

(2) $$u_x = v_y, \qquad u_y = -v_x.$$

Differentiating both sides of these equations with respect to x, we have

(3) $$u_{xx} = v_{yx}, \qquad u_{yx} = -v_{xx}.$$

Likewise, differentiation with respect to y yields

(4) $$u_{xy} = v_{yy}, \qquad u_{yy} = -v_{xy}.$$

Now, by a theorem in advanced calculus,* the continuity of the partial derivatives ensures that $u_{yx} = u_{xy}$ and $v_{yx} = v_{xy}$. It then follows from equations (3) and (4) that

---

*See, for instance, A. E. Taylor and W. R. Mann, "Advanced Calculus," 3d ed., pp. 199–201, 1983.

$$u_{xx}(x, y) + u_{yy}(x, y) = 0 \qquad \text{and} \qquad v_{xx}(x, y) + v_{yy}(x, y) = 0.$$

Thus, *if a function f(z) = u(x, y) + iv(x, y) is analytic in a domain D, its component functions u and v are harmonic in D.*

If two given functions u and v are harmonic in a domain D and their first-order partial derivatives satisfy the Cauchy-Riemann equations (2) throughout D, v is said to be a *harmonic conjugate* of u. The meaning of the word *conjugate* here is, of course, different from that in Sec. 3, where $\bar{z}$ is defined.

Evidently, if a function $f(z) = u(x, y) + iv(x, y)$ is analytic in a domain D, then v is a harmonic conjugate of u. Conversely, if v is a harmonic conjugate of u in a domain D, the function $f(z) = u(x, y) + iv(x, y)$ is analytic in D. This follows from the theorem in Sec. 17. Hence *a necessary and sufficient condition for a function f(z) = u(x, y) + iv(x, y) to be analytic in a domain D is that v be a harmonic conjugate of u in D.*

We note that if v is a harmonic conjugate of u in some domain, it is *not*, in general, true that u is a harmonic conjugate of v there. Example 1, below, shows this.

Example 1. Suppose that

$$u(x, y) = x^2 - y^2 \qquad \text{and} \qquad v(x, y) = 2xy.$$

Since these are the real and imaginary components, respectively, of the entire function $f(z) = z^2$, we know that v is a harmonic conjugate of u throughout the plane. But u cannot be a harmonic conjugate of v since, according to the theorem in Sec. 16, the function $2xy + i(x^2 - y^2)$ is not analytic anywhere. It is left as an exercise to show that if two functions u and v are to be harmonic conjugates of each other, both u and v must be constant functions.

It is, however, true that if v is a harmonic conjugate of u in a domain D, then $-u$ is a harmonic conjugate of v in D, and conversely. This is because the function $f(z) = u(x, y) + iv(x, y)$ is analytic in D if and only if the function

$$-if(z) = v(x, y) - iu(x, y)$$

is analytic there.

In Chap. 8 (Sec. 76) we show that a function u which is harmonic in a domain of a certain type always has a harmonic conjugate. Thus, in such domains, every harmonic function is the real part of an analytic function. It is also true that a harmonic conjugate, when it exists, is unique except for an additive constant.

Example 2. We now illustrate one method of obtaining a harmonic conjugate of a given harmonic function. The function

$$(5) \qquad\qquad u(x, y) = y^3 - 3x^2y$$

is readily seen to be harmonic throughout the entire xy plane. To find a harmonic conjugate $v(x, y)$, we note that

$$u_x(x, y) = -6xy.$$

So, in view of the condition $u_x = v_y$, we may conclude that

$$v_y(x, y) = -6xy.$$

Holding x fixed and integrating both sides of this equation with respect to y, we find that

(6) $$v(x, y) = -3xy^2 + \phi(x)$$

where ϕ is at present an arbitrary function of x. Since the condition $u_y = -v_x$ must hold, it follows from equations (5) and (6) that

$$3y^2 - 3x^2 = 3y^2 - \phi'(x).$$

So $\phi'(x) = 3x^2$, or $\phi(x) = x^3 + c$ where c is an arbitrary real number. Hence the function

$$v(x, y) = x^3 - 3xy^2 + c$$

is a harmonic conjugate of $u(x, y)$.

The corresponding analytic function is

(7) $$f(z) = (y^3 - 3x^2y) + i(x^3 - 3xy^2 + c).$$

It is easily verified that

$$f(z) = i(z^3 + c).$$

This form is suggested by noting that when $y = 0$, equation (7) becomes

$$f(x) = i(x^3 + c).$$

EXERCISES

1. Prove that each of these functions is entire:

   (a) $f(z) = 3x + y + i(3y - x)$;   (b) $f(z) = \sin x \cosh y + i \cos x \sinh y$;
   (c) $f(z) = e^{-y}e^{ix}$;   (d) $f(z) = (z^2 - 2)e^{-x}e^{-iy}$.

2. Show that each of these functions is nowhere analytic:

   (a) $f(z) = xy + iy$;   (b) $f(z) = e^y e^{ix}$.

3. In each case determine the singular points of the function and state why the function is analytic everywhere except at those points:

   (a) $\dfrac{2z + 1}{z(z^2 + 1)}$;   (b) $\dfrac{z^3 + i}{z^2 - 3z + 2}$;   (c) $\dfrac{z^2 + 1}{(z + 2)(z^2 + 2z + 2)}$.

   Ans.   (a) $z = 0, \pm i$;   (c) $z = -2, -1 \pm i$.

4. The composition $g[f(z)]$ of the entire function $f(z) = z^2$ and the function

   $$g(z) = z^{1/2} = \sqrt{r}\, e^{i\theta/2} \qquad (r > 0, -\pi < \theta < \pi)$$

   is shown in Sec. 19 to be analytic in a half plane, where $g[f(z)] = z$. Show that $f[g(z)]$ is analytic in a larger domain and that $f[g(z)] = z$ there.

5. The function $g(z) = z^{1/2} = \sqrt{r}\,e^{i\theta/2}$ $(r > 0, -\pi < \theta < \pi)$ in Sec. 19 is analytic in its domain of definition. Show that the composite function $g(z^2 + 1)$ is analytic in the quadrant $x > 0, y > 0$.

    Suggestion: Observe that $\mathrm{Im}(z^2 + 1) \neq 0$ when $x > 0, y > 0$.

6. Show that the function $g(z) = \mathrm{Log}\, r + i\theta\,(r > 0, -\pi/2 < \theta < \pi/2)$, where the natural logarithm is used, is analytic in the indicated domain of definition and show that $g'(z) = 1/z$ there. Then show why the composite function $g(2z - 2 + i)$ is an analytic function of z in the domain $x > 1$.

    Suggestion: Observe that $\mathrm{Re}(2z - 2 + i) > 0$ when $x > 1$.

7. State why a composition of two entire functions is entire. Also state why a *linear combination* $c_1 f_1(z) + c_2 f_2(z)$ of two entire functions, where c_1 and c_2 are complex constants, is entire.

8. Give an example illustrating that the compositions $g[f(z)]$ and $f[g(z)]$ of two entire functions are not, in general, the same.

9. Show that $u(x, y)$ is harmonic in some domain and find a harmonic conjugate $v(x, y)$ when

(a) $u(x, y) = 2x(1 - y)$; (b) $u(x, y) = 2x - x^3 + 3xy^2$;

(c) $u(x, y) = \sinh x \sin y$; (d) $u(x, y) = y/(x^2 + y^2)$.

    Ans. (a) $v(x, y) = x^2 - y^2 + 2y$; (c) $v(x, y) = -\cosh x \cos y$.

10. Show that if v is a harmonic conjugate of u in some domain and also u is a harmonic conjugate of v there, then u and v must be constant functions.

11. Let a function $f(z)$ be analytic in a domain D. Show that $f(z)$ is constant in D if

(a) the function $\overline{f(z)}$ is also analytic in D;

(b) $f(z)$ is real-valued for all z in D.

12. Suppose that in a domain D a function $f(z) = u(x, y) + iv(x, y)$ is analytic and its modulus $|f(z)|$ is constant. Show that $f(z)$ must also be constant in D.

    Suggestion: When $|f(z)| = c$, where $c \neq 0$, take appropriate derivatives of each side of the equation $u^2 + v^2 = c^2$ and refer to the Cauchy-Riemann equations to show that

$$u_x u - u_y v = 0 \qquad \text{and} \qquad u_y u + u_x v = 0.$$

Then, after noting that this linear homogeneous system in u and v must have a nontrivial solution, where not both u and v are identically zero, set the determinant of its coefficients equal to zero and show that $u_x = u_y = v_x = v_y = 0$.

13. Show that if v and V are harmonic conjugates of u in a domain D, then $v(x, y)$ and $V(x, y)$ can differ at most by an arbitrary additive constant.

14. Let the function $f(z) = u(r, \theta) + iv(r, \theta)$ be analytic in a domain D that does not include the origin. Using the Cauchy-Riemann equations in polar coordinates (Sec. 18) and assuming continuity of partial derivatives, show that throughout D the function $u(r, \theta)$ satisfies the partial differential equation

$$r^2 u_{rr}(r, \theta) + r u_r(r, \theta) + u_{\theta\theta}(r, \theta) = 0,$$

which is the *polar form of Laplace's equation*. Show that the same is true of the function $v(r, \theta)$.

15. Show that the function $u(r, \theta) = \mathrm{Log}\, r$, where the natural logarithm is used, is harmonic in the domain $r > 0, 0 < \theta < 2\pi$. (See Exercise 14.) Then find a harmonic conjugate $v(r, \theta)$.

                                                    Ans. $v(r, \theta) = \theta$.

16. Let the function $f(z) = u(x, y) + iv(x, y)$ be analytic in a domain D, and consider the families of *level curves* $u(x, y) = c_1$ and $v(x, y) = c_2$ where c_1 and c_2 are arbitrary real constants. Prove that these families are orthogonal. More precisely, show that if $z_0 = (x_0, y_0)$ is a point in D which is common to two particular curves $u(x, y) = c_1$ and $v(x, y) = c_2$ and if $f'(z_0) \neq 0$, then the lines tangent to those curves at (x_0, y_0) are perpendicular.

Suggestion: Note how it follows from the equation $u(x, y) = c_1$ that

$$u_x + u_y \frac{dy}{dx} = 0, \qquad \text{etc.}$$

17. Show that when $f(z) = z^2$, the level curves $u(x, y) = c_1$ and $v(x, y) = c_2$ of the component functions are those shown in Fig. 21. Note the orthogonality of the two families, as proved

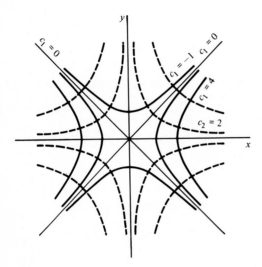

Figure 21

in Exercise 16. The curves $u(x, y) = 0$ and $v(x, y) = 0$ intersect at the origin but are not, however, orthogonal to each other. Why is this fact in agreement with the result of Exercise 16?

18. Sketch the families of level curves of the component functions u and v when $f(z) = 1/z$, and note the orthogonality proved in Exercise 16.

19. Do Exercise 18 using polar coordinates.

20. Sketch the families of level curves of the component functions u and v when

$$f(z) = \frac{z - 1}{z + 1},$$

and note how the results of Exercise 16 are illustrated here.

THREE

ELEMENTARY FUNCTIONS

We consider here various elementary functions studied in calculus and define corresponding functions of a complex variable. To be specific, we define analytic functions of a complex variable z which reduce to those elementary functions when $z = x + i0$. We start by defining the complex exponential function and then use it to develop the others.

21. The Exponential Function

If a function f of the complex variable $z = x + iy$ is to reduce to the familiar exponential function when z is real, we must require that

$$(1) \qquad\qquad f(x + i0) = e^x$$

for all real numbers x. Inasmuch as $d(e^x)/dx = e^x$ for all real x, it is also natural to impose the following conditions:

$$(2) \qquad\qquad f \text{ is entire} \quad \text{and} \quad f'(z) = f(z) \text{ for all } z.$$

As already pointed out in Example 1 of Sec. 17, the function

$$(3) \qquad\qquad f(z) = e^x(\cos y + i \sin y),$$

where y is to be taken in radians, is differentiable everywhere in the complex plane and $f'(z) = f(z)$. So conditions (1) and (2) are clearly satisfied by this function. It can be shown, moreover, that this is the only function satisfying conditions (1) and (2) (see Exercise 18, Sec. 22); and we write $f(z) = e^z$. Sometimes, for convenience, we use the notation $\exp z$ instead of e^z.

The exponential function of complex analysis is thus defined for all z by means of the equation

$$(4) \qquad\qquad e^z = e^x(\cos y + i \sin y).$$

As we have just seen, it reduces to the usual exponential function in calculus when $y = 0$, it is entire, and it satisfies the differentiation formula

(5) $$\frac{d}{dz}e^z = e^z$$

everywhere in the z plane.

Note that since it is customary to assign the value $\sqrt[n]{e}$, or the *positive* nth root of e, to e^x when $x = 1/n$ $(n = 2, 3, \ldots)$, the value of the complex exponential function e^z is also $\sqrt[n]{e}$ when $z = 1/n$ $(n = 2, 3, \ldots)$. This is an exception to the convention (Sec. 7) which would ordinarily require us to interpret $e^{1/n}$ as the set of all nth roots of e.

Note, too, that when z is the pure imaginary number $i\theta$, equation (4) becomes

$$e^{i\theta} = \cos\theta + i\sin\theta.$$

This is Euler's formula, which was introduced in Sec. 6. The definition of the symbol $e^{i\theta}$ given there thus agrees with definition (4).

The more compact form

(6) $$e^z = e^x e^{iy}$$

of equation (4) makes it especially easy to verify the additive property

(7) $$(\exp z_1)(\exp z_2) = \exp(z_1 + z_2).$$

To do this, we write $z_1 = x_1 + iy_1$ and $z_2 = x_2 + iy_2$. Then, in view of expression (6),

$$(\exp z_1)(\exp z_2) = (e^{x_1}e^{iy_1})(e^{x_2}e^{iy_2}) = (e^{x_1}e^{x_2})(e^{iy_1}e^{iy_2}).$$

But x_1 and x_2 are both real, and we also know that (Sec. 6)

$$e^{iy_1}e^{iy_2} = e^{i(y_1+y_2)}.$$

Hence

$$(\exp z_1)(\exp z_2) = e^{(x_1+x_2)}e^{i(y_1+y_2)};$$

and, since $(x_1 + x_2) + i(y_1 + y_2) = z_1 + z_2$, the right-hand side of this last equation becomes $\exp(z_1 + z_2)$ in accordance with expression (6). Property (7) is thus established.

Observe that since $e^{z+2\pi i} = e^z e^{2\pi i}$ and $e^{2\pi i} = 1$, *the exponential function is periodic with a pure imaginary period of* $2\pi i$:

(8) $$\exp(z + 2\pi i) = \exp z \qquad \text{for all } z.$$

It is left as an exercise to show that

(9) $$\frac{\exp z_1}{\exp z_2} = \exp(z_1 - z_2).$$

From this and the fact that $e^0 = 1$, it follows that $1/e^z = e^{-z}$. Another useful identity is

(10) $$(\exp z)^n = \exp(nz) \qquad (n = 1, 2, \ldots).$$

22. Other Properties of exp z

Let us write the expression $e^z = e^x e^{iy}$ in Sec. 21 as

(1) $\qquad\qquad e^z = \rho e^{i\phi} \qquad$ where $\qquad \rho = e^x$ and $\phi = y$.

The number $\rho = e^x$ is evidently positive for each value of x; and, according to statement (1), the modulus of e^z is e^x and an argument of e^z is y. That is,

(2) $\qquad |e^z| = e^x \qquad$ and $\qquad \arg(e^z) = y + 2n\pi \qquad (n = 0, \pm1, \pm2, \ldots).$

Note that since $|e^z|$ is always positive,

(3) $\qquad\qquad\qquad\qquad e^z \neq 0 \qquad\qquad\qquad$ for any complex number z.

A rephrasing of statement (1) is that when $w = e^z$, the number w can be written $w = \rho e^{i\phi}$ where

(4) $\qquad\qquad\qquad\qquad \rho = e^x \qquad$ and $\qquad \phi = y.$

Observe that to each positive value of ρ in the first of equations (4) there corresponds a value of x for which the equation is satisfied. That value is, of course, $x = \text{Log } \rho$, where the natural logarithm is used. Independently, for each angle ϕ there is a value of y, namely $y = \phi$, such that the second of equations (4) is satisfied. Thinking of the equation $w = e^z$ as a transformation from the z to the w plane, we thus find that any given *nonzero* point $w = \rho e^{i\phi}$ is the image of the point

(5) $\qquad\qquad\qquad\qquad z = \text{Log } \rho + i\phi.$

In view of statement (3), the point $w = 0$ cannot be the image of any point z under the transformation $w = e^z$. We may conclude, therefore, that *the range of the exponential function $w = e^z$ is the entire w plane except for the origin $w = 0$.*

Any nonzero point $w = \rho e^{i\phi}$ is actually the image of an infinite number of points in the z plane under the transformation $w = e^z$. For, in general, ϕ may have any one of the values $\phi = \Phi + 2n\pi$ $(n = 0, \pm1, \pm2, \ldots)$, where Φ denotes the principal value $(-\pi < \Phi \leq \pi)$ of $\arg w$. It then follows from equation (5) that w is the image of all the points

(6) $\qquad\qquad z = \text{Log } \rho + i(\Phi + 2n\pi) \qquad (n = 0, \pm1, \pm2, \ldots).$

Example. Values of z must exist, for instance, such that $e^z = -1$. To find them without relying on a formula, we write $e^z = e^x e^{iy}$ and $-1 = 1e^{i\pi}$, so that

$$e^x e^{iy} = 1e^{i\pi}.$$

According to the statement in italics in Sec. 6 regarding the equality of two complex numbers which are expressed in exponential form, this means that $e^x = 1$ and $y = \pi + 2n\pi$, where n is an integer. Since $x = \text{Log } 1 = 0$, then,

$$z = (2n + 1)\pi i \qquad\qquad (n = 0, \pm1, \pm2, \ldots).$$

Formula (6) could have been used, of course, with $\rho = 1$ and $\Phi = \pi$.

Finally, we note from equation (6) that under the transformation $w = e^z$, a nonzero point $w = \rho e^{i\phi}$ is, in particular, the image of the point

(7) $$z = \text{Log } \rho + i\Phi.$$

Since $-\pi < \Phi \leq \pi$, this point z lies in the horizontal strip $-\pi < y \leq \pi$; and the other points whose images are w lie outside that strip, equally spaced 2π units apart above and below z along the vertical line $x = \text{Log } \rho$. Such mapping properties of the exponential function are elaborated on in Chap. 7 (Sec. 70).

EXERCISES

1. Show that

(a) $\exp(2 \pm 3\pi i) = -e^2$; (b) $\exp\dfrac{2 + \pi i}{4} = \sqrt{\dfrac{e}{2}}(1 + i)$;

(c) $\exp(z + \pi i) = -\exp z$.

2. State why the function $2z^2 - 3 - ze^z + e^{-z}$ is entire.

3. Find all values of z such that

(a) $e^z = -2$; (b) $e^z = 1 + \sqrt{3}i$; (c) $\exp(2z - 1) = 1$.

$\quad$ Ans. (a) $z = \text{Log } 2 + (2n + 1)\pi i$ $(n = 0, \pm 1, \pm 2, \ldots)$;

$\quad\quad\quad$ (c) $z = \frac{1}{2} + n\pi i$ $(n = 0, \pm 1, \pm 2, \ldots)$.

4. Write $|\exp(2z + i)|$ and $|\exp(iz^2)|$ in terms of x and y. Then show that

$$|\exp(2z + i) + \exp(iz^2)| \leq e^{2x} + e^{-2xy}.$$

5. Show that $|\exp(z^2)| \leq \exp(|z|^2)$.

6. Prove that $|\exp(-2z)| < 1$ if and only if Re $z > 0$.

7. Let z be any nonzero complex number. Show that if $z = re^{i\theta}$, then

$$\exp(\text{Log } r + i\theta) = z.$$

8. Verify property (9), Sec. 21, of the exponential function.

9. Use mathematical induction to establish property (10), Sec. 21, of the exponential function.

10. Using the fact that property (10), Sec. 21, is valid when $n = 1, 2, \ldots$, prove that it also applies when $n = -1, -2, \ldots$.

$\quad$ *Suggestion*: Note that in view of the definition (Sec. 7) $z^n = (z^{-1})^{-n}$ of z^n when $z \neq 0$ and $n = -1, -2, \ldots$, one can write $(\exp z)^n = (1/\exp z)^m$ where $m = -n = 1, 2, \ldots$.

11. Show that

(a) $\exp \bar{z} = \overline{\exp z}$ for all z;

(b) $\exp(i\bar{z}) = \overline{\exp(iz)}$ if and only if $z = n\pi$, where $n = 0, \pm 1, \pm 2, \ldots$.

12. (a) Show that if e^z is real, then Im $z = n\pi$ $(n = 0, \pm 1, \pm 2, \ldots)$.

(b) If e^z is pure imaginary, what restriction is placed on z?

13. Describe the behavior of (a) $\exp(x + iy)$ as x tends to $-\infty$; (b) $\exp(2 + iy)$ as y tends to ∞.

14. Prove that the function $\exp \bar{z}$ is not analytic anywhere.

15. Show in two ways that the function $\exp(z^2)$ is entire. What is its derivative?

Ans. $2z \exp(z^2)$.

16. Write $\text{Re}(e^{1/z})$ in terms of x and y. Why is this function harmonic in every domain that does not contain the origin?

17. Let the function $f(z) = u(x, y) + iv(x, y)$ be analytic in some domain D. State why the functions

$$U(x, y) = e^{u(x, y)} \cos v(x, y), \qquad V(x, y) = e^{u(x, y)} \sin v(x, y)$$

are harmonic in D and why $V(x, y)$ is, in fact, a harmonic conjugate of $U(x, y)$.

18. Suppose that a function $f(z) = u(x, y) + iv(x, y)$ satisfies conditions (1) and (2) in Sec. 21. Follow the steps described below to show that $f(z)$ must be the function $f(z) = e^x(\cos y + i \sin y)$.

    (a) Obtain the equations $u_x = u$, $v_x = v$ and then show that there exist real-valued functions ϕ and ψ of the real variable y such that $u(x, y) = e^x \phi(y)$ and $v(x, y) = e^x \psi(y)$.

    (b) Use the fact that u is harmonic (Sec. 20) to obtain the differential equation $\phi''(y) + \phi(y) = 0$ and thus show that $\phi(y) = A \cos y + B \sin y$, where A and B are real numbers.

    (c) After pointing out why $\psi(y) = A \sin y - B \cos y$, use the fact that $u(x, 0) + iv(x, 0) = e^x$ to find A and B. Conclude that $u(x, y) = e^x \cos y$ and $v(x, y) = e^x \sin y$.

23. Trigonometric Functions

From the equations

$$e^{ix} = \cos x + i \sin x, \qquad e^{-ix} = \cos x - i \sin x$$

it follows that

$$\sin x = \frac{e^{ix} - e^{-ix}}{2i}, \qquad \cos x = \frac{e^{ix} + e^{-ix}}{2}$$

for every real number x. It is therefore natural to *define* the sine and cosine functions of a complex variable z by means of the equations

(1)
$$\sin z = \frac{e^{iz} - e^{-iz}}{2i}, \qquad \cos z = \frac{e^{iz} + e^{-iz}}{2}.$$

    The sine and cosine functions are entire since they are linear combinations (Exercise 7, Sec. 20) of the entire functions e^{iz} and e^{-iz}. Knowing the derivatives of those exponential functions, we find from equations (1) that

(2)
$$\frac{d}{dz} \sin z = \cos z, \qquad \frac{d}{dz} \cos z = -\sin z.$$

The expected trigonometric identities are valid with complex variables:

(3) $$\sin(-z) = -\sin z, \qquad \cos(-z) = \cos z,$$

(4) $$\sin^2 z + \cos^2 z = 1,$$

(5) $$\sin(z_1 + z_2) = \sin z_1 \cos z_2 + \cos z_1 \sin z_2,$$

(6) $$\cos(z_1 + z_2) = \cos z_1 \cos z_2 - \sin z_1 \sin z_2,$$

(7) $$\sin 2z = 2 \sin z \cos z, \qquad \cos 2z = \cos^2 z - \sin^2 z,$$

(8) $$\sin\left(z + \frac{\pi}{2}\right) = \cos z,$$

etc. Proofs may be based entirely on properties of the exponential function. According to the definition of $\sin z$,

$$\sin z = \frac{e^{i(x+iy)} - e^{-i(x+iy)}}{2i} = (\cos x + i \sin x)\frac{e^{-y}}{2i} - (\cos x - i \sin x)\frac{e^y}{2i}$$

$$= \sin x\left(\frac{e^y + e^{-y}}{2}\right) + i \cos x\left(\frac{e^y - e^{-y}}{2}\right),$$

where $z = x + iy$. Thus the real and imaginary parts of $\sin z$ are displayed as follows:

(9) $$\sin z = \sin x \cosh y + i \cos x \sinh y.$$

In the same way, or by using the first of formulas (2), one can show that

(10) $$\cos z = \cos x \cosh y - i \sin x \sinh y.$$

A number of important properties of $\sin z$ and $\cos z$ follow immediately from these last two expressions. For example,

(11) $$\sin(iy) = i \sinh y, \qquad \cos(iy) = \cosh y;$$

and $\sin \bar{z}$ and $\cos \bar{z}$ are the complex conjugates of $\sin z$ and $\cos z$, respectively. The periodic character of $\sin z$ and $\cos z$ is also evident:

(12) $$\sin(z + 2\pi) = \sin z, \qquad \sin(z + \pi) = -\sin z,$$

(13) $$\cos(z + 2\pi) = \cos z, \qquad \cos(z + \pi) = -\cos z.$$

With the aid of expressions (9) and (10), the reader can show that

(14) $$|\sin z|^2 = \sin^2 x + \sinh^2 y,$$

(15) $$|\cos z|^2 = \cos^2 x + \sinh^2 y.$$

It is clear from these two equations that $\sin z$ and $\cos z$ are not bounded in absolute value, whereas the absolute values of $\sin x$ and $\cos x$ are less than or equal to unity for all real values of x.

A specific value of z for which $f(z) = 0$ is called a *zero* of a given function f. The zeros of the sine and cosine functions are all real. In fact,

(16) $$\sin z = 0 \qquad \text{if and only if} \qquad z = n\pi \qquad (n = 0, \pm1, \pm2, \dots)$$

and

(17) $\cos z = 0$ if and only if $z = \left(n + \dfrac{1}{2}\right)\pi$ $(n = 0, \pm 1, \pm 2, \ldots)$.

Let us verify statement (16). The fact that $\sin z = 0$ if $z = n\pi$ is obvious since $\sin z$ becomes the usual sine function in calculus when z is real. To show that there are no other zeros of $\sin z$, we assume that $\sin z = 0$ and note how it follows from equation (14) that

$$\sin^2 x + \sinh^2 y = 0.$$

Thus

$$\sin x = 0 \quad \text{and} \quad \sinh y = 0.$$

Evidently, then, $x = n\pi$ and $y = 0$; that is, $z = n\pi$. Statement (17) can be verified in a similar way.

   The other four trigonometric functions are defined in terms of the sine and cosine functions by the usual relations:

(18)
$$\tan z = \frac{\sin z}{\cos z}, \qquad \cot z = \frac{\cos z}{\sin z},$$

$$\sec z = \frac{1}{\cos z}, \qquad \csc z = \frac{1}{\sin z}.$$

Observe that $\tan z$ and $\sec z$ are analytic everywhere except at the singularities (Sec. 19) $z = (n + \frac{1}{2})\pi$ $(n = 0, \pm 1, \pm 2, \ldots)$, which are the zeros of $\cos z$. A similar statement can be made regarding $\cot z$ and $\csc z$. By differentiating the right-hand sides of equations (18), we obtain the differentiation formulas

(19)
$$\frac{d}{dz}\tan z = \sec^2 z, \qquad \frac{d}{dz}\cot z = -\csc^2 z,$$

$$\frac{d}{dz}\sec z = \sec z \tan z, \qquad \frac{d}{dz}\csc z = -\csc z \cot z.$$

The periodicity of each of the trigonometric functions defined by equations (18) follows readily from equations (12) and (13). For example,

(20) $\tan(z + \pi) = \tan z.$

   Mapping properties of the transformation $w = \sin z$ are especially important in the applications later on. A reader who wishes at this time to learn some of those properties is sufficiently prepared to read Sec. 71 (Chap. 7), where they are discussed.

EXERCISES

1. Show that $e^{iz} = \cos z + i \sin z$ for every complex number z.
2. Establish trigonometric identities (4) and (5) in Sec. 23.
3. Verify identity (8), Sec. 23.

4. Derive expression (10), Sec. 23, for cos z.

5. Prove that   (*a*) $1 + \tan^2 z = \sec^2 z$;   (*b*) $1 + \cot^2 z = \csc^2 z$.

6. Establish differentiation formulas (19), Sec. 23.

7. Verify equation (14), Sec. 23, and then use it to show that

$$|\sin z| \geq |\sin x|.$$

8. Verify equation (15), Sec. 23, and then use it to show that

$$|\cos z| \geq |\cos x|.$$

9. With the aid of equations (14) and (15) in Sec. 23, show that

(*a*) $|\sinh y| \leq |\sin z| \leq \cosh y$;

(*b*) $|\sinh y| \leq |\cos z| \leq \cosh y$.

10. Prove statement (17), Sec. 23, regarding the zeros of cos z.

11. Establish the identities

(*a*) $2 \sin(z_1 + z_2) \sin(z_1 - z_2) = \cos 2z_2 - \cos 2z_1$;

(*b*) $2 \cos(z_1 + z_2) \sin(z_1 - z_2) = \sin 2z_1 - \sin 2z_2$.

12. With the aid of the identity in Exercise 11(*a*), show that if $\cos z_1 = \cos z_2$, then either $z_1 + z_2$ or $z_1 - z_2$ is an integral multiple of 2π. Obtain an analogous result when $\sin z_1 = \sin z_2$.

13. Show that (*a*) $\cos(i\bar{z}) = \overline{\cos(iz)}$ for all z; (*b*) $\sin(i\bar{z}) = \overline{\sin(iz)}$ if and only if $z = n\pi i$ ($n = 0, \pm1, \pm2, \ldots$).

14. Find all the roots of the equation $\sin z = \cosh 4$ by equating the real parts and the imaginary parts of $\sin z$ and $\cosh 4$.

$$\text{Ans.} \quad (2n + \tfrac{1}{2})\pi \pm 4i \ (n = 0, \pm1, \pm2, \ldots).$$

15. Find all the roots of the equation $\cos z = 2$.

$$\text{Ans.} \quad 2n\pi + i \cosh^{-1} 2; \text{ that is, } 2n\pi \pm i \operatorname{Log}(2 + \sqrt{3}) \ (n = 0, \pm1, \pm2, \ldots).$$

16. Show in two ways that each of the following functions is everywhere harmonic:

(*a*) $\sin x \sinh y$;   (*b*) $\cos 2x \sinh 2y$.

17. Let the function $f(z)$ be analytic in a domain D. State why the functions $\sin f(z)$ and $\cos f(z)$ are analytic there. Also, write $w = f(z)$ and state why

$$\frac{d}{dz} \sin w = \cos w \frac{dw}{dz}, \qquad \frac{d}{dz} \cos w = -\sin w \frac{dw}{dz}.$$

18. Show that neither $\sin \bar{z}$ nor $\cos \bar{z}$ is an analytic function of z anywhere.

24. Hyperbolic Functions

The hyperbolic sine and the hyperbolic cosine of a complex variable are defined as they are with a real variable; that is,

$$(1) \qquad \sinh z = \frac{e^z - e^{-z}}{2}, \qquad \cosh z = \frac{e^z + e^{-z}}{2}.$$

Since e^z and e^{-z} are entire, it follows from definitions (1) that $\sinh z$ and $\cosh z$ are entire. Furthermore,

(2) $$\frac{d}{dz} \sinh z = \cosh z, \qquad \frac{d}{dz} \cosh z = \sinh z.$$

Because of the way the exponential function appears in definitions (1) and in the definitions (Sec. 23)

$$\sin z = \frac{e^{iz} - e^{-iz}}{2i}, \qquad \cos z = \frac{e^{iz} + e^{-iz}}{2}$$

of $\sin z$ and $\cos z$, the hyperbolic sine and cosine functions are closely related to those trigonometric functions:

(3) $$-i \sinh(iz) = \sin z, \qquad -i \sin(iz) = \sinh z,$$

(4) $$\cosh(iz) = \cos z, \qquad \cos(iz) = \cosh z.$$

Various identities involving hyperbolic sines and cosines, which follow directly from definitions (1), are often more easily obtained from known properties of $\sin z$ and $\cos z$ with the aid of relations (3) and (4). Some of the most frequently used of those identities are

(5) $$\sinh(-z) = -\sinh z, \qquad \cosh(-z) = \cosh z,$$

(6) $$\cosh^2 z - \sinh^2 z = 1,$$

(7) $$\sinh(z_1 + z_2) = \sinh z_1 \cosh z_2 + \cosh z_1 \sinh z_2,$$

(8) $$\cosh(z_1 + z_2) = \cosh z_1 \cosh z_2 + \sinh z_1 \sinh z_2,$$

and

(9) $$\sinh z = \sinh x \cos y + i \cosh x \sin y,$$

(10) $$\cosh z = \cosh x \cos y + i \sinh x \sin y,$$

(11) $$|\sinh z|^2 = \sinh^2 x + \sin^2 y,$$

(12) $$|\cosh z|^2 = \sinh^2 x + \cos^2 y,$$

where $z = x + iy$.

Example. To illustrate the method of proof suggested, let us verify identity (11). According to the second of relations (3), $|\sinh z|^2 = |\sin(iz)|^2$. That is,

(13) $$|\sinh z|^2 = |\sin(-y + ix)|^2$$

where $z = x + iy$. But from equation (14), Sec. 23, we know that

$$|\sin(x + iy)|^2 = \sin^2 x + \sinh^2 y;$$

and this enables us to write equation (13) in the desired form (11).

In view of the periodicity of $\sin z$ and $\cos z$, it follows immediately from relations (3) and (4) that $\sinh z$ and $\cosh z$ are periodic with period $2\pi i$. Also, the zeros of sinh z are the numbers $z = n\pi i$ $(n = 0, \pm 1, \pm 2, \ldots)$ and the zeros of cosh z are $z = (n + \frac{1}{2})\pi i$ $(n = 0, \pm 1, \pm 2, \ldots)$.

The hyperbolic tangent of z is defined by the equation

$$(14) \qquad\qquad \tanh z = \frac{\sinh z}{\cosh z}$$

and is analytic in every domain in which $\cosh z \neq 0$. The functions coth z, sech z, and csch z are the reciprocals of tanh z, cosh z, and sinh z, respectively. It is straightforward to verify the following differentiation formulas, which are the same as those established in calculus for the corresponding functions of a real variable:

$$(15) \qquad \frac{d}{dz}\tanh z = \operatorname{sech}^2 z, \qquad\qquad \frac{d}{dz}\coth z = -\operatorname{csch}^2 z,$$

$$(16) \qquad \frac{d}{dz}\operatorname{sech} z = -\operatorname{sech} z \tanh z, \qquad \frac{d}{dz}\operatorname{csch} z = -\operatorname{csch} z \coth z.$$

EXERCISES

1. Derive differentiation formulas (2), Sec. 24.

2. Prove that $\sinh 2z = 2 \sinh z \cosh z$ by (a) starting with definitions (1), Sec. 24, of sinh z and cosh z; (b) appealing to the identity $\sin 2z = 2 \sin z \cos z$ (Sec. 23) and using certain of relations (3) and (4) in Sec. 24.

3. Show how identities (6) and (8) in Sec. 24 follow from identities (4) and (6), respectively, in Sec. 23.

4. Write $\sinh z = \sinh(x + iy)$ and $\cosh z = \cosh(x + iy)$, and show how expressions (9) and (10) in Sec. 24 follow from identities (7) and (8), respectively, in that section.

5. Show that $\sinh(z + \pi i) = -\sinh z$ and $\cosh(z + \pi i) = -\cosh z$. Then show that
$$\tanh (z + \pi i) = \tanh z.$$

6. Verify expression (12), Sec. 24, for $|\cosh z|^2$.

7. Show that $|\sinh x| \leq |\cosh z| \leq \cosh x$ by using (a) identity (12), Sec. 24; (b) the inequalities in Exercise 9(b), Sec. 23.

8. Prove that the zeros of sinh z and cosh z are as stated in Sec. 24.

9. Using the results proved in Exercise 8, locate all the zeros and singularities of the hyperbolic tangent function.

10. Derive differentiation formulas (15), Sec. 24.

11. Find all the roots of the equations
    (a) $\cosh z = 1/2$; (b) $\sinh z = i$; (c) $\cosh z = -2$.
    $\qquad\qquad$ *Ans.* (a) $(2n \pm \frac{1}{3})\pi i$ $(n = 0, \pm 1, \pm 2, \ldots)$;
    $\qquad\qquad\qquad$ (b) $(2n + \frac{1}{2})\pi i$ $(n = 0, \pm 1, \pm 2, \ldots)$.

12. Why is the function $\sinh(e^z)$ entire? Write its real part as a function of x and y and state why that function must be harmonic everywhere.

25. The Logarithmic Function and Its Branches

Let Log r denote the natural logarithm of a positive real number r, as used in calculus.* We now define the logarithmic function of a *nonzero* complex variable $z = re^{i\theta}$ by means of the equation

$$\text{(1)} \qquad\qquad \log z = \text{Log } r + i\theta.$$

That is,

$$\text{(2)} \qquad\qquad \log z = \text{Log } |z| + i \arg z \qquad\qquad (z \neq 0).$$

Definition (1) is suggested by using familiar properties of the natural logarithm of a positive real number to formally expand $\log(re^{i\theta})$.

The value of θ used in the expression $z = re^{i\theta}$ for a specific nonzero number z is, of course, not unique; if Θ denotes the principal value $(-\pi < \Theta \leq \pi)$ of θ, we can write

$$\theta = \Theta + 2n\pi \qquad\qquad (n = 0, \pm 1, \pm 2, \ldots).$$

Definition (1) then takes the form

$$\text{(3)} \qquad\qquad \log z = \text{Log } r + i(\Theta + 2n\pi) \qquad (n = 0, \pm 1, \pm 2, \ldots),$$

and it is clear that the function $\log z$ is multiple-valued with infinitely many values. Those values all have the same real part, and their imaginary parts differ by integral multiples of 2π.

Examples. Note that

$$\log 1 = 2n\pi i \qquad\qquad (n = 0, \pm 1, \pm 2, \ldots)$$

and

$$\log(-1) = (2n + 1)\pi i \qquad\qquad (n = 0, \pm 1, \pm 2, \ldots).$$

The *principal value* of $\log z$ is the value obtained from equation (3) when $n = 0$ and is denoted by Log z. Thus

$$\text{(4)} \qquad\qquad \text{Log } z = \text{Log } r + i\Theta,$$

or

$$\text{(5)} \qquad\qquad \text{Log } z = \text{Log } |z| + i \text{ Arg } z \qquad\qquad (z \neq 0).$$

This logarithmic function is a *single-valued* function, still defined on the set of all nonzero complex numbers. It reduces to the usual natural logarithm in calculus when z is a positive real number $z = r$. To see this, one need only write $z = re^{i0}$, in which case equation (4) becomes Log $z = $ Log r.

The single-valued function Log z, whose component functions are

---

*The notation Ln r, rather than Log r, is also frequently used.

(6) $u(r, \Theta) = \text{Log } r$ and $v(r, \Theta) = \Theta$,

is *not* continuous, and therefore not analytic, throughout its domain of definition $r > 0, -\pi < \Theta \leq \pi$. For the value of the function v at any point on the negative real axis is π, whereas just below and arbitrarily close to that point the values of v are near $-\pi$. The component functions u and v of Log z are, however, each continuous throughout the domain $r > 0, -\pi < \Theta < \pi$, consisting of all points in the plane except $z = 0$ and points on the negative real axis. The partial derivatives

$$u_r = \frac{1}{r}, \qquad u_\Theta = 0, \qquad v_r = 0, \qquad v_\Theta = 1$$

are, in fact, continuous everywhere in that domain, and they satisfy the polar form (Sec. 18)

$$u_r = \frac{1}{r} v_\Theta, \qquad \frac{1}{r} u_\Theta = -v_r$$

of the Cauchy-Riemann equations there. We thus conclude from the theorem in Sec. 18 that *the function* Log z *is analytic in the domain* $r > 0, -\pi < \Theta < \pi$. Moreover, if $z = re^{i\Theta}$,

$$\frac{d}{dz} \text{Log } z = e^{-i\Theta}(u_r + iv_r) = e^{-i\Theta}\left(\frac{1}{r} + i0\right) = \frac{1}{re^{i\Theta}};$$

that is,

(7) $$\frac{d}{dz} \text{Log } z = \frac{1}{z} \qquad\qquad (|z| > 0, -\pi < \text{Arg } z < \pi).$$

More generally, if we restrict the value of θ in definition (1) so that $\alpha < \theta < \alpha + 2\pi$ for a fixed but arbitrary α, then the function

(8) $$\log z = \text{Log } r + i\theta \qquad (r > 0, \alpha < \theta < \alpha + 2\pi)$$

is single-valued and continuous everywhere in the indicated domain. Throughout that domain the component functions possess continuous first-order partial derivatives with respect to r and θ which satisfy the polar form of the Cauchy-Riemann equations. Hence the function defined by equation (8) is analytic throughout its domain of definition; furthermore,

(9) $$\frac{d}{dz} \log z = \frac{1}{z} \qquad\qquad (|z| > 0, \alpha < \text{arg } z < \alpha + 2\pi).$$

A *branch* of a multiple-valued function f is any single-valued function F which is analytic in some domain at each point z of which the value $F(z)$ is one of the values $f(z)$. The requirement of analyticity, of course, prevents F from taking on a random selection of the values of f. Observe that the function Log z, when restricted to the domain $r > 0, -\pi < \Theta < \pi$, is a branch of the logarithmic function (1). This branch is called the *principal branch*. For each fixed α the function defined by equation (8) is a branch of the same multiple-valued function.

Each point on the negative real axis $\Theta = \pi$, as well as the origin, is a singular point of the principal branch Log z, according to the definition (Sec. 19) of a singular point. The ray $\Theta = \pi$ is the *branch cut* for the principal branch, a branch cut being a portion of a line or a curve, consisting of singular points, which is introduced in order to define a branch of a multiple-valued function. The ray $\theta = \alpha$ is the branch cut for the branch (8) of the logarithmic function. A singular point common to all branch cuts for a multiple-valued function is called a *branch point*. The origin is the branch point for the logarithmic function.

26. Further Properties of Logarithms

Various identities involving natural logarithms of positive real numbers carry over to complex analysis, with certain modifications.

First of all,

$$(1) \qquad\qquad \exp(\log z) = z \qquad\qquad (z \neq 0),$$

regardless of what value of log z we select after z is specified. To see this, write $z = re^{i\theta}$ and log $z = \text{Log } r + i\theta$, where θ is any particular value of arg z. Then

$$\exp(\log z) = \exp(\text{Log } r + i\theta) = e^{\text{Log } r}e^{i\theta} = re^{i\theta} = z.$$

It is *not* true, however, that $\log(e^z)$ is always equal to z. This is evident from the fact that $\log(e^z)$ has an infinite number of values for any given z. Specifically, since (Sec. 22)

$$|e^z| = e^x \qquad \text{and} \qquad \arg(e^z) = y + 2n\pi \qquad (n = 0, \pm1, \pm2, \ldots)$$

when $z = x + iy$, we see that

$$\log(e^z) = \text{Log } |e^z| + i \arg(e^z) = x + i(y + 2n\pi),$$

or

$$\log(e^z) = z + 2n\pi i \qquad\qquad (n = 0, \pm1, \pm2, \ldots).$$

But suppose that the number $z = x + iy$ is confined to the horizontal strip $-\pi < y \leq \pi$ and that principal values of the logarithmic function are taken. Since $\text{Arg}(e^z) = y$, the above expression for $\log(e^z)$ becomes

$$(2) \qquad\qquad \text{Log}(e^z) = z \qquad\qquad (-\pi < \text{Im } z \leq \pi).$$

For another property of logarithms, we let z_1 and z_2 be any two nonzero complex numbers. It is straightforward to show that

$$(3) \qquad\qquad \log(z_1 z_2) = \log z_1 + \log z_2.$$

This statement involving a multiple-valued function is to be interpreted in the same way that the statement

$$(4) \qquad\qquad \arg(z_1 z_2) = \arg z_1 + \arg z_2$$

was in Sec. 5. That is, any value of log z_1 plus any value of log z_2 is a value of $\log(z_1 z_2)$;

and any value of $\log(z_1 z_2)$ can be expressed as any value of $\log z_1$ plus some value of $\log z_2$.

The proof of statement (3) can be based on statement (4) in the following way. Since $|z_1 z_2| = |z_1||z_2|$ and since these moduli are all positive real numbers, we know from experience with logarithms of such numbers in calculus that

$$\text{Log } |z_1 z_2| = \text{Log } |z_1| + \text{Log } |z_2|.$$

So it follows from this and equation (4) that

(5) $\quad \text{Log } |z_1 z_2| + i \arg(z_1 z_2) = (\text{Log } |z_1| + i \arg z_1) + (\text{Log } |z_2| + i \arg z_2).$

Finally, because of the way equations (3) and (4) are to be interpreted, equation (5) is the same as equation (3).

In like manner, it can be shown that

(6) $$\log\left(\frac{z_1}{z_2}\right) = \log z_1 - \log z_2.$$

Example. To illustrate statement (3), write $z_1 = z_2 = -1$ and note that $z_1 z_2 = 1$. The statement is clearly valid when $\log(z_1 z_2) = 0$, $\log z_1 = \pi i$, and $\log z_2 = -\pi i$. Observe that for the same numbers z_1 and z_2,

$$\text{Log}(z_1 z_2) = 0 \quad \text{and} \quad \text{Log } z_1 + \text{Log } z_2 = 2\pi i.$$

Thus statement (3) is not, in general, valid when *log* is replaced everywhere by *Log*. A similar remark applies to statement (6).

We include here two other properties of $\log z$ which will be of special interest in Sec. 27. If z is a nonzero complex number, then

(7) $$z^n = \exp(n \log z) \qquad (n = 0, \pm 1, \pm 2, \ldots)$$

for any value of $\log z$ that is taken. This identity is, of course, a generalization of identity (1), and the proof is similar.

It is also true that when $z \neq 0$,

(8) $$z^{1/n} = \exp\left(\frac{1}{n} \log z\right) \qquad (n = 1, 2, \ldots).$$

That is, the term on the right here has n distinct values and those values are the nth roots of z. To prove this, write $z = r \exp(i\Theta)$, where Θ is the principal value of $\arg z$. Then, in view of expression (3), Sec. 25, for $\log z$,

$$\exp\left(\frac{1}{n} \log z\right) = \exp\left[\frac{1}{n} \text{Log } r + \frac{i(\Theta + 2k\pi)}{n}\right],$$

where $k = 0, \pm 1, \pm 2, \ldots$. That is,

(9) $$\exp\left(\frac{1}{n} \log z\right) = \sqrt[n]{r} \exp\left[i\left(\frac{\Theta}{n} + \frac{2k\pi}{n}\right)\right] \qquad (k = 0, \pm 1, \pm 2, \ldots).$$

Because $\exp(i2k\pi/n)$ has distinct values only when $k = 0, 1, \ldots, n - 1$, the right-hand side of equation (9) has only n values. That right-hand side is, in fact, an expression for the nth roots of z (Sec. 7), and so it can be written $z^{1/n}$. This establishes property (8), which is actually valid when n is a negative integer too (see Exercise 18).

EXERCISES

1. Show that

   (a) $\text{Log}(-ei) = 1 - \dfrac{\pi}{2}i$;   (b) $\text{Log}(1 - i) = \dfrac{1}{2}\text{Log }2 - \dfrac{\pi}{4}i$.

2. Show that when $n = 0, \pm 1, \pm 2, \ldots$,

   (a) $\log e = 1 + 2n\pi i$;   (b) $\log i = (2n + \frac{1}{2})\pi i$;

   (c) $\log(-1 + \sqrt{3}\,i) = \text{Log }2 + 2(n + \frac{1}{3})\pi i$.

3. Show that $\text{Log}[(1 + i)^2] = 2\,\text{Log}(1 + i)$ but $\text{Log}[(-1 + i)^2] \neq 2\,\text{Log}(-1 + i)$.

4. Show that

   (a) if $\log z = \text{Log }r + i\theta$ $(r > 0, \pi/4 < \theta < 9\pi/4)$, then $\log(i^2) = 2\log i$;

   (b) if $\log z = \text{Log }r + i\theta$ $(r > 0, 3\pi/4 < \theta < 11\pi/4)$, then $\log(i^2) \neq 2\log i$.

5. Show that (a) the set of values of $\log(i^{1/2})$ is $(n + \frac{1}{4})\pi i$ $(n = 0, \pm 1, \pm 2, \ldots)$ and that the same is true of $(1/2)\log i$; (b) the set of values of $\log(i^2)$ is *not* the same as the set of values of $2\log i$.

6. Find all the roots of the equation $\log z = (\pi/2)i$.   *Ans.* $z = i$.

7. Given that the branch (8), Sec. 25, of $\log z$ is analytic at each point z in the stated domain, show that its derivative at such a point is $1/z$. Do this by differentiating each side of identity (1), Sec. 26, with respect to z.

8. Suppose that the point $z = x + iy$ lies in the horizontal strip $\alpha < y < \alpha + 2\pi$. Show that when the branch

   $$\log z = \text{Log }r + i\theta \qquad\qquad (r > 0, \alpha < \theta < \alpha + 2\pi)$$

   of the logarithmic function is used, $\log(e^z) = z$.

9. Show that if $\text{Re }z_1 > 0$ and $\text{Re }z_2 > 0$, then

   $$\text{Log}(z_1 z_2) = \text{Log }z_1 + \text{Log }z_2.$$

10. Show that for any two nonzero complex numbers z_1 and z_2,

    $$\text{Log}(z_1 z_2) = \text{Log }z_1 + \text{Log }z_2 + 2N\pi i$$

    where N has one of the values $0, \pm 1$.

11. Using the fact that $\arg(z_1/z_2) = \arg z_1 - \arg z_2$ (Sec. 5), verify expression (6), Sec. 26, for $\log(z_1/z_2)$.

12. By choosing specific nonzero values of z_1 and z_2, show that statement (6), Sec. 26, is not always valid when *log* is replaced everywhere by *Log*.

13. Verify identity (7), Sec. 26.

14. Show that for all points $z = x + iy$ in the right half plane $x > 0$, the function Log z can be written

$$\text{Log } z = \frac{1}{2} \text{Log}(x^2 + y^2) + i \arctan\left(\frac{y}{x}\right),$$

where $-\pi/2 < \arctan(y/x) < \pi/2$. Use this expression for Log z, together with results in Sec. 17, to give another proof that the principal branch Log z is analytic in the domain $x > 0$ and that $d(\text{Log } z)/dz = 1/z$ there. (Some complications arise with the inverse tangent and its derivative in the remaining part of the full domain of analyticity of Log z, especially on the line $x = 0$.)

15. Show that (a) the function $\text{Log}(z - i)$ is analytic everywhere except on the half line $y = 1$ $(x \leq 0)$; (b) the function

$$\frac{\text{Log}(z + 4)}{z^2 + i}$$

is analytic everywhere except at the points $\pm(1 - i)/\sqrt{2}$ and on the portion $x \leq -4$ of the real axis.

16. Show in two ways that the function $\text{Log}(x^2 + y^2)$ is harmonic in every domain that does not contain the origin.

17. Show that

$$\text{Re}[\log(z - 1)] = \frac{1}{2} \text{Log}[(x - 1)^2 + y^2] \qquad (z \neq 1).$$

Why must this function satisfy Laplace's equation when $z \neq 1$?

18. Show that property (8), Sec. 26, also holds when n is a negative integer. Do this by writing $z^{1/n} = (z^{1/m})^{-1}$ $(m = -n)$, where n has any one of the negative values $n = -1, -2, \ldots$ [see Exercise 11(b), Sec. 7], and using the fact that the property is already known to be valid for positive integers.

19. Let z denote any nonzero complex number, written $z = re^{i\Theta}$ $(-\pi < \Theta \leq \pi)$, and let n denote any fixed positive integer $(n = 1, 2, \ldots)$. Show that all the values of $\log(z^{1/n})$ are given by the equation

$$\log(z^{1/n}) = \frac{1}{n} \text{Log } r + i\frac{\Theta + 2(pn + k)\pi}{n},$$

where $p = 0, \pm 1, \pm 2, \ldots$ and $k = 0, 1, 2, \ldots, n - 1$. Then, after writing

$$\frac{1}{n} \log z = \frac{1}{n} \text{Log } r + i\frac{\Theta + 2q\pi}{n},$$

where $q = 0, \pm 1, \pm 2, \ldots$, show that the set of values of $\log(z^{1/n})$ is the same as the set of values of $(1/n) \log z$. Thus show that $\log(z^{1/n}) = (1/n) \log z$, where corresponding to a value of $\log(z^{1/n})$ taken on the left the appropriate value of $\log z$ is to be selected on the right, and conversely. [The result in Exercise 5(a) is a special case of this one.]

   Suggestion: Use the fact that the remainder upon dividing an integer by a positive integer n is always an integer between 0 and $n - 1$, inclusive; that is, when a positive integer n is specified, any integer q can be written $q = pn + k$, where p is an integer and k has one of the values $k = 0, 1, 2, \ldots, n - 1$.

27. Complex Exponents

When $z \neq 0$ and *the exponent c is any complex number*, z^c is defined by means of the equation

$$(1) \qquad z^c = \exp(c \log z)$$

where $\log z$ denotes the multiple-valued logarithmic function. Equation (1) provides a consistent definition of z^c in the sense that the equation is already known to be valid (Sec. 26) when $c = n$ $(n = 0, \pm 1, \pm 2, \ldots)$ and $c = 1/n$ $(n = \pm 1, \pm 2, \ldots)$. The definition of z^c is, in fact, suggested by those particular choices of c.

Example 1. Powers of z are, in general, multiple-valued, as illustrated by writing

$$i^{-2i} = \exp(-2i \log i) = \exp\left[-2i\left(2n + \frac{1}{2}\right)\pi i\right] = \exp[(4n + 1)\pi]$$

$$(n = 0, \pm 1, \pm 2, \ldots).$$

Note that in view of the relation $1/e^z = e^{-z}$, the two sets of numbers $1/z^c$ and z^{-c} are the same. So we can write

$$(2) \qquad \frac{1}{z^c} = z^{-c};$$

and, in particular,

$$\frac{1}{i^{2i}} = \exp[(4n + 1)\pi] \qquad (n = 0, \pm 1, \pm 2, \ldots).$$

If $z = re^{i\theta}$ and α is any real number, the function

$$\log z = \text{Log } r + i\theta \qquad (r > 0, \alpha < \theta < \alpha + 2\pi)$$

is single-valued and analytic in the indicated domain (Sec. 25). When that branch of $\log z$ is used, it follows that the function $z^c = \exp(c \log z)$ is single-valued and analytic in the same domain. The derivative of such a *branch* of z^c is found by writing

$$\frac{d}{dz} z^c = \frac{d}{dz} \exp(c \log z) = [\exp(c \log z)]\frac{c}{z}$$

$$= c\frac{\exp(c \log z)}{\exp(\log z)} = c \exp[(c - 1) \log z]$$

and observing that the final term here is the single-valued function cz^{c-1} $(|z| > 0, \alpha < \arg z < \alpha + 2\pi)$. That is,

$$(3) \qquad \frac{d}{dz} z^c = cz^{c-1} \qquad (|z| > 0, \alpha < \arg z < \alpha + 2\pi).$$

The *principal value* of z^c occurs when $\log z$ is replaced by $\text{Log } z$ in definition (1), so that

(4) $$z^c = \exp(c \text{ Log } z).$$

Equation (4) also serves to define the *principal branch* of the function z^c on the domain $|z| > 0, -\pi < \text{Arg } z < \pi$.

Example 2. The principal value of $(-i)^i$ is

$$\exp[i \text{ Log}(-i)] = \exp\left[i\left(-\frac{\pi}{2}i\right)\right] = \exp\frac{\pi}{2}.$$

Example 3. The principal branch of $z^{2/3}$ can be written

$$\exp\left(\frac{2}{3} \text{ Log } z\right) = \exp\left(\frac{2}{3} \text{ Log } r + \frac{2}{3}i\Theta\right) = \sqrt[3]{r^2} \exp\left(\frac{i2\Theta}{3}\right).$$

It is analytic in the domain $r > 0, -\pi < \Theta < \pi$, as one can also show directly from the theorem in Sec. 18.

According to definition (1), *the exponential function with base c,* where c is any nonzero complex constant, is written

(5) $$c^z = \exp(z \log c).$$

Note that although e^z is, in general, multiple-valued according to definition (5), the usual interpretation of e^z occurs when the principal value of the logarithm is taken. For $\exp(z \text{ Log } e) = e^z$, the principal value of $\log e$ being 1.

When a value of $\log c$ is specified, c^z is an entire function of z. In fact,

$$\frac{d}{dz}c^z = \frac{d}{dz}\exp(z \log c) = [\exp(z \log c)] \log c;$$

and this shows that

(6) $$\frac{d}{dz}c^z = c^z \log c.$$

28. Inverse Trigonometric and Hyperbolic Functions

Inverses of the trigonometric and hyperbolic functions can be described in terms of logarithms.

To define the inverse sine function $\sin^{-1} z$, we write $w = \sin^{-1} z$ when $z = \sin w$. That is, $w = \sin^{-1} z$ when

$$z = \frac{e^{iw} - e^{-iw}}{2i}.$$

Let us write this equation in the form

$$(e^{iw})^2 - 2iz(e^{iw}) - 1 = 0,$$

which is quadratic in e^{iw}. Solving for e^{iw} [see Exercise 20(a), Sec. 7], we find that

(1) $$e^{iw} = iz + (1 - z^2)^{1/2}$$

where $(1 - z^2)^{1/2}$ is, of course, a double-valued function of z. Taking logarithms of each side of equation (1) and recalling that $w = \sin^{-1} z$, we arrive at the expression

(2) $$\sin^{-1} z = -i \log[iz + (1 - z^2)^{1/2}].$$

Because of this logarithm, $\sin^{-1} z$ is a multiple-valued function with infinitely many values at each point z.

Example. Expression (2) tells us that

$$\sin^{-1}(-i) = -i \log(1 \pm \sqrt{2}).$$

But

$$\log(1 + \sqrt{2}) = \text{Log}(1 + \sqrt{2}) + 2n\pi i \qquad (n = 0, \pm 1, \pm 2, \ldots)$$

and

$$\log(1 - \sqrt{2}) = \text{Log}(\sqrt{2} - 1) + (2n + 1)\pi i \qquad (n = 0, \pm 1, \pm 2, \ldots).$$

Since

$$\text{Log}(\sqrt{2} - 1) = \text{Log} \frac{1}{1 + \sqrt{2}} = -\text{Log}(1 + \sqrt{2}),$$

then, the numbers

$$(-1)^n \text{Log}(1 + \sqrt{2}) + n\pi i \qquad (n = 0, \pm 1, \pm 2, \ldots)$$

constitute the set of values of $\log(1 \pm \sqrt{2})$. Thus

$$\sin^{-1}(-i) = n\pi + i(-1)^{n+1} \text{Log}(1 + \sqrt{2}) \qquad (n = 0, \pm 1, \pm 2, \ldots).$$

One can apply the technique used to derive expression (2) for $\sin^{-1} z$ to show that

(3) $$\cos^{-1} z = -i \log[z + i(1 - z^2)^{1/2}]$$

and

(4) $$\tan^{-1} z = \frac{i}{2} \log \frac{i + z}{i - z}.$$

The functions $\cos^{-1} z$ and $\tan^{-1} z$ are also multiple-valued. When specific branches of the square root and logarithmic functions are used, all three inverse functions become single-valued and analytic because they are then compositions of analytic functions.

The derivatives of these three functions are readily obtained from the above expressions. The derivatives of the first two depend on the values chosen for the square roots:

(5) $$\frac{d}{dz} \sin^{-1} z = \frac{1}{(1 - z^2)^{1/2}},$$

(6) $$\frac{d}{dz} \cos^{-1} z = \frac{-1}{(1 - z^2)^{1/2}}.$$

The derivative of the last one,

$$(7) \qquad \frac{d}{dz} \tan^{-1} z = \frac{1}{1 + z^2},$$

does not, however, depend on the manner in which the function is made single-valued.

Inverse hyperbolic functions can be treated in a corresponding manner. It turns out that

$$(8) \qquad \sinh^{-1} z = \log[z + (z^2 + 1)^{1/2}],$$

$$(9) \qquad \cosh^{-1} z = \log[z + (z^2 - 1)^{1/2}],$$

and

$$(10) \qquad \tanh^{-1} z = \frac{1}{2} \log \frac{1 + z}{1 - z}.$$

Finally, we remark that common alternative notation for all these inverse functions is arcsin z, etc.

EXERCISES

1. Show that when $n = 0, \pm 1, \pm 2, \ldots,$

(a) $(1 + i)^i = \exp[(-\pi/4) + 2n\pi] \exp[(i/2) \, \text{Log} \, 2]$; (b) $(-1)^{1/\pi} = \exp[(2n + 1)i]$.

2. Find the principal value of

(a) i^i; (b) $[(e/2)(-1 - \sqrt{3}i)]^{3\pi i}$; (c) $(1 - i)^{4i}$.

> *Ans.* (a) $\exp(-\pi/2)$; (b) $-\exp(2\pi^2)$.

3. Use definition (1), Sec. 27, of z^c to show that $(-1 + \sqrt{3}i)^{3/2} = \pm 2\sqrt{2}$.

4. Show that the result in Exercise 3 could have been obtained by writing

(a) $(-1 + \sqrt{3}i)^{3/2} = [(-1 + \sqrt{3}i)^{1/2}]^3$ and first finding the square roots of $-1 + \sqrt{3}i$;

(b) $(-1 + \sqrt{3}i)^{3/2} = [(-1 + \sqrt{3}i)^3]^{1/2}$ and first cubing $-1 + \sqrt{3}i$.

5. Show that if $z \neq 0$ and a is a real number, then $|z^a| = \exp(a \, \text{Log} \, |z|) = |z|^a$.

6. Let c be a fixed nonzero complex number, and note that i^c is multiple-valued. What restriction must be placed on the constant c so that the values of $|i^c|$ are all the same?

> *Ans.* The number c is real.

7. Let $c, d,$ and z denote complex numbers, where $z \neq 0$. Prove that if all the powers involved are principal values, then

(a) $1/z^c = z^{-c}$; (b) $(z^c)^n = z^{cn}$ $(n = 1, 2, \ldots)$; (c) $z^c z^d = z^{c+d}$;

(d) $z^c/z^d = z^{c-d}$.

8. Using the principal branch of z^i, find the functions $u(r, \theta)$ and $v(r, \theta)$ when

$$z^i = u(r, \theta) + iv(r, \theta).$$

9. Assuming that $f'(z)$ exists, find the differentiation formula for $d[c^{f(z)}]/dz$.

10. Find all the values of

(a) $\tan^{-1}(2i)$;   (b) $\tan^{-1}(1 + i)$;   (c) $\cosh^{-1}(-1)$;   (d) $\tanh^{-1} 0$.

$$\textit{Ans.} \quad (a)\ \left(n + \frac{1}{2}\right)\pi + \frac{i}{2}\ \text{Log}\ 3\ (n = 0, \pm1, \pm2, \ldots);$$

$$(d)\ n\pi i\ (n = 0, \pm1, \pm2, \ldots).$$

11. Solve the equation $\sin z = 2$ for z by (a) identifying real components and imaginary components; (b) using expression (2), Sec. 28, for $\sin^{-1} z$.

$$\textit{Ans.} \quad (2n + \tfrac{1}{2})\pi \pm i\ \text{Log}(2 + \sqrt{3})\ (n = 0, \pm1, \pm2, \ldots).$$

12. Solve the equation $\cos z = \sqrt{2}$ for z.

13. Derive formula (5), Sec. 28, for the derivative of $\sin^{-1} z$.

14. Derive expression (4), Sec. 28, for $\tan^{-1} z$.

15. Derive formula (7), Sec. 28, for the derivative of $\tan^{-1} z$.

16. Derive expression (9), Sec. 28, for $\cosh^{-1} z$.

FOUR

INTEGRALS

Integrals are extremely important in the study of functions of a complex variable. The theory of integration, to be developed in this chapter, is noted for its mathematical elegance. The theorems are generally concise and powerful, and most of the proofs are simple. The theory is also noted for its great utility in applied mathematics.

The reader could at this time pass directly to Chap. 7, which is concerned with the manner in which various curves and regions are mapped from the z plane into the w plane by elementary functions. That would lead more quickly to the theory of a special type of mapping (Chap. 8) which is useful in solving certain physical problems (Chap. 9). It must be pointed out, however, that in Chap. 8 there are a few references to this chapter. For example, we have not yet established the continuity of the first- and second-order partial derivatives of the components of an analytic function; and the reader would need to assume that continuity, the verification of which appears in our treatment of integrals here.

29. Definite Integrals of $w(t)$

To introduce integrals of $f(z)$ in a fairly simple way, we first define the definite integral of a complex-valued function of a *real* variable t over a given interval $a \leq t \leq b$.

We write

$$(1) \qquad\qquad w(t) = u(t) + iv(t) \qquad\qquad (a \leq t \leq b).$$

Here u and v are real-valued piecewise continuous functions of t defined on a closed bounded interval $a \leq t \leq b$. That is, each of those two functions is real-valued and continuous everywhere in the stated interval except possibly for a finite number of points where the function, although discontinuous, has finite left-hand and right-hand limits. Of course, only the right-hand limit is required at a; and only the left-hand limit is required at b. We then say that w is *piecewise continuous* on the interval $a \leq t \leq b$ and define the definite integral of w from a to b in terms of two definite integrals of the type studied in calculus:

$$(2) \qquad \int_a^b w(t)\, dt = \int_a^b u(t)\, dt + i \int_a^b v(t)\, dt.$$

The conditions on u and v stated above are sufficient to ensure the existence of their integrals. In many instances that we shall encounter w is actually continuous on the interval $a \leqq t \leqq b$, the functions u and v being continuous at every point in it.

An improper integral of w over an *unbounded* interval is similarly defined in terms of improper integrals of u and v, and it exists when the improper integrals of u and v both exist.

Example. To illustrate definition (2), we observe that

$$\int_0^{\pi/6} e^{i2t}\, dt = \int_0^{\pi/6} \cos 2t\, dt + i \int_0^{\pi/6} \sin 2t\, dt = \frac{\sqrt{3}}{4} + \frac{i}{4}.$$

Another way of evaluating such integrals is given in Exercise 7, Sec. 30. There it is shown that the technique of finding an antiderivative, learned in calculus, can be applied to *complex-valued* functions of t.

From definition (2) it follows that

$$(3) \qquad \operatorname{Re} \int_a^b w(t)\, dt = \int_a^b \operatorname{Re}[w(t)]\, dt.$$

Also, for each complex constant $z_0 = x_0 + iy_0$,

$$\int_a^b z_0 w(t)\, dt = \int_a^b (x_0 + iy_0)(u + iv)\, dt$$

$$= \int_a^b (x_0 u - y_0 v)\, dt + i \int_a^b (y_0 u + x_0 v)\, dt$$

$$= (x_0 + iy_0) \left(\int_a^b u\, dt + i \int_a^b v\, dt \right);$$

that is,

$$(4) \qquad \int_a^b z_0 w(t)\, dt = z_0 \int_a^b w(t)\, dt.$$

Many other of the usual rules, such as those for integrating a sum and for interchanging the limits of integration, apply just as they do for real-valued functions of t. As was the case with property (4), the proofs follow readily from the corresponding rules in calculus.

To establish a basic property of absolute values of integrals, let us assume that $a < b$ and that the value of the integral defined in equation (2) is a nonzero complex number. If r_0 is the modulus and θ_0 is an argument of that number, then

$$r_0 e^{i\theta_0} = \int_a^b w\, dt.$$

Solving for r_0 and using property (4), we write

$$(5) \qquad\qquad r_0 = \int_a^b e^{-i\theta_0} w \, dt.$$

Now the left-hand side of this equation is a real number, and therefore the right-hand side is too. Thus, using the fact that the real part of a real number is the number itself and referring to property (3), we see that the right-hand side of equation (5) can be rewritten in the following way:

$$\int_a^b e^{-i\theta_0} w \, dt = \operatorname{Re} \int_a^b e^{-i\theta_0} w \, dt = \int_a^b \operatorname{Re}(e^{-i\theta_0} w) \, dt.$$

Equation (5) then takes the form

$$(6) \qquad\qquad r_0 = \int_a^b \operatorname{Re}(e^{-i\theta_0} w) \, dt.$$

But

$$\operatorname{Re}(e^{-i\theta_0} w) \leqq |e^{-i\theta_0} w| = |e^{-i\theta_0}| \, |w| = |w|;$$

and so

$$r_0 \leqq \int_a^b |w| \, dt.$$

That is,

$$(7) \qquad\qquad \left| \int_a^b w(t) \, dt \right| \leqq \int_a^b |w(t)| \, dt \qquad\qquad (a < b).$$

This property is clearly also valid when the value of the integral on the left is zero, in particular when $a = b$.

With only minor modifications, the above discussion yields inequalities such as

$$(8) \qquad\qquad \left| \int_a^\infty w(t) \, dt \right| \leqq \int_a^\infty |w(t)| \, dt,$$

provided both improper integrals exist.

30. Contours

Integrals of complex-valued functions of a *complex* variable are defined on curves in the complex plane, rather than on just intervals of the real line. Classes of curves that are adequate for the study of such integrals are introduced in this section.

An *arc* C in the complex plane is a set of points $z = (x, y)$ such that

$$(1) \qquad\qquad x = x(t), \qquad y = y(t) \qquad\qquad (a \leqq t \leqq b),$$

where $x(t)$ and $y(t)$ are continuous functions of the real parameter t. This definition establishes a continuous mapping of the interval $a \leqq t \leqq b$ into the xy, or z, plane; and the image points are ordered according to increasing values of t. It is convenient to describe the points of C by means of the equation

$$z = z(t) \qquad (a \leq t \leq b)$$ (2)

where

$$z(t) = x(t) + iy(t).$$ (3)

The arc C is a *simple arc,* or a Jordan arc, if it does not cross itself; that is, C is simple if $z(t_1) \neq z(t_2)$ when $t_1 \neq t_2$. When the arc C is simple except for the fact that $z(b) = z(a)$, we say that C is a *simple closed curve,* or a Jordan curve.

Example 1. The polygonal line

$$(4) \qquad z = \begin{cases} t + it & \text{when} \quad 0 \leq t \leq 1, \\ t + i & \text{when} \quad 1 \leq t \leq 2, \end{cases}$$

consisting of a line segment from 0 to $1 + i$ followed by one from $1 + i$ to $2 + i$, is an example of a simple arc.

Example 2. The unit circle

$$z = e^{i\theta} \qquad (0 \leq \theta \leq 2\pi)$$ (5)

about the origin is, on the other hand, a simple closed curve oriented in the counterclockwise direction. So is the circle

$$z = z_0 + Re^{i\theta} \qquad (0 \leq \theta \leq 2\pi),$$ (6)

centered at the point z_0 and with radius R (see Sec. 6).

The derivative $z'(t)$, or $d[z(t)]/dt$, of the function (3) is defined as

$$z'(t) = x'(t) + iy'(t),$$ (7)

provided the derivatives $x'(t)$ and $y'(t)$ both exist. Those derivatives are, of course, understood to be the appropriate one-sided derivatives at the end points $t = a$ and $t = b$. Anticipated rules for differentiating sums of functions of the type $z(t)$ in equation (3) and for differentiating a complex constant times such a function are valid. Note, too, that $z'(t) = 0$ if $z(t) = z_0$, where z_0 is any fixed complex number.

Suppose that the derivatives $x'(t)$ and $y'(t)$ of the components of the function $z(t)$ used to describe an arc C exist and are continuous throughout the entire interval $a \leq t \leq b$. Under these conditions C is called a *differentiable arc*. It follows that the real-valued function

$$|z'(t)| = \sqrt{[x'(t)]^2 + [y'(t)]^2}$$

is integrable over the interval $a \leq t \leq b$ and that the arc C has length

$$L = \int_a^b |z'(t)| \, dt.$$ (8)

Expression (8) arises from the definition of arc length in calculus.

To examine the effect on expression (8) when the parameter for C is changed, write

(9) $t = \phi(\tau)$ $(\alpha \leqq \tau \leqq \beta)$

where ϕ is a real-valued function mapping the interval $\alpha \leqq \tau \leqq \beta$ onto the interval $a \leqq t \leqq b$. We assume that ϕ is continuous with a continuous derivative. We also assume that $\phi'(\tau) > 0$ for each τ; this ensures that t increases with τ. Observe that with the change of variable indicated in equation (9), expression (8) for the length of C takes the form

$$L = \int_{\alpha}^{\beta} |z'[\phi(\tau)]| \phi'(\tau) \, d\tau.$$

But C may be described in terms of the new parametric representation

(10) $z = Z(\tau) = z[\phi(\tau)]$ $(\alpha \leqq \tau \leqq \beta)$.

Then, because (see Exercise 10)

(11) $Z'(\tau) = z'[\phi(\tau)]\phi'(\tau),$

we have the result

$$L = \int_{\alpha}^{\beta} |Z'(\tau)| \, d\tau.$$

Thus the number L given by expression (8) is invariant under such changes in the parametric representation for C, as one would expect.

If the equation $z = z(t)$ $(a \leqq t \leqq b)$ represents a differentiable arc and $z'(t)$ is never zero, the unit tangent vector

$$\mathbf{T} = \frac{z'(t)}{|z'(t)|},$$

whose angle of inclination is arg $z'(t)$, turns continuously as the parameter t varies from $t = a$ to $t = b$. This expression for $\mathbf{T}$ is the one learned in calculus when $z(t)$ is interpreted as a radius vector. Such an arc is said to be *smooth*. In referring to a smooth arc $z = z(t)$ $(a \leqq t \leqq b)$, then, we agree that the derivative $z'(t)$ is both continuous and nonzero throughout the entire interval $a \leqq t \leqq b$.

A *contour*, or piecewise smooth arc, is an arc consisting of a finite number of smooth arcs joined end to end. Hence if equation (2) represents a contour, $z(t)$ is continuous, whereas its derivative $z'(t)$ is piecewise continuous (and never zero). The polygonal path (4) is, for example, a contour. When only the initial and final values of $z(t)$ are the same, a contour C is called a *simple closed contour*. Examples are the circles (5) and (6), as well as the boundary of a triangle or a rectangle taken in a specific direction. The length of a contour or a simple closed contour is the sum of the lengths of the smooth arcs which make up the contour.

Associated with any simple closed curve or simple closed contour C are two domains each of which has the points of C as its only boundary points. One of these domains, called the interior of C, is bounded; the other, which is the exterior of C, is unbounded. It will be convenient to accept this statement, known as the *Jordan curve theorem*, as geometrically evident; the proof is not easy.*

EXERCISES

1. Apply definition (2), Sec. 29, or its adaptation to improper integrals, to evaluate

$$(a)\ \int_0^1 (1 + it^2)\, dt; \quad (b)\ \int_0^{\pi/4} e^{it}\, dt; \quad (c)\ \int_0^\infty e^{-zt}\, dt.$$

$$\textit{Ans.} \quad (a)\ 1 + \frac{i}{3}; \quad (b)\ \frac{1}{\sqrt{2}} + i\left(1 - \frac{1}{\sqrt{2}}\right); \quad (c)\ \frac{1}{z}\ (\text{Re } z > 0).$$

2. Show that if m and n are integers,

$$\int_0^{2\pi} e^{im\theta} e^{-in\theta}\, d\theta = \begin{cases} 0 & \text{when} & m \neq n, \\ 2\pi & \text{when} & m = n. \end{cases}$$

3. Show that if $w(t)$ is a function of the type (1), Sec. 29, then

$$\int_{-b}^{-a} w(-t)\, dt = \int_a^b w(\tau)\, d\tau.$$

4. Apply inequality (7), Sec. 29, to show that for all values of x in the interval $-1 \leqq x \leqq 1$, the functions

$$P_n(x) = \frac{1}{\pi} \int_0^\pi \left(x + i\sqrt{1 - x^2}\ \cos \theta\right)^n d\theta \qquad (n = 0, 1, 2, \ldots)$$

satisfy the inequality $|P_n(x)| \leqq 1$. †

5. Let $w(t) = u(t) + iv(t)$ denote a continuous complex-valued function defined on an interval $-a \leqq t \leqq a$.

   (a) Suppose that $w(t)$ is *even*; that is, $w(-t) = w(t)$ for each point t in the given interval. Show that

$$\int_{-a}^a w(t)\, dt = 2 \int_0^a w(t)\, dt.$$

   (b) Show that if $w(t)$ is an *odd* function, one where $w(-t) = -w(t)$ for each point t in the interval, then

$$\int_{-a}^a w(t)\, dt = 0.$$

   Suggestion: In each part of this exercise use the corresponding property of integrals of *real-valued* functions of t, which is graphically evident.

*See pp. 115–116 of the book by Newman or sec. 13 of the one by Thron, both of which are cited in Appendix 1.

†These functions are actually polynomials in x. They are known as *Legendre polynomials* and are important in applied mathematics. See the authors' "Fourier Series and Boundary Value Problems," 3d ed., chap. 9, 1978.

6. Let $z_0 = x_0 + iy_0$ be a fixed complex number. Verify the following differentiation rules in the manner indicated.

   (a) Using the corresponding rules in calculus, show that

   $$\frac{d}{dt}[z_0 z(t)] = z_0 z'(t), \qquad \frac{d}{dt}[z(t)]^2 = 2z(t)z'(t)$$

   when $z(t) = x(t) + iy(t)$ is a complex-valued function of a real variable t and $z'(t)$ exists.

   (b) Use the expression

   $$e^{z_0 t} = e^{x_0 t} \cos y_0 t + ie^{x_0 t} \sin y_0 t$$

   to show that

   $$\frac{d}{dt} e^{z_0 t} = z_0 e^{z_0 t}.$$

7. (a) Let $w(t) = u(t) + iv(t)$ and $W(t) = U(t) + iV(t)$ be complex-valued functions of a real variable t which are continuous on an interval $a \leq t \leq b$. Show that if $W'(t) = w(t)$ when $a < t < b$, then

   $$\int_a^b w(t)\,dt = W(t)\bigg]_a^b = W(b) - W(a).$$

   Do this by appealing to the corresponding result from calculus for real-valued functions of t.

   (b) Use the result in part (a), as well as differentiation rules obtained in Exercise 6, to evaluate the integral

   $$\int_0^{\pi/6} e^{i2t}\,dt,$$

   which was evaluated in Sec. 29 directly from the definition of an integral of a complex-valued function of t.

8. According to definition (2), Sec. 29,

   $$\int_0^{\pi} e^{(1+i)x}\,dx = \int_0^{\pi} e^x \cos x\,dx + i\int_0^{\pi} e^x \sin x\,dx.$$

   Evaluate the two integrals on the right here by evaluating the single integral on the left with the aid of results in Exercises 6 and 7(a) and then taking the real and imaginary parts of the value that is found.

   $$\text{Ans.} \quad -(1 + e^{\pi})/2, \ (1 + e^{\pi})/2.$$

9. Let $w(t)$ be a continuous complex-valued function of t defined on an interval $a \leq t \leq b$. By considering the special case $w(t) = e^{it}$ on the interval $0 \leq t \leq 2\pi$, show that it is *not* always true that there is a number c in the interval $a < t < b$ such that $w(c)$ times $b - a$ equals the value of the definite integral of $w(t)$ from a to b. Thus show that the *mean-value theorem for definite integrals* in calculus does not apply to such functions.

10. Verify expression (11), Sec. 30, for the derivative of $Z(\tau) = z[\phi(\tau)]$.

    Suggestion: Write $Z(\tau) = x[\phi(\tau)] + iy[\phi(\tau)]$ and apply the chain rule for real-valued functions of a real variable.

11. Suppose that a function $f(z)$ is analytic at a point $z_0 = z(t_0)$ on a differentiable arc $z = z(t)$ $(a \leqq t \leqq b)$. Show that if $w(t) = f[z(t)]$, then

$$w'(t) = f'[z(t)]z'(t)$$

when $t = t_0$.

    Suggestion: Write $f(z) = u(x, y) + iv(x, y)$ and $z(t) = x(t) + iy(t)$, so that

$$w(t) = u[x(t), y(t)] + iv[x(t), y(t)].$$

Then apply the chain rule in calculus for functions of two variables to write

$$w' = u_x x' + u_y y' + i(v_x x' + v_y y'),$$

and use the Cauchy-Riemann equations.

31. Line Integrals

We turn now to integrals of complex-valued functions f of the complex variable z. Such an integral is defined in terms of the values $f(z)$ along a given contour C extending from a point $z = z_1$ to a point $z = z_2$ in the complex plane. It is therefore a line integral, and its value depends, in general, on the contour C as well as the function f. It is written

$$\int_C f(z)\,dz \qquad \text{or} \qquad \int_{z_1}^{z_2} f(z)\,dz,$$

the latter notation often being used when the value of the integral is independent of the choice of the contour taken between the two end points. While the integral may be defined directly as the limit of a sum, we choose to define it in terms of a definite integral of the type introduced in Sec. 29.

    Suppose that the equation

$$(1) \qquad\qquad\qquad\qquad z = z(t) \qquad\qquad\qquad\qquad (a \leqq t \leqq b)$$

represents a contour C, extending from a point $z_1 = z(a)$ to a point $z_2 = z(b)$. Let the function $f(z) = u(x, y) + iv(x, y)$ be piecewise continuous on C; that is, if $z(t) = x(t) + iy(t)$, the function

$$f[z(t)] = u[x(t), y(t)] + iv[x(t), y(t)]$$

is piecewise continuous on the interval $a \leqq t \leqq b$. We define the line integral, or *contour integral*, of f along C as follows:

$$(2) \qquad\qquad\qquad \int_C f(z)\,dz = \int_a^b f[z(t)]z'(t)\,dt.$$

Note that since C is a contour, $z'(t)$ is also piecewise continuous on the interval $a \leqq t \leqq b$; and so the existence of integral (2) is ensured.

    The integrand on the right-hand side in equation (2) is the product of the complex-valued functions

$$u[x(t), y(t)] + iv[x(t), y(t)], \qquad x'(t) + iy'(t)$$

of the real variable t. Thus, by definition (2), Sec. 29,

$$(3) \qquad \int_C f(z)\,dz = \int_a^b (ux' - vy')\,dt + i\int_a^b (vx' + uy')\,dt.$$

In terms of line integrals of real-valued functions of two real variables, then,

$$(4) \qquad \int_C f(z)\,dz = \int_C u\,dx - v\,dy + i\int_C v\,dx + u\,dy.$$

Observe that expression (4) can be obtained formally by replacing f by $u + iv$ and dz by $dx + i\,dy$ and then expanding the product.

It also follows from equation (3) that the value of a contour integral is invariant under a change in the representation of its contour when the change is of the type described by equation (9), Sec. 30. This can be seen by writing the integrals on the right in equation (3) in terms of the new parameter τ and following the same general procedure used in Sec. 30 to show the invariance of arc length.

Associated with the contour C used in integral (2) is the contour $-C$, consisting of the same set of points but with the order reversed so that the new contour extends from the point z_2 to the point z_1. The contour $-C$ has the parametric representation $z = z(-t)$ $(-b \leq t \leq -a)$. Thus

$$\int_{-C} f(z)\,dz = \int_{-b}^{-a} f[z(-t)][-z'(-t)]\,dt$$

where $z'(-t)$ denotes the derivative of $z(t)$ with respect to t evaluated at $-t$. After a change of variable in the last integral (see Exercise 3, Sec. 30), we find that

$$(5) \qquad \int_{-C} f(z)\,dz = -\int_C f(z)\,dz.$$

Suppose that C consists of a contour C_1 from z_1 to z_2 followed by a contour C_2 from z_2 to z_3, the initial point of C_2 being the terminal point of C_1. It is clear from equation (4) and rules for line integrals learned in calculus that

$$(6) \qquad \int_C f(z)\,dz = \int_{C_1} f(z)\,dz + \int_{C_2} f(z)\,dz.$$

Sometimes the contour C is called the *sum* of its legs C_1 and C_2 and is denoted by $C_1 + C_2$. The sum of two contours C_1 and $-C_2$ is well defined when C_1 and C_2 have the same terminal points, and it is written $C_1 - C_2$. Two further properties of line integrals follow immediately from definition (2) and properties of integrals of complex-valued functions $w(t)$ mentioned in Sec. 29. Namely,

$$(7) \qquad \int_C z_0 f(z)\,dz = z_0 \int_C f(z)\,dz,$$

for any complex constant z_0, and

$$(8) \qquad \int_C [f(z) + g(z)]\,dz = \int_C f(z)\,dz + \int_C g(z)\,dz.$$

According to definition (2) above and property (7), Sec. 29,

$$\left| \int_C f(z)\, dz \right| \leqq \int_a^b |f[z(t)]z'(t)|\, dt.$$

So for any nonnegative constant M such that the values of f on C satisfy the inequality $|f(z)| \leqq M$,

$$\left| \int_C f(z)\, dz \right| \leqq M \int_a^b |z'(t)|\, dt.$$

Now the integral on the right here represents the length L of the contour (see Sec. 30). Hence the modulus of the value of the integral of f along C does not exceed ML:

(9) $$\left| \int_C f(z)\, dz \right| \leqq ML.$$

This is, of course, a strict inequality when the values of f on C are such that $|f(z)| < M$.

Note that since all the paths of integration to be considered here are contours and the integrands are piecewise continuous functions defined on those contours, a number M such as the one appearing in inequality (9) will always exist. This is because the real-valued function $|f[z(t)]|$ is continuous on the closed bounded interval $a \leqq t \leqq b$ when f is *continuous* on C; and such a function always reaches a maximum value M on that interval.* Hence $|f(z)|$ has a maximum value on C when f is continuous on it. It now follows immediately that the same is true when f is *piecewise* continuous on C.

Definite integrals in calculus can be interpreted as areas, and they have other interpretations as well. Except in special cases, no corresponding helpful interpretation, geometric or physical, is available for integrals in the complex plane. Nonetheless, as noted earlier, the theory of integration in the complex plane is remarkably useful in both pure and applied mathematics.

32. Examples

In this section we give examples illustrating the theory of line integrals just developed. In Sec. 37 we extend the theory to include antiderivatives of the integrands $f(z)$ in line integrals. Some of the integrals in the examples to follow here can be evaluated more easily by that method. Special care must be taken, however, in the use of such antiderivatives, and we defer that topic until the later section. Our main goal at this time is to reinforce the basic definition of line integrals and to illustrate those properties that we have seen.

Example 1. Let us find the value of the integral

$$I_1 = \int_{C_1} z^2\, dz$$

---

*See, for instance, A. E. Taylor and W. R. Mann, "Advanced Calculus," 3d ed., pp. 86–90, 1983.

when C_1 is the line segment from $z = 0$ to $z = 2 + i$ (Fig. 22). Observe that points of C_1 lie on the line $y = x/2$, or $x = 2y$. Hence, if the coordinate y is used as the parameter, a parametric equation for C_1 is

$$z = 2y + iy \qquad\qquad (0 \leqq y \leqq 1).$$

Also, on C_1 the integrand z^2 becomes

$$z^2 = (2y + iy)^2 = 3y^2 + i4y^2.$$

Therefore,

$$I_1 = \int_0^1 (3y^2 + i4y^2)(2 + i)\,dy = (3 + 4i)(2 + i)\int_0^1 y^2\,dy = \frac{2}{3} + \frac{11}{3}i.$$

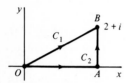

Figure 22

Example 2. We next let C_2 denote the contour OAB shown in Fig. 22 and evaluate the integral

$$I_2 = \int_{C_2} z^2\,dz = \int_{OA} z^2\,dz + \int_{AB} z^2\,dz.$$

A parametric equation for the path OA is $z = x + i0$ $(0 \leqq x \leqq 2)$; and for the path AB one can write $z = 2 + iy$ $(0 \leqq y \leqq 1)$. Hence

$$I_2 = \int_0^2 x^2\,dx + \int_0^1 (2 + iy)^2 i\,dy$$

$$= \frac{8}{3} + i\left[\int_0^1 (4 - y^2)\,dy + 4i\int_0^1 y\,dy\right] = \frac{2}{3} + \frac{11}{3}i.$$

Incidentally, a parametric representation for the entire contour OAB is

$$z = \begin{cases} t & \text{when } 0 \leqq t \leqq 2, \\ 2 + i(t - 2) & \text{when } 2 \leqq t \leqq 3. \end{cases}$$

Observe that $I_2 = I_1$, where I_1 is the integral evaluated in Example 1. Thus the integral of z^2 over the simple closed contour $OABO$, or $C_2 - C_1$, has the value $I_2 - I_1 = 0$. We shall soon see that this is a consequence of the fact that the integrand z^2 is analytic interior to and on that contour.

Example 3. For another example, we evaluate the integral

$$I_3 = \int_{C_3} \bar{z}\,dz$$

where the path of integration C_3 is the upper half of the circle $|z| = 1$ from $z = -1$ to $z = 1$ (Fig. 23). A parametric representation for $-C_3$ is

$$z = e^{i\theta} \qquad\qquad (0 \leqq \theta \leqq \pi);$$

and since $d(e^{i\theta})/d\theta = ie^{i\theta}$ [see Exercise 6(b), Sec. 30],

$$I_3 = -\int_{-C_3} \bar{z} \, dz = -\int_0^\pi e^{-i\theta} ie^{i\theta} \, d\theta = -\pi i.$$

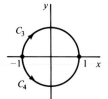

Figure 23

Example 4. Let us now evaluate the integral

$$I_4 = \int_{C_4} \bar{z} \, dz$$

between the same two points as in Example 3 but along the lower semicircle C_4 shown in Fig. 23. The contour is represented by the equation

$$z = e^{i\theta} \qquad\qquad (\pi \leqq \theta \leqq 2\pi),$$

and the integral is readily found to have the value

$$I_4 = \int_\pi^{2\pi} e^{-i\theta} ie^{i\theta} \, d\theta = \pi i.$$

Note that $I_4 \neq I_3$, where I_3 is the integral in Example 3. Thus the value of the integral I_0 of $\bar{z}$ around the entire circle $C_0 = C_4 - C_3$ in the counterclockwise direction is not zero:

$$I_0 = \int_{C_0} \bar{z} \, dz = I_4 - I_3 = 2\pi i.$$

When z is on the unit circle $|z| = 1$, it follows that $z\bar{z} = 1$, or

$$\bar{z} = \frac{1}{z};$$

hence the integrands of the integrals I_3, I_4, and I_0 can be replaced by $1/z$. In particular,

$$I_0 = \int_{C_0} \frac{dz}{z} = 2\pi i.$$

Example 5. Let C denote the semicircular path

$$z = 3e^{i\theta} \qquad\qquad (0 \leq \theta \leq \pi)$$

from the point $z = 3$ to the point $z = -3$ (Fig. 24). Although the branch (Sec. 27)

$$f(z) = \sqrt{r}\, e^{i\theta/2} \qquad\qquad (r > 0,\ 0 < \theta < 2\pi)$$

of the multiple-valued function $z^{1/2}$ is not defined at the initial point $z = 3$ of the contour C, the integral

$$I = \int_C z^{1/2}\, dz$$

of that branch nevertheless exists. For the integrand is piecewise continuous on C.
More precisely, when

$$z(\theta) = 3e^{i\theta},$$

the right-hand limits of the real and imaginary components of the function

$$f[z(\theta)] = \sqrt{3}\, e^{i\theta/2} = \sqrt{3}\, \cos\frac{\theta}{2} + i\sqrt{3}\, \sin\frac{\theta}{2} \qquad (0 < \theta \leq \pi)$$

at $\theta = 0$ are $\sqrt{3}$ and 0, respectively. Hence $f[z(\theta)]$ is continuous on the closed interval $0 \leq \theta \leq \pi$ when its value at $\theta = 0$ is defined as $\sqrt{3}$. Consequently,

$$I = \int_0^\pi \sqrt{3}\, e^{i\theta/2}\, 3ie^{i\theta}\, d\theta = 3\sqrt{3}\, i \int_0^\pi e^{i3\theta/2}\, d\theta;$$

and, in view of Exercise 7(a), Sec. 30,

$$\int_0^\pi e^{i3\theta/2}\, d\theta = \frac{2}{3i} e^{i3\theta/2} \bigg]_0^\pi = -\frac{2}{3i}(1 + i).$$

Finally, then,

$$I = -2\sqrt{3}\,(1 + i).$$

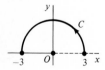

Figure 24

Example 6. Let C be the same contour and $z^{1/2}$ the same branch of the square root function as in Example 5. Without actually knowing the value of the integral, one can easily show that

$$\left| \int_C \frac{z^{1/2}}{z^2 + 1}\, dz \right| \leq \frac{3\sqrt{3}\, \pi}{8}.$$

For when z is a point on C, $|z| = 3$; and it follows that

$$|z^{1/2}| = |\sqrt{3}\, e^{i\theta/2}| = \sqrt{3}$$

and

$$|z^2 + 1| \geq ||z|^2 - 1| = 8$$

at the points on C where the integrand is defined. Hence, at those points,

$$\left| \frac{z^{1/2}}{z^2 + 1} \right| \leq M$$

where $M = \sqrt{3}/8$. Since the length of the contour is the number $L = 3\pi$, the stated upper bound for the modulus of the value of the integral is now a consequence of property (9), Sec. 31.

EXERCISES

For each arc C and function f in Exercises 1 through 5, find the value of

$$\int_C f(z) \, dz$$

after observing that C is a contour and that f is piecewise continuous on C.

1. $f(z) = y - x - i3x^2$ and C
  (a) is the line segment from $z = 0$ to $z = 1 + i$;
  (b) consists of two line segments, one from $z = 0$ to $z = i$ and the other from $z = i$ to $z = 1 + i$.

> *Ans.* (a) $1 - i$; (b) $(1 - i)/2$.

2. $f(z) = (z + 2)/z$ and C is
  (a) the semicircle $z = 2e^{i\theta}$ $(0 \leq \theta \leq \pi)$;
  (b) the semicircle $z = 2e^{i\theta}$ $(\pi \leq \theta \leq 2\pi)$;
  (c) the circle $z = 2e^{i\theta}$ $(0 \leq \theta \leq 2\pi)$.

> *Ans.* (a) $-4 + 2\pi i$; (b) $4 + 2\pi i$; (c) $4\pi i$.

3. $f(z) = z - 1$ and C is the arc from $z = 0$ to $z = 2$ consisting of
  (a) the semicircle $z = 1 + e^{i\theta}$ $(\pi \leq \theta \leq 2\pi)$;
  (b) the segment $0 \leq x \leq 2$ of the real axis.

> *Ans.* (a) 0; (b) 0.

4. C is the arc from $z = -1 - i$ to $z = 1 + i$ along the curve $y = x^3$ and

$$f(z) = \begin{cases} 4y & \text{when } y > 0, \\ 1 & \text{when } y < 0. \end{cases}$$

> *Ans.* $2 + 3i$.

5. $f(z) = e^z$ and C is the arc from $z = \pi i$ to $z = 1$ consisting of
  (a) the line segment joining those points;
  (b) the portion of the coordinate axes joining those points.

> *Ans.* (a) $1 + e$; (b) $1 + e$.

6. With the aid of the result in Exercise 2, Sec. 30, evaluate the integral

$$\int_C z^m \bar{z}^n \, dz$$

where m and n are integers and C is the circle $|z| = 1$ taken counterclockwise.

7. Let C be the contour in Exercise 6. Evaluate the integral

$$\int_C z^{-1+i} \, dz,$$

where the integrand denotes the branch

$$z^{-1+i} = \exp[(-1 + i) \log z] \qquad (|z| > 0, \ 0 < \arg z < 2\pi)$$

of the indicated power function.

$$\text{Ans.} \quad i(1 - e^{-2\pi}).$$

8. Show that if C is the boundary of the square with vertices at the points $z = 0$, $z = 1$, $z = 1 + i$, $z = i$ and the orientation of C is counterclockwise, then

$$\int_C (3z + 1) \, dz = 0.$$

9. Let C be the contour in Exercise 8 and evaluate the integral

$$\int_C \pi \exp(\pi \bar{z}) \, dz.$$

$$\text{Ans.} \quad 4(e^\pi - 1).$$

10. Evaluate the integral I_3 in Example 3, Sec. 32, using this representation for C_3:

$$z = x + i\sqrt{1 - x^2} \qquad (-1 \leq x \leq 1).$$

11. Let C be the arc of the circle $|z| = 2$ from $z = 2$ to $z = 2i$ that lies in the first quadrant. Without evaluating the integral, show that

$$\left| \int_C \frac{dz}{z^2 - 1} \right| \leq \frac{\pi}{3}.$$

12. Let C denote the line segment from $z = i$ to $z = 1$. By observing that of all the points on that line segment the midpoint is the closest to the origin, show that

$$\left| \int_C \frac{dz}{z^4} \right| \leq 4\sqrt{2}$$

without evaluating the integral.

13. Show that if C is the boundary of the triangle with vertices at the points $z = 0$, $z = 3i$, and $z = -4$, oriented in the counterclockwise direction, then

$$\left| \int_C (e^z - \bar{z}) \, dz \right| \leq 60.$$

14. (a) By referring to definition (2), Sec. 31, of the integral of a function $f(z)$ along a contour C and writing the following integrals in terms of integrals of complex-valued functions of a real variable t, show that

$$\int_{z_1}^{z_2} dz = z_2 - z_1, \qquad \int_{z_1}^{z_2} z\, dz = \frac{1}{2}(z_2^2 - z_1^2)$$

whenever the path of integration from the point $z = z_1$ to the point $z = z_2$ is a smooth arc $z = z(t)\,(a \le t \le b)$.

(b) Point out why each of the results in part (a) remains valid when the path is a contour. In particular, point out why

$$\int_C dz = 0, \qquad \int_C z\, dz = 0$$

whenever C is a closed contour which need not be simple and is allowed to intersect itself any number of times.

Suggestion: Results obtained in Exercises 6(a) and 7(a) of Sec. 30 are useful in evaluating the integrals arising in part (a) here.

15. (a) Show that if C_0 is a circle $|z - z_0| = R$, oriented in the counterclockwise direction, then

$$\int_{C_0} f(z)\, dz = iR \int_{-\pi}^{\pi} f(z_0 + Re^{i\theta})e^{i\theta}\, d\theta$$

when f is piecewise continuous on C_0.

(b) Let C denote the circle centered at the origin and with the same radius and orientation as the circle C_0 in part (a). With the aid of the result in part (a), show that

$$\int_C f(z)\, dz = \int_{C_0} f(z - z_0)\, dz$$

when f is piecewise continuous on C.

16. Let C_0 denote the positively oriented circle $|z - z_0| = R$, and use the result obtained in Exercise 15(a) to show that

(a) $\displaystyle\int_{C_0} \frac{dz}{z - z_0} = 2\pi i$;   (b) $\displaystyle\int_{C_0} (z - z_0)^{n-1}\, dz = 0 \quad (n = \pm 1, \pm 2, \ldots)$;

(c) $\displaystyle\int_{C_0} (z - z_0)^{a-1}\, dz = i\frac{2R^a}{a}\sin(a\pi)$, where a is any real number other than zero and the principal branch of the integrand is taken.

17. Let C_R be the circle $|z| = R$, described in the counterclockwise direction, where $R > 1$. Show that

$$\left| \int_{C_R} \frac{\mathrm{Log}\, z}{z^2}\, dz \right| < 2\pi \left(\frac{\pi + \mathrm{Log}\, R}{R} \right)$$

and hence that the value of the integral approaches zero as R tends to infinity.

18. Let C_ρ denote the circle $|z| = \rho\ (0 < \rho < 1)$, oriented in the counterclockwise direction, and suppose that $f(z)$ is analytic in the disk $|z| \le 1$. Show that if $z^{-1/2}$ represents any particular branch of that power of z, there is a nonnegative constant M, *independent of* ρ, such that

$$\left| \int_{C_\rho} z^{-1/2} f(z)\, dz \right| \le 2\pi M \sqrt{\rho}.$$

Thus show that the value of the integral here approaches 0 as ρ tends to 0.

Suggestion: Note that since $f(z)$ is analytic, and therefore continuous, throughout the disk $|z| \leq 1$, it is bounded there (Sec. 13).

19. Let C_N denote the boundary of the square formed by the lines

$$x = \pm(N + \tfrac{1}{2})\pi \quad \text{and} \quad y = \pm(N + \tfrac{1}{2})\pi$$

where N is a positive integer, and let the orientation of C_N be counterclockwise.

(a) With the aid of the inequalities

$$|\sin z| \geq |\sin x| \quad \text{and} \quad |\sin z| \geq |\sinh y|,$$

obtained in Exercises 7 and 9(a) of Sec. 23, show that $|\sin z| \geq 1$ on the vertical sides of the square and that $|\sin z| > \sinh(\pi/2)$ on the horizontal sides. Thus show that there is a positive constant A, *independent of N*, such that $|\sin z| \geq A$ for all points z lying on the contour C_N.

(b) Using the final result in part (a), show that

$$\left| \int_{C_N} \frac{dz}{z^2 \sin z} \right| \leq \frac{8}{(N + \tfrac{1}{2})\pi A}$$

and hence that the value of this integral tends to zero as N tends to infinity.

33. Cauchy-Goursat Theorem

Suppose that two real-valued functions $P(x, y)$ and $Q(x, y)$, together with their partial derivatives of the first order, are continuous throughout a closed region R consisting of points interior to and on a simple closed contour C in the xy plane. The contour is described in the *positive sense* (counterclockwise), so that the interior points of R lie to the left of C. According to Green's theorem for line integrals in advanced calculus,

$$\int_C P\, dx + Q\, dy = \iint_R (Q_x - P_y)\, dx\, dy.*$$

We now consider a function

$$f(z) = u(x, y) + iv(x, y)$$

which is analytic throughout such a region R in the xy, or z, plane. According to Sec. 31, the line integral of f along C can be written

(1) $$\int_C f(z)\, dz = \int_C u\, dx - v\, dy + i \int_C v\, dx + u\, dy.$$

Since f is continuous in R, the functions u and v are also continuous there; and if the derivative f' of f is continuous in R, so are the first-order partial derivatives of u and v. Green's theorem then enables us to rewrite equation (1) as

(2) $$\int_C f(z)\, dz = \iint_R (-v_x - u_y)\, dx\, dy + i \iint_R (u_x - v_y)\, dx\, dy.$$

*See, for example, A. E. Taylor and W. R. Mann, "Advanced Calculus," 3d ed., p. 457, 1983.

But, in view of the Cauchy-Riemann equations

$$u_x = v_y, \qquad u_y = -v_x,$$

the integrands of these two double integrals are zero throughout R. So *when f is analytic in R and f' is continuous there,*

(3) $$\int_C f(z)\, dz = 0.$$

This result was obtained by Cauchy in the early part of the last century.

Note that once it has been established that the value of this integral is zero, the orientation of C is immaterial. That is, statement (3) is also true if C is taken in the clockwise direction, since

$$\int_C f(z)\, dz = -\int_{-C} f(z)\, dz = 0.$$

Examples. To illustrate statement (3), we observe that if C is a simple closed contour,

$$\int_C dz = 0, \qquad \int_C z\, dz = 0, \qquad \text{and} \qquad \int_C z^2\, dz = 0$$

because the functions 1, z, and z^2 are entire and their derivatives are everywhere continuous.

Goursat* was the first to prove that *the condition of continuity on f' can be omitted.* Its removal is important and will allow us to show, for example, that the derivative f' of an analytic function f is analytic without having to assume the continuity of f', which follows as a consequence. We now state the revised form of Cauchy's result, known as the *Cauchy-Goursat theorem.*

> **Theorem.** *If a function f is analytic at all points interior to and on a simple closed contour C, then*
>
> $$\int_C f(z)\, dz = 0.$$

The proof is presented in the next two sections where, to be specific, we assume that C is positively oriented. It will be a simple matter to extend the theorem to more general paths, such as the entire boundary of the region between two concentric circles. The reader who wishes to accept this theorem without proof may pass directly to Sec. 36.

34. Preliminary Lemma

We start by forming subsets of the region R which consists of the points on a positively oriented simple closed contour C together with the points interior to C. To do this, we

---

*E. Goursat (1858–1936), pronounced *gour-sah'*.

draw equally spaced lines parallel to the real and imaginary axes such that the distance between adjacent vertical lines is the same as that between adjacent horizontal lines. We thus form a finite number of closed square subregions where each point of R lies in at least one such subregion and each subregion contains points of R.

We refer to these square subregions simply as squares, always keeping in mind that by a square we mean a boundary together with the points interior to it. If a particular square contains points that are not in R, we remove those points and call what remains a partial square. We thus *cover* the region R with a finite number of squares and partial squares (Fig. 25), and our proof of the following lemma starts with this covering.

Lemma. *Let f be analytic throughout a closed region R consisting of the points interior to a positively oriented simple closed contour C together with the points on C itself. For any positive number ε, the region R can be covered with a finite number of squares and partial squares, indexed by $j = 1, 2, \ldots, n$, such that in each one there is a fixed point z_j for which the inequality*

$$(1) \qquad \left| \frac{f(z) - f(z_j)}{z - z_j} - f'(z_j) \right| < \varepsilon \qquad (z \neq z_j)$$

is satisfied by all other points in that square or partial square.

To start the proof, we consider the possibility that in the covering constructed just prior to the statement of the lemma there is some square or partial square in which no point z_j exists such that inequality (1) holds for all other points z in it. If that subregion is a square, construct four smaller squares by drawing line segments joining the midpoints of its opposite sides (Fig. 25). If the subregion is a partial square, treat the whole square in the same manner and then let the portions that lie outside R be discarded. If in any one of these smaller subregions no point z_j exists such that inequality (1) holds for all other points z in it, construct still smaller squares and partial squares, etc.

Doing this to each of the original subregions that requires it, we may find that after a finite number of steps we can cover the region R with a collection of squares and partial squares such that the lemma is true.

Suppose, on the other hand, that needed points z_j do not exist after subdividing one of the original subregions a finite number of times. Let σ_0 denote that subregion if it is a square; if it is a partial square, let σ_0 denote the entire square of which it is a part. After we subdivide σ_0, at least one of the four smaller squares, denoted by σ_1, must contain points of R but no appropriate point z_j. We then subdivide σ_1 and continue in

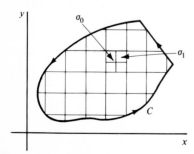

Figure 25

this manner. It may be that after a square σ_{k-1} $(k = 1, 2, \ldots)$ has been subdivided, more than one of the four smaller squares constructed from it can be chosen. To make a specific choice, we take σ_k to be the one lowest and then farthest to the left.

In view of the manner in which the nested infinite sequence

(2) $\sigma_0, \sigma_1, \sigma_2, \ldots, \sigma_{k-1}, \sigma_k, \ldots$

of squares is constructed, it is easily shown (Exercise 12, Sec. 37) that there is a point z_0 common to each σ_k; also, each of these squares contains points of R other than possibly z_0. Recall how the sizes of the squares in the sequence are decreasing, and note that any δ neighborhood $|z - z_0| < \delta$ of z_0 contains such squares when their diagonals have lengths less than δ. Every δ neighborhood $|z - z_0| < \delta$ therefore contains points of R distinct from z_0, and this means that z_0 is an accumulation point of R. Since the region R is a closed set, it follows that z_0 is a point in R. (See Sec. 8.)

The function f is analytic throughout R and, in particular, at z_0. Consequently, $f'(z_0)$ exists. According to the definition of derivative (Sec. 14), there is for each positive number ε a δ neighborhood $|z - z_0| < \delta$ such that the inequality

$$\left| \frac{f(z) - f(z_0)}{z - z_0} - f'(z_0) \right| < \varepsilon$$

is satisfied by all points distinct from z_0 in that neighborhood. But the neighborhood $|z - z_0| < \delta$ contains a square σ_K when the integer K is large enough that the length of a diagonal of that square is less than δ (Fig. 26). Consequently, z_0 serves as the

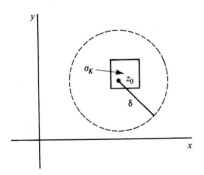

Figure 26

point z_j in inequality (1) for the subregion consisting of the square σ_K or a part of σ_K. Contrary to the way sequence (2) was formed, then, it is not necessary to subdivide σ_K. We thus arrive at a contradiction, and the proof of the lemma is complete.

35. Proof of the Cauchy-Goursat Theorem

Continuing with a function f which is analytic throughout a region R consisting of a positively oriented simple closed contour C and points interior to it, we are now ready to prove that

(1) $$\int_C f(z)\, dz = 0.$$

Given an arbitrary positive number ε, consider the covering of R in the lemma of Sec. 34. Let us define on the jth square or partial square the following function, where z_j is the fixed point lying in that subregion and appearing in the statement of the lemma:

$$(2) \qquad \delta_j(z) = \begin{cases} \dfrac{f(z) - f(z_j)}{z - z_j} - f'(z_j) & \text{when } z \neq z_j, \\ 0 & \text{when } z = z_j. \end{cases}$$

According to inequality (1) in the lemma,

$$(3) \qquad |\delta_j(z)| < \varepsilon$$

at all points z in the subregion on which $\delta_j(z)$ is defined. Also, the function $\delta_j(z)$ is continuous throughout the subregion since $f(z)$ is continuous there and

$$\lim_{z \to z_j} \delta_j(z) = f'(z_j) - f'(z_j) = 0.$$

Next, let C_j $(j = 1, 2, \ldots, n)$ denote the positively oriented boundaries of the above squares or partial squares covering R; and observe that in view of definition (2), the value of f at a point z on any particular C_j can be written

$$(4) \qquad f(z) = f(z_j) - z_j f'(z_j) + f'(z_j)z + (z - z_j)\delta_j(z).$$

Hence, integrating $f(z)$ counterclockwise around C_j and using the fact that

$$\int_{C_j} dz = 0 \qquad \text{and} \qquad \int_{C_j} z\, dz = 0$$

[see Exercise 14(b), Sec. 32], we can write

$$(5) \qquad \int_{C_j} f(z)\, dz = \int_{C_j} (z - z_j)\, \delta_j(z)\, dz \qquad (j = 1, 2, \ldots, n).$$

If we add all n integrals on the left in equations (5), we find that

$$\sum_{j=1}^{n} \int_{C_j} f(z)\, dz = \int_C f(z)\, dz.$$

This is true because the two integrals along the common boundary of every pair of adjacent subregions cancel each other, the integral being taken in one sense along that line segment in one subregion and in the opposite sense in the other (Fig. 27). Only

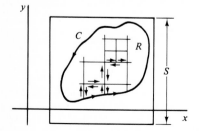

Figure 27

the integrals along the arcs that are parts of C remain. Thus, in view of equations (5),

$$\int_C f(z)\,dz = \sum_{j=1}^{n} \int_{C_j} (z - z_j)\,\delta_j(z)\,dz;$$

and so

(6)
$$\left| \int_C f(z)\,dz \right| \leqq \sum_{j=1}^{n} \left| \int_{C_j} (z - z_j)\,\delta_j(z)\,dz \right|.$$

Let us now use property (9), Sec. 31, to find an upper bound for each absolute value on the right in inequality (6). To do this, we first recall that each C_j coincides either entirely or partially with the boundary of a square. In either case, we let s_j denote the length of a side of the square. Since in the jth integral both the variable z and the point z_j lie in that square,

$$|z - z_j| \leqq \sqrt{2}\, s_j.$$

In view of inequality (3), then, we know that each integrand on the right in inequality (6) satisfies the condition

(7)
$$|(z - z_j)\,\delta_j(z)| < \sqrt{2}\, s_j\, \varepsilon.$$

As for the length of the path C_j, it is $4s_j$ if C_j is the boundary of a square. In that case we let A_j denote the area of the square and observe that

(8)
$$\left| \int_{C_j} (z - z_j)\,\delta_j(z)\,dz \right| < \sqrt{2}\, s_j\, \varepsilon 4 s_j = 4\sqrt{2}\, A_j\, \varepsilon.$$

If C_j is the boundary of a partial square, its length does not exceed $4s_j + L_j$, where L_j is the length of that part of C_j which is also a part of C. Again letting A_j denote the area of the full square, we find that

(9)
$$\left| \int_{C_j} (z - z_j)\,\delta_j(z)\,dz \right| < \sqrt{2}\, s_j\, \varepsilon(4 s_j + L_j) < 4\sqrt{2}\, A_j\, \varepsilon + \sqrt{2}\, S L_j\, \varepsilon$$

where S is the length of a side of some square that encloses the entire contour C as well as all the squares originally used in covering R (Fig. 27). Note that the sum of all the A_j's does not exceed S^2.

If L denotes the length of C, it now follows from inequalities (6), (8), and (9) that

$$\left| \int_C f(z)\,dz \right| < (4\sqrt{2}\, S^2 + \sqrt{2}\, SL)\varepsilon.$$

Since the value of the positive number ε is arbitrary, we can choose it so that the right-hand side of this last inequality is as small as we please. The left-hand side, which is independent of ε, must therefore be equal to zero; and statement (1) follows. This completes the proof of the Cauchy-Goursat theorem.

36. Simply and Multiply Connected Domains

A *simply connected* domain D is a domain such that every simple closed contour within it encloses only points of D. The set of points interior to a simple closed contour is an example. The annular domain between two concentric circles is, however, not simply connected. A domain that is not simply connected is said to be *multiply connected*.

The Cauchy-Goursat theorem can be stated in the following alternative form.

If a function f is analytic throughout a simply connected domain D, then

$$\text{(1)} \qquad\qquad \int_C f(z)\, dz = 0$$

for every simple closed contour C lying in D.

The simple closed contour here can be replaced by an arbitrary closed contour C which is not necessarily simple. For if C intersects itself a finite number of times, it consists of a finite number of simple closed contours, as illustrated in Fig. 28. By

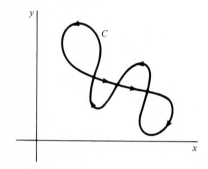

Figure 28

applying the Cauchy-Goursat theorem to each of those simple closed contours, we obtain the desired result for C. Also, a portion of C may be traversed twice in opposite directions since the integrals along that portion in the two directions cancel each other. Subtleties arise if the closed contour has an infinite number of self-intersection points. One method that can often be used to show that the theorem still applies is illustrated in Exercise 13, Sec. 37.*

Note that if two contours C_1 and C_2 each lie in a simply connected domain D and have the same initial and terminal points z_1 and z_2, respectively, then C_1 and $-C_2$ together form a closed contour (Fig. 29). Since the Cauchy-Goursat theorem is now seen to hold for any closed contour in a simply connected domain, one can write

$$\int_{C_1} f(z)\, dz + \int_{-C_2} f(z)\, dz = 0,$$

*For a proof of the theorem involving more general paths, see, for example, secs. 63–65 in vol. I of the book by Markushevich cited in Appendix 1.

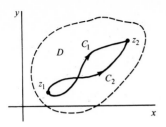

z_2

C_1

D

C_2

z_1

x **Figure 29**

or

$$\int_{C_1} f(z)\, dz = \int_{C_2} f(z)\, dz,$$

when f is analytic in D. Values of line integrals from z_1 to z_2 are therefore independent of the contours taken so long as those contours lie entirely within the simply connected domain D and the integrand is analytic in D. This observation is developed more fully in Sec. 37.

The Cauchy-Goursat theorem can also be modified in the following way, involving the boundary B of a multiply connected domain.

Theorem. *Let C be a simple closed contour and let C_j $(j = 1, 2, \ldots, n)$ be a finite number of simple closed contours inside C such that the regions interior to each C_j have no points in common. Let R be the closed region consisting of all points within and on C except for points interior to each C_j (Fig. 30). Let B denote the*

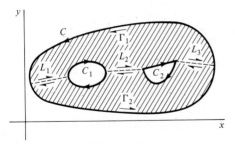

C

Γ_1

L_3

L_1

L_2

C_1

C_2

Γ_2

x **Figure 30**

entire oriented boundary of R consisting of C and all the contours C_j, described in a direction such that the interior points of R lie to the left of B. Then, if f is analytic throughout R,

(2)
$$\int_B f(z)\, dz = 0.$$

To establish this result, we introduce a polygonal path L_1, consisting of a finite number of line segments joined end to end, to connect the outer contour C to the inner contour C_1. We introduce another polygonal path L_2 which connects C_1 to C_2; and we continue in this manner, with L_{n+1} connecting C_n to C. As indicated by the single-barbed arrows in Fig. 30, two simple closed contours Γ_1 and Γ_2 can be formed, each consisting of polygonal paths L_j or $-L_j$ and pieces of C and C_j and each described in such a direction that points enclosed by them lie to the left. The Cauchy-Goursat theorem can now be applied to f on Γ_1 and Γ_2, and the sum of the integrals over those contours is found to be zero. Since the integrals in opposite directions along each path L_j cancel, only the integral along B remains; and statement (2) follows.

Example. We note that

$$\int_B \frac{dz}{z^2(z^2 + 9)} = 0$$

where B consists of the circle $|z| = 2$, described in the positive direction, together with the circle $|z| = 1$, described in the negative direction (Fig. 31). The integrand is analytic except at the points $z = 0$ and $z = \pm 3i$, all of which lie outside the annular region with boundary B.

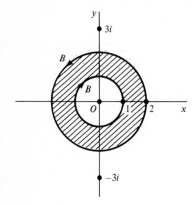

Figure 31

37. Antiderivatives and Independence of Path

Let f be a function which is continuous throughout a domain D, and suppose that there is an analytic function F such that $F'(z) = f(z)$ at each point in D. The function F is said to be an *antiderivative* of f in the domain D. A line integral

(1)
$$\int_C f(z)\,dz = \int_a^b f[z(t)]z'(t)\,dt$$

along a contour C, with parametric representation $z = z(t)$ ($a \leq t \leq b$), is readily evaluated with the aid of the antiderivative F when C lies entirely within D.

To show how this is done, we first consider C to be a smooth arc and observe that since $F'(z) = f(z)$,

$$\frac{d}{dt}F[z(t)] = F'[z(t)]z'(t) = f[z(t)]z'(t) \qquad (a < t < b).$$

(See Exercise 11, Sec. 30.) Then, in view of Exercise 7(a), Sec. 30, expression (1) becomes

$$\int_C f(z)\, dz = F[z(t)]\Big]_a^b = F[z(b)] - F[z(a)].$$

Hence, if z_1 is the initial point of C and z_2 is the terminal point, the value of the line integral is simply $F(z_2) - F(z_1)$. This method is, of course, also valid when C is a contour since a contour consists of a finite number of smooth arcs joined end to end.

Evidently, then, *if a continuous function f has an antiderivative F everywhere in a domain D, then the value of a line integral of f from a point z_1 in D to a point z_2 in D is independent of the contour taken, provided the contour lies entirely in D.* Furthermore, that value is found by writing

$$(2) \qquad\qquad \int_{z_1}^{z_2} f(z)\, dz = F(z)\Big]_{z_1}^{z_2} = F(z_2) - F(z_1).$$

Observe that when $z_1 = z_2$, the contour is closed and the value of integral (2) is zero.

When, as in the statement in italics, the value of a line integral of a given function f between two specified points in a domain D is always independent of the contour in D that is taken, we say that line integrals of f are *independent of path* in D. Let us now assume only that f is continuous throughout a domain D and that integrals of f are independent of path in D. By taking any fixed point z_0 in D and agreeing that all paths of integration are to lie entirely in D, we can define the new function

$$(3) \qquad\qquad F(z) = \int_{z_0}^{z} f(s)\, ds$$

on D. It is straightforward to show that $F'(z) = f(z)$ at each point z in D and hence that F is actually an antiderivative of f in D.

To do this, we let $z + \Delta z$ be any point, distinct from z, lying in some neighborhood of z where that neighborhood is small enough to be contained in D. Then

$$F(z + \Delta z) - F(z) = \int_{z_0}^{z+\Delta z} f(s)\, ds - \int_{z_0}^{z} f(s)\, ds = \int_{z}^{z+\Delta z} f(s)\, ds,$$

where the path of integration from z to $z + \Delta z$ may be selected as a line segment (Fig. 32). Since

$$\int_{z}^{z+\Delta z} ds = \Delta z$$

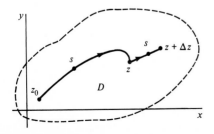

Figure 32

[see Exercise 14(*a*), Sec. 32], we can write

$$f(z) = \frac{1}{\Delta z} \int_z^{z+\Delta z} f(z)\, ds;$$

and it follows that

$$\frac{F(z + \Delta z) - F(z)}{\Delta z} - f(z) = \frac{1}{\Delta z} \int_z^{z+\Delta z} [f(s) - f(z)]\, ds.$$

But f is continuous at the point z. Hence for each positive number ε, a positive number δ exists such that

$$|f(s) - f(z)| < \varepsilon \qquad \text{whenever} \qquad |s - z| < \delta.$$

Consequently, if the point $z + \Delta z$ is close enough to z that $|\Delta z| < \delta$, then

$$\left| \frac{F(z + \Delta z) - F(z)}{\Delta z} - f(z) \right| < \frac{1}{|\Delta z|} \varepsilon |\Delta z| = \varepsilon;$$

that is,

$$\lim_{\Delta z \to 0} \frac{F(z + \Delta z) - F(z)}{\Delta z} = f(z),$$

or $F'(z) = f(z)$.

So we have a converse of the statement in italics earlier in this section. That is, *if line integrals of a continuous function f are independent of path in a domain D, then f has an antiderivative everywhere in D.* While the function $F(z)$ defined by equation (3) is such an antiderivative, it should be noted that if $G(z)$ is any analytic function other than $F(z)$ such that $G'(z) = f(z)$ everywhere in D, then the derivative of the analytic function $H(z) = F(z) - G(z)$ is zero. But, as shown in Sec. 19, an analytic function is constant when its derivative is zero throughout a domain; hence any two anti-derivatives of a given function f can differ at most by an additive complex constant.

For an important application of these results, consider a function f which is *analytic* throughout a *simply connected* domain D. As pointed out in Sec. 36, line integrals of f are independent of path in such a domain. This means that f has an antiderivative F throughout D and that expression (2) can be used to evaluate any line integral of f over a contour in D from a point z_1 to a point z_2 in that domain.

Example 1. The continuous function $f(z) = z^2$ has the antiderivative $F(z) = z^3/3$ throughout the plane. Hence

$$\int_0^{1+i} z^2\, dz = \frac{z^3}{3} \bigg]_0^{1+i} = \frac{1}{3}(1 + i)^3 = \frac{2}{3}(-1 + i)$$

for every contour between the points $z = 0$ and $z = 1 + i$. Note that the existence of an antiderivative is ensured by the fact that f is entire.

Example 2. The function $1/z^2$, which is continuous everywhere except at the origin, has the antiderivative $-1/z$ in the multiply connected domain $|z| > 0$. Consequently,

$$\int_{z_1}^{z_2} \frac{dz}{z^2} = -\frac{1}{z} \Bigg]_{z_1}^{z_2} = \frac{1}{z_1} - \frac{1}{z_2} \qquad\qquad (z_1 \neq 0,\, z_2 \neq 0)$$

for any contour from z_1 to z_2 which does not pass through the origin. In particular,

$$\int_C \frac{dz}{z^2} = 0$$

when C is the circle $z = 2e^{i\theta}$ $(-\pi \leq \theta \leq \pi)$ about the origin.

Note that the integral of the function $f(z) = 1/z$ around the same circle *cannot* be evaluated in a similar way. Although the derivative of any branch $F(z)$ of log z is $1/z$ (Sec. 25), $F(z)$ is not differentiable, or even defined, along its branch cut. In particular, if the branch cut is a ray $\theta = \alpha$ from the origin, $F'(z)$ fails to exist at the point where that ray intersects the circle C. So C does not lie in a domain throughout which $F'(z) = 1/z$, and we cannot make direct use of an antiderivative (see Exercise 8, however).

Example 3. The fact that the domain $|z| > 0$, $-\pi < \text{Arg } z < \pi$ is simply connected ensures that the analytic function $1/z$ has an antiderivative in that domain. The principal branch Log z of the logarithmic function serves as the antiderivative, and we can write

$$\int_{-2i}^{2i} \frac{dz}{z} = \text{Log } z \Bigg]_{-2i}^{2i} = \text{Log}(2i) - \text{Log}(-2i) = \pi i$$

when the path of integration from $-2i$ to $2i$ is, for instance, the arc

$$z = 2e^{i\theta} \qquad\qquad \left(-\frac{\pi}{2} \leq \theta \leq \frac{\pi}{2}\right)$$

of the circle in Example 2.

Example 4. Let us use an antiderivative to evaluate the integral

(4)
$$\int_{-3}^{3} z^{1/2}\, dz$$

over a given contour C_1 where

(5)
$$z^{1/2} = \sqrt{r}\, e^{i\theta/2} \qquad\qquad (r > 0,\, 0 < \theta < 2\pi)$$

and where, except for the end points $z = \pm 3$, C_1 lies above the real axis (Fig. 33).

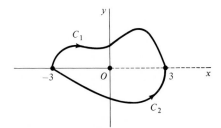

Figure 33

Although the integrand is piecewise continuous on C_1, and the integral therefore exists, that branch of $z^{1/2}$ is not defined on the ray $\theta = 0$, in particular at the point $z = 3$. But another branch,

$$f_1(z) = \sqrt{r}\, e^{i\theta/2} \qquad \left(r > 0, -\frac{\pi}{2} < \theta < \frac{3\pi}{2} \right),$$

is defined and continuous everywhere except on the ray $\theta = -\pi/2$. The values of $f_1(z)$ at all points on C_1 except $z = 3$ coincide with those of our integrand (5); so the integrand can be replaced by $f_1(z)$. Since an antiderivative of $f_1(z)$ is the function

$$F_1(z) = \frac{2}{3} z^{3/2} = \frac{2}{3} r\sqrt{r}\, e^{i3\theta/2} \qquad \left(r > 0, -\frac{\pi}{2} < \theta < \frac{3\pi}{2} \right),$$

we can now write

$$\int_{-3}^{3} z^{1/2}\, dz = \int_{-3}^{3} f_1(z)\, dz = F_1(z)\Big]_{-3}^{3} = 2\sqrt{3}\,(e^{i0} - e^{i3\pi/2}) = 2\sqrt{3}\,(1 + i).$$

(Compare Example 5 in Sec. 32.)

Integral (4) over any contour C_2 below the real axis has another value. There we can replace the integrand by the branch

$$f_2(z) = \sqrt{r}\, e^{i\theta/2} \qquad \left(r > 0, \frac{\pi}{2} < \theta < \frac{5\pi}{2} \right),$$

whose values coincide with those of the branch (5) in the lower half plane. The analytic function

$$F_2(z) = \frac{2}{3} z^{3/2} = \frac{2}{3} r\sqrt{r}\, e^{i3\theta/2} \qquad \left(r > 0, \frac{\pi}{2} < \theta < \frac{5\pi}{2} \right)$$

is an antiderivative of $f_2(z)$. Thus

$$\int_{-3}^{3} z^{1/2}\, dz = \int_{-3}^{3} f_2(z)\, dz = F_2(z)\Big]_{-3}^{3} = 2\sqrt{3}\,(e^{i3\pi} - e^{i3\pi/2}) = 2\sqrt{3}\,(-1 + i).$$

The integral of the function (5) in the positive sense around a closed contour consisting of a path of the first type, in the opposite direction, combined with one of the second type therefore has the value

$$-2\sqrt{3}\,(1 + i) + 2\sqrt{3}\,(-1 + i) = -4\sqrt{3}.$$

EXERCISES

1. Determine the domain of analyticity of the function f, and apply the Cauchy-Goursat theorem to show that

$$\int_C f(z)\, dz = 0.$$

when the simple closed contour C is the circle $|z| = 1$ and when

(a) $f(z) = \dfrac{z^2}{z - 3}$;   (b) $f(z) = ze^{-z}$;   (c) $f(z) = \dfrac{1}{z^2 + 2z + 2}$;

(d) $f(z) = \operatorname{sech} z$;   (e) $f(z) = \tan z$;   (f) $f(z) = \operatorname{Log}(z + 2)$.

2. Let B denote the boundary of the domain between the circle $|z| = 4$ and the square whose sides lie along the lines $x = \pm 1, y = \pm 1$. Assuming that B is oriented so that the points of the domain lie to the left of B, state why

$$\int_B f(z)\, dz = 0$$

when

(a) $f(z) = \dfrac{1}{3z^2 + 1}$;   (b) $f(z) = \dfrac{z + 2}{\sin(z/2)}$;   (c) $f(z) = \dfrac{z}{1 - e^z}$.

3. Let C be a simple closed contour interior to a simple closed contour C_0, where both C and C_0 are oriented in the positive direction. Show how it follows from the Cauchy-Goursat theorem applied to boundaries of multiply connected domains that if a function f is analytic in the closed region bounded by C and C_0, then

$$\int_C f(z)\, dz = \int_{C_0} f(z)\, dz.$$

4. According to Exercise 16, Sec. 32, the integral of $1/(z - z_0)$ around a circle $|z - z_0| = R$ in the positive sense always has the value $2\pi i$, and the integral of $(z - z_0)^{n-1}$ ($n = \pm 1, \pm 2, \ldots$) around the same circle always has the value 0. Use those results and the one in Exercise 3 above to show that

$$\int_C \frac{dz}{z - 2 - i} = 2\pi i, \qquad \int_C (z - 2 - i)^{n-1}\, dz = 0 \qquad (n = \pm 1, \pm 2, \ldots)$$

when C is the boundary of the rectangle $0 \leq x \leq 3, 0 \leq y \leq 2$, described in the positive sense.

5. Use an antiderivative to show that for every contour C extending from a point z_1 to a point z_2,

$$\int_C z^n\, dz = \frac{1}{n + 1}(z_2^{n+1} - z_1^{n+1}) \qquad (n = 0, 1, 2, \ldots).$$

6. Evaluate each of these integrals where the path is an arbitrary contour between the limits of integration:

(a) $\displaystyle\int_i^{i/2} e^{\pi z}\, dz$;   (b) $\displaystyle\int_0^{\pi + 2i} \cos\left(\frac{z}{2}\right) dz$;   (c) $\displaystyle\int_1^3 (z - 2)^3\, dz$.

$$\textit{Ans.} \quad (a)\ (1 + i)/\pi; \quad (b)\ e + (1/e); \quad (c)\ 0.$$

7. With the aid of an antiderivative, show that

$$\int_{C_0} (z - z_0)^{n-1}\, dz = 0 \qquad (n = \pm 1, \pm 2, \ldots)$$

when C_0 is any closed contour which does not pass through the point z_0. [Compare Exercise 16(b), Sec. 32.]

8. (a) By using the branch

$$\log z = \text{Log } r + i\theta \qquad\qquad (r > 0,\, 0 < \theta < 2\pi)$$

of the logarithmic function as an antiderivative of $1/z$, show that

$$\int_{-2i}^{2i} \frac{dz}{z} = -\pi i$$

when the path of integration from $-2i$ to $2i$ is the left-hand half of the circle $|z| = 2$.

(b) Point out how it follows from the results in part (a) and Example 3, Sec. 37, that

$$\int_C \frac{dz}{z} = 2\pi i$$

when C is the entire positively oriented circle $|z| = 2$. [Compare Exercise 16(a), Sec. 32].

9. By finding an antiderivative of another branch of the indicated power function, show that

$$\int_{-1}^{1} z^i\, dz = \frac{1 + e^{-\pi}}{2}(1 - i)$$

where z^i denotes the principal branch

$$z^i = \exp(i \text{ Log } z) \qquad\qquad (|z| > 0,\, -\pi < \text{Arg } z < \pi)$$

and where the path of integration is any contour that, except for its end points, lies above the real axis.

10. Let C denote the entire positively oriented boundary of the half disk $0 \leq r \leq 1$, $0 \leq \theta \leq \pi$, and let $f(z)$ be a continuous function defined on that half disk by writing $f(0) = 0$ and using the branch

$$f(z) = z^{1/2} = \sqrt{r}\, e^{i\theta/2} \qquad\qquad \left(r > 0,\, -\frac{\pi}{2} < \theta < \frac{3\pi}{2}\right)$$

of the multiple-valued function $z^{1/2}$. Show that

$$\int_C f(z)\, dz = 0$$

by evaluating separately the integrals of $f(z)$ over the semicircle and the two radii which constitute C. Why does the Cauchy-Goursat theorem not apply here?

11. *Nested Intervals.* An infinite sequence of closed intervals $a_n \leq x \leq b_n$ ($n = 0, 1, 2, \ldots$) is formed in the following way. The interval $a_1 \leq x \leq b_1$ is either the left-hand or right-hand half of the first interval $a_0 \leq x \leq b_0$, and the interval $a_2 \leq x \leq b_2$ is then one of the two halves of $a_1 \leq x \leq b_1$, etc. Prove that there is a point x_0 which belongs to every one of the closed intervals $a_n \leq x \leq b_n$.

   Suggestion: Note that the left-hand end points a_n represent a bounded nondecreasing sequence of numbers since $a_0 \leq a_n \leq a_{n+1} < b_0$; hence they have a limit A as n tends to infinity. Show that the end points b_n also have a limit B; then show that $A = B$ and write $x_0 = A = B$.

12. *Nested Squares.* A square σ_0: $a_0 \leq x \leq b_0, c_0 \leq y \leq d_0$, where $b_0 - a_0 = d_0 - c_0$, is

divided into four equal squares by line segments parallel to the coordinate axes. One of those four smaller squares σ_1: $a_1 \leqq x \leqq b_1, c_1 \leqq y \leqq d_1$, where $b_1 - a_1 = d_1 - c_1$, is selected according to some rule. It, in turn, is divided into four equal squares one of which, called σ_2, is selected, etc. (see Sec. 34). Prove that there is a point (x_0, y_0) which belongs to each of the closed regions of the infinite sequence $\sigma_0, \sigma_1, \sigma_2, \ldots$.

  Suggestion: Apply the result in Exercise 11 to each of the sequences of closed intervals $a_n \leqq x \leqq b_n$ and $c_n \leqq y \leqq d_n$ $(n = 0, 1, 2, \ldots)$.

13. (a) Accept the fact that the path C_1 from the origin to the point $z = 1$ along the graph (Fig. 34) of the function

$$y = y(x) = x^3 \sin\left(\frac{\pi}{x}\right) \qquad\qquad (0 < x \leqq 1),$$

where $y(0) = y'(0) = 0$, is a smooth arc. Let C denote the closed contour consisting of

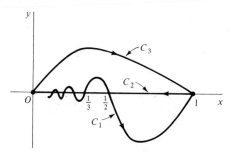

Figure 34

the path C_1 followed by the line segment C_2 along the real axis back to the origin. Verify that C intersects itself at the points $z = 1/n$ $(n = 2, 3, \ldots)$.

(b) Let f be a function which is analytic in a simply connected domain D that contains the paths C_1 and C_2 in part (a), and let C_3 denote a smooth arc in D from the origin to the point $z = 1$ which does not intersect itself and which has only its end points in common with C_1 and C_2. Apply the statement of the Cauchy-Goursat theorem in Sec. 36 to show that

$$\int_{C_1} f(z)\,dz = \int_{C_3} f(z)\,dz \qquad \text{and} \qquad \int_{C_2} f(z)\,dz = -\int_{C_3} f(z)\,dz.$$

(c) Conclude from the results in part (b) that

$$\int_{C} f(z)\,dz = 0,$$

even though the closed contour C, described in part (a), has an infinite number of self-intersection points.

38. Cauchy Integral Formula

Another fundamental result will now be established.

  Theorem. *Let f be analytic everywhere within and on a simple closed contour C, taken in the positive sense. If z_0 is any point interior to C, then*

$$(1) \qquad f(z_0) = \frac{1}{2\pi i} \int_C \frac{f(z)\,dz}{z - z_0}.$$

Formula (1) is called the *Cauchy integral formula*. It says that if a function f is to be analytic within and on a simple closed contour C, then the values of f interior to C are completely determined by the values of f on C.

When the Cauchy integral formula is written

$$(2) \qquad \int_C \frac{f(z)\,dz}{z - z_0} = 2\pi i f(z_0),$$

it can be used to evaluate certain integrals along simple closed contours.

Example. Let C be the positively oriented circle $|z| = 2$. Since the function $f(z) = z/(9 - z^2)$ is analytic within and on C and the point $z_0 = -i$ is interior to C, formula (2) shows that

$$\int_C \frac{z\,dz}{(9 - z^2)(z + i)} = \int_C \frac{z/(9 - z^2)}{z - (-i)}\,dz = 2\pi i \left(\frac{-i}{10}\right) = \frac{\pi}{5}.$$

Our proof of the theorem begins with the fact (Sec. 13) that since f is continuous at z_0, there corresponds to any positive number ε, however small, a positive number δ such that

$$|f(z) - f(z_0)| < \varepsilon \qquad \text{whenever} \qquad |z - z_0| < \delta.$$

Let us now choose a positive number ρ which is less than δ and is so small that the positively oriented circle $|z - z_0| = \rho$, denoted by C_0 in Fig. 35, is interior to C. Then

$$(3) \qquad |f(z) - f(z_0)| < \varepsilon \qquad \text{whenever} \qquad |z - z_0| = \rho.$$

This statement plays an important role in what follows.

Observe that the function $f(z)/(z - z_0)$ is analytic at all points within and on C except at the point z_0. Hence, by the Cauchy-Goursat theorem for multiply connected domains (Sec. 36), its integral around the oriented boundary of the region between C and C_0 has value zero:

$$\int_C \frac{f(z)\,dz}{z - z_0} - \int_{C_0} \frac{f(z)\,dz}{z - z_0} = 0.$$

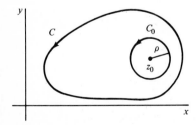

Figure 35

That is,

$$\int_C \frac{f(z)\,dz}{z - z_0} = \int_{C_0} \frac{f(z)\,dz}{z - z_0}.$$

This allows us to write

(4) $$\int_C \frac{f(z)\,dz}{z - z_0} - f(z_0) \int_{C_0} \frac{dz}{z - z_0} = \int_{C_0} \frac{f(z) - f(z_0)}{z - z_0}\,dz.$$

But [see Exercise 16(*a*), Sec. 32]

$$\int_{C_0} \frac{dz}{z - z_0} = 2\pi i;$$

and so equation (4) becomes

(5) $$\int_C \frac{f(z)\,dz}{z - z_0} - 2\pi i f(z_0) = \int_{C_0} \frac{f(z) - f(z_0)}{z - z_0}\,dz.$$

Referring now to statement (3) and noting that the length of C_0 is $2\pi\rho$, we may apply property (9), Sec. 31, of integrals:

$$\left| \int_{C_0} \frac{f(z) - f(z_0)}{z - z_0}\,dz \right| < \frac{\varepsilon}{\rho} 2\pi\rho = 2\pi\varepsilon.$$

In view of equation (5), then,

$$\left| \int_C \frac{f(z)\,dz}{z - z_0} - 2\pi i f(z_0) \right| < 2\pi\varepsilon.$$

Since the left-hand side of this inequality is a nonnegative constant which is less than an arbitrarily small positive number, it must be equal to zero. Hence equation (2) is valid, and the theorem is proved.

39. Derivatives of Analytic Functions

We are now ready to prove that if a function is analytic at a point, its derivatives of all orders exist at that point and are themselves analytic there.

We begin by assuming that f is analytic within and on a positively oriented simple closed contour C, and we let z be any point interior to C. Letting s denote points on C and using the Cauchy integral formula

(1) $$f(z) = \frac{1}{2\pi i} \int_C \frac{f(s)\,ds}{s - z},$$

we shall show that the derivative of f at z exists and has the integral representation

(2) $$f'(z) = \frac{1}{2\pi i} \int_C \frac{f(s)\,ds}{(s - z)^2}.$$

Note that expression (2) can be obtained formally by differentiating the integrand in equation (1) with respect to z.

To verify expression (2), we observe that according to formula (1),

$$\frac{f(z + \Delta z) - f(z)}{\Delta z} = \frac{1}{2\pi i} \int_C \left(\frac{1}{s - z - \Delta z} - \frac{1}{s - z} \right) \frac{f(s)}{\Delta z} \, ds$$

$$= \frac{1}{2\pi i} \int_C \frac{f(s) \, ds}{(s - z - \Delta z)(s - z)}$$

when $0 < |\Delta z| < d$, where d is the shortest distance from z to points s on C. We then use the fact that f is continuous on C to show that the value of the last integral here approaches the value of the integral

$$\int_C \frac{f(s) \, ds}{(s - z)^2}$$

as Δz approaches zero. To accomplish this, we first write the difference

$$\int_C \left[\frac{1}{(s - z - \Delta z)(s - z)} - \frac{1}{(s - z)^2} \right] f(s) \, ds$$

of these two integrals in the form

$$\Delta z \int_C \frac{f(s) \, ds}{(s - z - \Delta z)(s - z)^2}.$$

Next we let M denote the maximum value of $|f(s)|$ on C, and we let L be the length of C. After noting that $|s - z| \geqq d$ and

$$|s - z - \Delta z| \geqq ||s - z| - |\Delta z|| \geqq d - |\Delta z|,$$

we readily obtain the inequality

$$\left| \Delta z \int_C \frac{f(s) \, ds}{(s - z - \Delta z)(s - z)^2} \right| \leqq \frac{|\Delta z| M L}{(d - |\Delta z|) d^2},$$

where the last fraction approaches zero as Δz approaches zero. Consequently,

$$\lim_{\Delta z \to 0} \frac{f(z + \Delta z) - f(z)}{\Delta z} = \frac{1}{2\pi i} \int_C \frac{f(s) \, ds}{(s - z)^2};$$

and expression (2) is now verified.

If we apply the same technique to expression (2), we find that

(3) $$f''(z) = \frac{1}{\pi i} \int_C \frac{f(s) \, ds}{(s - z)^3}.$$

More precisely, when $0 < |\Delta z| < d$,

$$\frac{f'(z + \Delta z) - f'(z)}{\Delta z} = \frac{1}{2\pi i} \int_C \left[\frac{1}{(s - z - \Delta z)^2} - \frac{1}{(s - z)^2} \right] \frac{f(s) \, ds}{\Delta z}$$

$$= \frac{1}{2\pi i} \int_C \frac{2(s - z) - \Delta z}{(s - z - \Delta z)^2 (s - z)^2} f(s) \, ds;$$

and, again since f is continuous on C, it can be shown that the value of the integral

$$\int_C\left[\frac{2(s-z)-\Delta z}{(s-z-\Delta z)^2(s-z)^2}-\frac{2}{(s-z)^3}\right]f(s)\,ds=\int_C\frac{3(s-z)\Delta z-2(\Delta z)^2}{(s-z-\Delta z)^2(s-z)^3}f(s)\,ds$$

approaches zero as Δz approaches zero. (See Exercise 7, Sec. 40.)

Expression (3) establishes the existence of the second derivative of the function f at each point z interior to C. In fact, it shows that if a function f is analytic at a point, its derivative f' is also analytic at that point. For if f is analytic at a point z, there must exist a circle about z such that f is analytic within and on the circle. Then, in view of expression (3), $f''(z)$ exists at each point interior to the circle; and the derivative f' is therefore analytic at z. One can apply the same argument to the analytic function f' to conclude that its derivative f'' is analytic, etc. The following fundamental result is thus obtained.

Theorem. *If a function f is analytic at a point, then its derivatives of all orders are also analytic functions at that point.*

In particular, when a function

$$f(z)=u(x,y)+iv(x,y)$$

is analytic at a point $z=(x,y)$, the analyticity of f' ensures the continuity of f' there. Then, since

$$f'(z)=u_x(x,y)+iv_x(x,y)=v_y(x,y)-iu_y(x,y),$$

we may conclude that the first-order partial derivatives of u and v are continuous at the point. Furthermore, since f'' is analytic and continuous at z and since

$$f''(z)=u_{xx}(x,y)+iv_{xx}(x,y)=v_{yx}(x,y)-iu_{yx}(x,y),$$

etc., it follows that *the partial derivatives of u and v of all orders are continuous at any point where f is analytic*. This result was anticipated in Sec. 20 for partial derivatives of the first and second order in the discussion of harmonic functions.

With the understanding that $f^{(0)}(z)$ denotes $f(z)$ and that $0!=1$, we can use mathematical induction to verify the remarkable formula

(4) $$f^{(n)}(z)=\frac{n!}{2\pi i}\int_C\frac{f(s)\,ds}{(s-z)^{n+1}}\qquad(n=0,1,2,\ldots).$$

When $n=0$, this is just the Cauchy integral formula (1); if we assume that formula (4) holds for any particular nonnegative integer $n=m$, we can, by proceeding as we did in obtaining formulas (2) and (3), show that it is valid when $n=m+1$. The details of the proof are left as an exercise.

Formula (4), and the Cauchy integral formula in particular, can be extended to the case in which the simple closed contour C is replaced by *the oriented boundary B of a multiply connected domain,* described in the theorem of Sec. 36. This can be done when z is a point in the domain and f is analytic in the region consisting of the domain and its boundary B. The method is illustrated in Sec. 46, where such an extension of the Cauchy integral formula is needed.

When written in the form

$$(5) \qquad f^{(n)}(z_0) = \frac{n!}{2\pi i} \int_C \frac{f(z)\,dz}{(z - z_0)^{n+1}} \qquad (n = 0, 1, 2, \ldots),$$

where z_0 is a fixed point interior to C, expression (4) is an extension of the Cauchy integral formula in the notation of Sec. 38, where that formula was derived. This provides the useful formula

$$(6) \qquad \int_C \frac{f(z)\,dz}{(z - z_0)^{n+1}} = \frac{2\pi i}{n!} f^{(n)}(z_0) \qquad (n = 0, 1, 2, \ldots)$$

for evaluating certain integrals. It has already been illustrated in Sec. 38 when $n = 0$.

Example. Let z_0 be any point interior to a positively oriented simple closed contour C. When $f(z) = 1$, formula (6) shows that

$$\int_C \frac{dz}{z - z_0} = 2\pi i, \qquad \int_C \frac{dz}{(z - z_0)^{n+1}} = 0 \qquad (n = 1, 2, \ldots).$$

(Compare Exercise 16, Sec. 32.)

40. Morera's Theorem

Let D denote a domain, which need not be simply connected, and let f be a function that is continuous at each point in D. In addition, suppose that integrals of f around closed contours in D always have value zero. By an elementary argument already used in Sec. 36 when the domain was simply connected, it follows that line integrals of f are independent of path in D. But, according to Sec. 37, this means that f has an antiderivative in D; that is, there is an analytic function F such that $F'(z) = f(z)$ at each point in D.

Now, we know from Sec. 39 that the derivative of an analytic function is also analytic; and, since f is the derivative of F, it follows that f is analytic in D. A theorem due to E. Morera (1856–1909) is now established.

    Theorem. If a function f is continuous throughout a domain D and if

$$(1) \qquad \int_C f(z)\,dz = 0$$

for every closed contour C lying in D, then f is analytic throughout D.

In particular, when D is *simply connected,* we have for the class of continuous functions on D a converse of the statement of the Cauchy-Goursat theorem in Sec. 36, applied to arbitrary closed contours.

EXERCISES

1. Let C denote the boundary of the square whose sides lie along the lines $x = \pm 2$ and $y = \pm 2$, where C is described in the positive sense. Evaluate each of these integrals:

(a) $\displaystyle\int_C \frac{e^{-z}\,dz}{z - (\pi i/2)}$; (b) $\displaystyle\int_C \frac{\cos z}{z(z^2 + 8)}\,dz$; (c) $\displaystyle\int_C \frac{z\,dz}{2z + 1}$;

(d) $\displaystyle\int_C \frac{\tan(z/2)}{(z - x_0)^2}\,dz$ $(-2 < x_0 < 2)$; (e) $\displaystyle\int_C \frac{\cosh z}{z^4}\,dz$.

      Ans. (a) 2π; (b) $\pi i/4$; (c) $-\pi i/2$; (d) $i\pi\,\sec^2(x_0/2)$; (e) 0.

2. Find the value of the integral of $g(z)$ around the circle $|z - i| = 2$ in the positive sense when

(a) $g(z) = \dfrac{1}{z^2 + 4}$; (b) $g(z) = \dfrac{1}{(z^2 + 4)^2}$.

      Ans. (a) $\pi/2$; (b) $\pi/16$.

3. Let C be the circle $|z| = 3$, described in the positive sense. Show that if

$$g(z) = \int_C \frac{2s^2 - s - 2}{s - z}\,ds \qquad (|z| \ne 3),$$

then $g(2) = 8\pi i$. What is the value of $g(z)$ when $|z| > 3$?

4. Let C be a simple closed contour described in the positive sense, and write

$$g(z) = \int_C \frac{s^3 + 2s}{(s - z)^3}\,ds.$$

Show that $g(z) = 6\pi i z$ when z is inside C and that $g(z) = 0$ when z is outside C.

5. Show that if f is analytic within and on a simple closed contour C and z_0 is not on C, then

$$\int_C \frac{f'(z)\,dz}{z - z_0} = \int_C \frac{f(z)\,dz}{(z - z_0)^2}.$$

6. Let f denote a function that is *continuous* on a simple closed contour C. Following the procedure used in Sec. 39, prove that the function

$$g(z) = \frac{1}{2\pi i}\int_C \frac{f(s)\,ds}{s - z}$$

is *analytic* at each point z interior to C and that

$$g'(z) = \frac{1}{2\pi i}\int_C \frac{f(s)\,ds}{(s - z)^2}$$

at such a point.

7. Give details in the derivation of integral representation (3), Sec. 39, for $f''(z)$.

    Suggestion: In the algebraic simplifications, retain the difference $s - z$ as a single term. Also, let D denote the *largest* distance from z to points s on C.

8. Carry out the induction argument used to establish integral representation (4), Sec. 39, for $f^{(n)}(z)$ $(n = 0, 1, 2, \ldots)$.

Suggestion: Use the binomial formula (Exercise 17, Sec. 2) and the suggestion in Exercise 7.

9. Let C be the unit circle $z = e^{i\theta}$ $(-\pi \leq \theta \leq \pi)$. First show that for any real constant a,

$$\int_C \frac{e^{az}}{z} \, dz = 2\pi i.$$

Then write the integral in terms of θ to derive the integration formula

$$\int_0^\pi e^{a \cos \theta} \cos(a \sin \theta) \, d\theta = \pi.$$

10. (a) With the aid of the binomial formula (Exercise 17, Sec. 2), show that for each value of n the function

$$P_n(z) = \frac{1}{n! 2^n} \frac{d^n}{dz^n} (z^2 - 1)^n \qquad (n = 0, 1, 2, \ldots)$$

is a polynomial of degree n.*

(b) Let C denote any positively oriented simple closed contour surrounding a fixed point z. With the aid of the integral representation (4), Sec. 39, for the nth derivative of an analytic function, show that the polynomials in part (a) can be expressed in the form

$$P_n(z) = \frac{1}{2^{n+1} \pi i} \int_C \frac{(s^2 - 1)^n}{(s - z)^{n+1}} \, ds \qquad (n = 0, 1, 2, \ldots).$$

(c) Point out how the integrand in the representation for $P_n(z)$ in part (b) can be written $(s + 1)^n / (s - 1)$ if $z = 1$. Then apply the Cauchy integral formula to show that $P_n(1) = 1$ $(n = 0, 1, 2, \ldots)$. Similarly, show that $P_n(-1) = (-1)^n (n = 0, 1, 2, \ldots)$.

41. Maximum Moduli of Functions

Let f be a function which is analytic in a neighborhood $|z - z_0| < \varepsilon$ of a given point z_0. If C is any positively oriented circle $|z - z_0| = \rho$ where $0 < \rho < \varepsilon$ (Fig. 36),

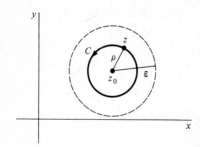

Figure 36

the Cauchy integral formula tells us that

(1)
$$f(z_0) = \frac{1}{2\pi i} \int_C \frac{f(z) \, dz}{z - z_0}.$$

---

*These are the Legendre polynomials which appear in Exercise 4, Sec. 30, when $z = x$. See the footnote to that exercise.

A parametric representation for C is $z = z_0 + \rho e^{i\theta}$ $(0 \leq \theta \leq 2\pi)$, and so expression (1) can be written in the form

$$(2) \qquad\qquad f(z_0) = \frac{1}{2\pi} \int_0^{2\pi} f(z_0 + \rho e^{i\theta})\, d\theta \qquad\qquad (0 < \rho < \varepsilon).$$

We note from this expression that when a function is analytic within and on a given circle, its value at the center is the arithmetic mean of its values on the circle.

From expression (2) we obtain the inequality

$$(3) \qquad\qquad |f(z_0)| \leq \frac{1}{2\pi} \int_0^{2\pi} |f(z_0 + \rho e^{i\theta})|\, d\theta \qquad\qquad (0 < \rho < \varepsilon).$$

On the other hand, if we *assume* that

$$(4) \qquad\qquad |f(z)| \leq |f(z_0)| \qquad \text{whenever} \qquad |z - z_0| < \varepsilon,$$

then

$$(5) \qquad\qquad \frac{1}{2\pi} \int_0^{2\pi} |f(z_0 + \rho e^{i\theta})|\, d\theta \leq |f(z_0)| \qquad\qquad (0 < \rho < \varepsilon);$$

and we see from inequalities (3) and (5) that

$$|f(z_0)| = \frac{1}{2\pi} \int_0^{2\pi} |f(z_0 + \rho e^{i\theta})|\, d\theta,$$

or

$$\int_0^{2\pi} [|f(z_0 + \rho e^{i\theta})| - |f(z_0)|]\, d\theta = 0 \qquad\qquad (0 < \rho < \varepsilon).$$

The integrand in this last integral is continuous in the variable θ and, in view of our assumption (4), it is less than or equal to zero on the entire interval $0 \leq \theta \leq 2\pi$. Because the value of the integral is zero, it follows that the integrand is identically zero; that is,

$$|f(z_0 + \rho e^{i\theta})| = |f(z_0)| \qquad (0 < \rho < \varepsilon,\ 0 \leq \theta \leq 2\pi).$$

In terms of z, this equation takes the form

$$(6) \qquad\qquad |f(z)| = |f(z_0)| \qquad\qquad (|z - z_0| < \varepsilon),$$

where the special case $z = z_0$ is included.

Now, according to Exercise 12, Sec. 20, when the modulus of an analytic function is constant in some domain, the function itself is constant there. In view of equation (6), then, assumption (4) has led us to the conclusion that our function f is constant throughout the neighborhood $|z - z_0| < \varepsilon$. We have thus shown that *if a function f is analytic and not constant in a neighborhood of a point z_0, there exists at least one point z in that neighborhood such that*

$$(7) \qquad\qquad |f(z)| > |f(z_0)|.$$

An important consequence of this result is the following *maximum principle*.

Theorem. *If f is analytic and not constant in a domain, then $|f(z)|$ has no maximum value in that domain. That is, there is no point z_0 in the domain such that $|f(z)| \leq |f(z_0)|$ for all points z in it.*

To show this, we also need to use Corollary 2 in Sec. 103, Chap. 12, of the theorem to be proved there; namely, if an analytic function is not constant throughout a domain D, then it is not constant over any neighborhood $|z - z_0| < \varepsilon$ contained in D. Suppose now that $|f(z)|$ assumes a maximum value at a point z_0 in D. It follows that $|f(z)| \leq |f(z_0)|$ for each point z in a neighborhood $|z - z_0| < \varepsilon$ which is contained in D. But this contradicts statement (7), since f is analytic and not constant in that neighborhood; and the proof is complete.

If a function f that is analytic at each point in the interior of a closed bounded region R is also continuous throughout R, the modulus $|f(z)|$ has a maximum value somewhere in R (Sec. 13). That is, there exists a nonnegative constant M such that $|f(z)| \leq M$ for all points z in R, and equality holds for at least one such point. If f is a constant function, then $|f(z)| = M$ for all z in R. If, however, $f(z)$ is not constant, then, according to the maximum principle, $|f(z)| \neq M$ for any point z in the interior of R. Thus, *if a function f is continuous in a closed bounded region R and is analytic and not constant in the interior of R, the modulus $|f(z)|$ assumes its maximum value on the boundary of R and never in the interior.*

When the function f in this last statement is written $f(z) = u(x, y) + iv(x, y)$, *the component function $u(x, y)$ also has a maximum value in R which is assumed on the boundary of R and never in the interior,* where it is harmonic. For the composite function $g(z) = \exp[f(z)]$ is continuous in R and analytic and not constant in the interior. Consequently, its modulus $|g(z)| = \exp[u(x, y)]$, which is continuous in R, must assume its maximum value in R on the boundary. Because of the increasing nature of the exponential function, it follows that the maximum value of $u(x, y)$ also occurs on the boundary.

Properties of *minimum* values of $|f(z)|$ and $u(x, y)$ are treated in the exercises.

42. Liouville's Theorem and the Fundamental Theorem of Algebra

When f is analytic within and on a circle $|z - z_0| = R$, taken in the positive sense and denoted by C, we know from Sec. 39 that

$$f^{(n)}(z_0) = \frac{n!}{2\pi i} \int_C \frac{f(z)\, dz}{(z - z_0)^{n+1}} \qquad (n = 1, 2, \ldots).$$

Now the maximum value of $|f(z)|$ on the circle C depends, in general, on the radius of C; and if we let M_R denote that maximum value, *Cauchy's inequality* follows:

(1)
$$|f^{(n)}(z_0)| \leq \frac{n! M_R}{R^n} \qquad (n = 1, 2, \ldots).$$

In particular, when $n = 1$,

(2)
$$|f'(z_0)| \leq \frac{M_R}{R}.$$

From this it is easy to show that no entire function except a constant is bounded for all z, as stated below in *Liouville's theorem* in a slightly different way.

Theorem. *If f is entire and bounded for all values of z in the complex plane, then $f(z)$ is constant throughout the plane.*

To start the proof, we note that since f is entire, inequality (2) holds for any choices of z_0 and R. The boundedness condition tells us that a nonnegative constant M exists such that $|f(z)| \leq M$ for all z; and because the constant M_R in inequality (2) is always less than or equal to M, it follows that

$$(3) \qquad\qquad |f'(z_0)| \leq \frac{M}{R}$$

where z_0 is any fixed point in the plane and R is arbitrarily large. Now the number M in inequality (3) is independent of the value of R that is taken. Hence that inequality can hold for arbitrarily large values of R only if $f'(z_0) = 0$. Since the choice of z_0 was arbitrary, this means that $f'(z) = 0$ everywhere in the complex plane. Consequently, f a constant function (Sec. 19).

The following theorem, known as the *fundamental theorem of algebra*, follows readily from Liouville's theorem. Its proof by purely algebraic methods is difficult.

Theorem. *Any polynomial*

$$P(z) = a_0 + a_1 z + a_2 z^2 + \cdots + a_n z^n \qquad\qquad (a_n \neq 0)$$

of degree n ($n \geq 1$) has at least one zero. That is, there exists at least one point z_0 such that $P(z_0) = 0$.

The proof here is by contradiction. Suppose that $P(z)$ is not zero for any value of z. Then the function

$$f(z) = \frac{1}{P(z)}$$

is entire, and it is also bounded for all z. To show that it is bounded, we need only refer to Exercise 8 below to see that there is a positive number R_0 such that

$$(4) \qquad\qquad |f(z)| = \frac{1}{|P(z)|} < \frac{2}{|a_n| |z|^n} < \frac{2}{|a_n| R_0^n}$$

whenever $|z| > R_0$. That is, f is bounded in the region *exterior to* the disk $|z| \leq R_0$. But f is continuous in that closed disk, and this means that f is bounded there too. So f is bounded in the entire plane. It then follows from Liouville's theorem that $f(z)$, and consequently $P(z)$, is constant. But $P(z)$ is not constant, and we have reached a contradiction.*

---

*For an interesting proof of the fundamental theorem using the Cauchy-Goursat theorem, see R. P. Boas, Jr., *Amer. Math. Monthly*, vol. 71, no. 2, p. 180, 1964.

In elementary algebra courses the fundamental theorem is often stated without proof. Then, as a consequence, it is shown that a polynomial of degree n ($n \geq 1$) has no more than n distinct zeros.

EXERCISES

1. Let a function f be continuous in a closed bounded region R, and let it be analytic and not constant throughout the interior of R. Assuming that $f(z) \neq 0$ anywhere in R, prove that $|f(z)|$ has a *minimum value m* in R which occurs on the boundary of R and never in the interior. Do this by applying the corresponding result for maximum values (Sec. 41) to the function $F(z) = 1/f(z)$.

2. Use the function $f(z) = z$ to show that in Exercise 1 the condition $f(z) \neq 0$ anywhere in R is necessary to prove the result of that exercise. That is, show that $|f(z)|$ *can* reach its minimum value at an interior point when that minimum value is zero.

3. Consider the function $f(z) = (z + 1)^2$ and the closed triangular region R with vertices at the points $z = 0$, $z = 2$, and $z = i$. Find points in R where $|f(z)|$ has its maximum and minimum values, thus illustrating results in Sec. 41 and Exercise 1.

   Suggestion: Interpret $|f(z)|$ as the square of the distance between z and -1.

   Ans. $z = 2, z = 0.$

4. Let $f(z) = u(x, y) + iv(x, y)$ be a function which is continuous in a closed bounded region R and analytic and not constant throughout the interior of R. Prove that the component function $u(x, y)$ has a minimum value in R which occurs on the boundary of R and never in the interior. (See Exercise 1.)

5. Let f be the function $f(z) = e^z$ and R the rectangular region $0 \leq x \leq 1, 0 \leq y \leq \pi$. Illustrate results in Sec. 41 and Exercise 4 by finding points in R where the component function $u(x, y) = \text{Re}[f(z)]$ reaches its maximum and minimum values.

   Ans. $z = 1, z = 1 + \pi i.$

6. Suppose that $f(z)$ is entire and that the harmonic function $u(x, y) = \text{Re}[f(z)]$ has an upper bound; that is, $u(x, y) \leq u_0$ for all points (x, y) in the xy plane. Show that $u(x, y)$ must be constant throughout the plane.

   Suggestion: Apply Liouville's theorem (Sec. 42) to the function $g(z) = \exp[f(z)]$.

7. Let the function $f(z) = u(x, y) + iv(x, y)$ be continuous in a closed bounded region R, and let it be analytic and not constant in the interior of R. Show that the component function $v(x, y)$ has maximum and minimum values in R which are reached on the boundary of R and never in the interior, where it is harmonic.

   Suggestion: Apply results in Sec. 41 and Exercise 4 to the function $g(z) = -if(z)$.

8. Show that if

$$P(z) = a_0 + a_1 z + a_2 z^2 + \cdots + a_n z^n \qquad (a_n \neq 0)$$

is a polynomial of degree n ($n \geq 1$), there exists a positive number R_0 such that

$$|P(z)| > \frac{|a_n||z|^n}{2}$$

for each value of z such that $|z| > R_0$.

Suggestion: Observe first that for some positive number R_0 each of the numbers $|a_{n-1}/z|$, $|a_{n-2}/z^2|$, ..., $|a_1/z^{n-1}|$, $|a_0/z^n|$ is less than $|a_n|/(2n)$ when $|z| > R_0$. Hence

$$\left| \frac{a_{n-1}}{z} + \frac{a_{n-2}}{z^2} + \cdots + \frac{a_1}{z^{n-1}} + \frac{a_0}{z^n} \right| < \frac{|a_n|}{2}$$

when $|z| > R_0$. Use this result and the inequality $|z_1 + z_2| \geq ||z_1| - |z_2||$ to show that when $|z| > R_0$,

$$\left| a_n + \left(\frac{a_{n-1}}{z} + \frac{a_{n-2}}{z^2} + \cdots + \frac{a_1}{z^{n-1}} + \frac{a_0}{z^n} \right) \right| > \frac{|a_n|}{2}.$$

The desired result is obtained by multiplying both sides of this inequality by $|z|^n$.

9. Let f be an entire function such that $|f(z)| \leq A|z|$ for all z, where A is a fixed positive number. Show that $f(z) = a_1 z$ where a_1 is a complex constant.

   Suggestion: Use Cauchy's inequality (Sec. 42) to show that the second derivative $f''(z)$ is zero everywhere in the plane. Note that the constant M_R in Cauchy's inequality is less than or equal to $A(|z_0| + R)$.

FIVE

SERIES

This chapter is devoted mainly to series representations of analytic functions. We present theorems which guarantee the existence of such representations, and we develop some facility in manipulating series.

43. Convergence of Sequences and Series

An infinite *sequence* of complex numbers,

$$z_1, z_2, \ldots, z_n, \ldots,$$

has a *limit* z if for each positive number ε there exists a positive integer n_0 such that

$$(1) \qquad\qquad |z_n - z| < \varepsilon \qquad \text{whenever} \qquad n > n_0.$$

Geometrically, this means that for sufficiently large values of n, the points z_n lie in any given ε neighborhood of z (Fig. 37). Since we can choose ε as small as we please, it follows that the points z_n become arbitrarily close to z as their subscripts increase. Note that the value of n_0 that is needed will, in general, depend on the value of ε.

It is left as an exercise to show that a sequence can have at most one limit. That is, a limit is unique if it exists. When the limit z exists, the sequence is said to *converge*

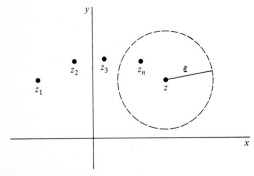

Figure 37

to z; and we write

$$\lim_{n \to \infty} z_n = z.$$

If the sequence has no limit, it *diverges*.

Theorem 1. *Suppose that* $z_n = x_n + iy_n$ $(n = 1, 2, \ldots)$ *and* $z = x + iy$. *Then*

(2) $$\lim_{n \to \infty} z_n = z$$

if and only if

(3) $$\lim_{n \to \infty} x_n = x \quad \text{and} \quad \lim_{n \to \infty} y_n = y.$$

To prove the theorem, we first assume that condition (2) holds and obtain conditions (3) from it. According to condition (2), for each positive number ε there exists a positive integer n_0 such that

$$\left|(x_n - x) + i(y_n - y)\right| < \varepsilon \quad \text{whenever} \quad n > n_0.$$

But

$$\left|x_n - x\right| \leqq \left|(x_n - x) + i(y_n - y)\right|$$

and

$$\left|y_n - y\right| \leqq \left|(x_n - x) + i(y_n - y)\right|.$$

Consequently,

$$\left|x_n - x\right| < \varepsilon \quad \text{and} \quad \left|y_n - y\right| < \varepsilon \quad \text{whenever} \quad n > n_0;$$

that is, conditions (3) are satisfied.

Conversely, if we start with conditions (3), we know that for each positive number ε there are positive integers n_1 and n_2 such that

$$\left|x_n - x\right| < \frac{\varepsilon}{2} \quad \text{whenever} \quad n > n_1$$

and

$$\left|y_n - y\right| < \frac{\varepsilon}{2} \quad \text{whenever} \quad n > n_2.$$

Hence if n_0 is the larger of the two integers n_1 and n_2, then

$$\left|x_n - x\right| < \frac{\varepsilon}{2} \quad \text{and} \quad \left|y_n - y\right| < \frac{\varepsilon}{2} \quad \text{whenever} \quad n > n_0.$$

But

$$\left|(x_n - x) + i(y_n - y)\right| \leqq \left|x_n - x\right| + \left|y_n - y\right|,$$

and so

$$\left|z_n - z\right| < \varepsilon \quad \text{whenever} \quad n > n_0.$$

Condition (2) thus holds.

An infinite *series* of complex numbers,

$$\sum_{n=1}^{\infty} z_n = z_1 + z_2 + \cdots + z_n + \cdots,$$

converges to a number S, called the *sum* of the series, if the sequence

$$S_N = \sum_{n=1}^{N} z_n = z_1 + z_2 + \cdots + z_N \qquad (N = 1, 2, \ldots)$$

of *partial sums* converges to S; we then write

$$\sum_{n=1}^{\infty} z_n = S.$$

Note that since a sequence can have at most one limit, a series can have at most one sum. When a series does not converge, we say that it *diverges*.

Theorem 2. *Suppose that* $z_n = x_n + iy_n$ $(n = 1, 2, \ldots)$ *and* $S = X + iY$. *Then*

(4)
$$\sum_{n=1}^{\infty} z_n = S$$

if and only if

(5)
$$\sum_{n=1}^{\infty} x_n = X \qquad and \qquad \sum_{n=1}^{\infty} y_n = Y.$$

The proof is based on Theorem 1. Let S_N denote the partial sum consisting of the first N terms of the series in condition (4), and observe that

(6)
$$S_N = X_N + iY_N$$

where

$$X_N = \sum_{n=1}^{N} x_n, \qquad Y_N = \sum_{n=1}^{N} y_n.$$

Now condition (4) holds if and only if

$$\lim_{N \to \infty} S_N = S;$$

and, in view of relation (6) and Theorem 1, this condition holds if and only if

(7)
$$\lim_{N \to \infty} X_N = X \qquad and \qquad \lim_{N \to \infty} Y_N = Y.$$

Condition (4) therefore implies conditions (7), and conversely. Since X_N and Y_N are the partial sums of the series in conditions (5), Theorem 2 is proved.

In establishing the fact that the sum of a series is a given number S, it is often convenient to define the *remainder* ρ_N after N terms:

(8)
$$\rho_N = S - S_N.$$

Thus $S = S_N + \rho_N$. Observe that since $|S_N - S| = |\rho_N - 0|$, *a series converges to a number* S *if and only if the sequence of remainders converges to zero.*

Especially important in the theory of complex variables are *power series*. They are series of the form

$$\sum_{n=0}^{\infty} a_n(z - z_0)^n = a_0 + a_1(z - z_0) + a_2(z - z_0)^2 + \cdots + a_n(z - z_0)^n + \cdots$$

where z_0 and the coefficients a_n are complex constants and z may be any point in a stated region containing z_0. In such series involving a variable z we shall denote sums, partial sums, and remainders by $S(z)$, $S_N(z)$, and $\rho_N(z)$, respectively.

EXERCISES

1. Show in two ways that the sequence

$$z_n = -2 + i\frac{(-1)^n}{n^2} \qquad\qquad (n = 1, 2, \ldots)$$

converges to -2.

2. Let r_n denote the moduli and Θ_n the principal values of the arguments of the complex numbers z_n in Exercise 1. Show that the sequence r_n $(n = 1, 2, \ldots)$ converges but that the sequence Θ_n $(n = 1, 2, \ldots)$ does not.

3. Show that

$$\text{if}\quad \lim_{n\to\infty} z_n = z, \qquad \text{then}\qquad \lim_{n\to\infty} |z_n| = |z|.$$

4. By considering the remainders $\rho_N(z)$, verify that

$$\sum_{n=1}^{\infty} z^n = \frac{z}{1-z} \qquad \text{when}\qquad |z| < 1.$$

Suggestion: Use the identity (see Exercise 18, Sec. 7)

$$1 + z + z^2 + \cdots + z^N = \frac{1 - z^{N+1}}{1 - z} \qquad\qquad (z \neq 1)$$

to show that $\rho_N(z) = z^{N+1}/(1 - z)$.

5. Write $z = re^{i\theta}$, where $0 < r < 1$, in the summation formula obtained in Exercise 4. Then, with the aid of Theorem 2 in Sec. 43, show that

$$\sum_{n=1}^{\infty} r^n \cos n\theta = \frac{r\cos\theta - r^2}{1 - 2r\cos\theta + r^2}, \qquad \sum_{n=1}^{\infty} r^n \sin n\theta = \frac{r\sin\theta}{1 - 2r\cos\theta + r^2}$$

when $0 < r < 1$. (Note that these formulas are also valid when $r = 0$.)

6. Show that a limit of a sequence of complex numbers is unique by appealing to the corresponding result for a sequence of real numbers.

7. Show that

$$\text{if}\quad \sum_{n=1}^{\infty} z_n = S, \qquad \text{then}\qquad \sum_{n=1}^{\infty} \bar{z}_n = \bar{S}.$$

8. Let c be any complex number and show that

$$\text{if}\quad \sum_{n=1}^{\infty} z_n = S, \qquad \text{then}\qquad \sum_{n=1}^{\infty} cz_n = cS.$$

9. By recalling the corresponding result for series of real numbers and referring to Theorem 2 in Sec. 43, show that if

$$\sum_{n=1}^{\infty} z_n = S \quad \text{and} \quad \sum_{n=1}^{\infty} w_n = T,$$

then

$$\sum_{n=1}^{\infty} (z_n + w_n) = S + T.$$

10. Let a sequence z_n ($n = 1, 2, \ldots$) converge to a number z. Show that there exists a positive number M such that the inequality $|z_n| \leq M$ holds for all n. Do this by (a) noting that there exists a positive integer n_0 such that

$$|z_n| = |z + (z_n - z)| < |z| + 1$$

whenever $n > n_0$; (b) writing $z_n = x_n + iy_n$ and recalling from the theory of sequences of real numbers that the convergence of x_n and y_n ($n = 1, 2, \ldots$) implies that $|x_n| \leq M_1$ and $|y_n| \leq M_2$ ($n = 1, 2, \ldots$) for some positive numbers M_1 and M_2.

44. Taylor Series

We turn now to *Taylor's theorem,* which is one of the most important results of the chapter.

Theorem. *Let f be analytic everywhere inside a circle C with center at z_0 and radius R. Then at each point z inside C (Fig. 38)*

(1) $$f(z) = f(z_0) + \frac{f'(z_0)}{1!}(z - z_0) + \frac{f''(z_0)}{2!}(z - z_0)^2 + \cdots$$

$$+ \frac{f^{(n)}(z_0)}{n!}(z - z_0)^n + \cdots;$$

that is, the power series here converges to f(z) when $|z - z_0| < R$.

This is the expansion of $f(z)$ into a *Taylor series* about the point z_0. It is the familiar Taylor series from calculus, adapted to functions of a complex variable.

We first prove the theorem when $z_0 = 0$; the proof when z_0 is arbitrary will follow as an immediate consequence. To begin, let z be any fixed point inside the circle C,

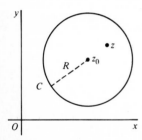

Figure 38

centered now at the origin (Fig. 39). Then write $|z| = r$ and note that $r < R$ where R is the radius of C, as stated in the theorem. Also, let s denote points lying on a positively oriented circle C_1 about the origin with radius R_1 where $r < R_1 < R$; thus $|s| = R_1$. Since z is interior to C_1 and f is analytic within and on that circle, the Cauchy integral formula (Sec. 38) applies:

(2) $$f(z) = \frac{1}{2\pi i} \int_{C_1} \frac{f(s)\,ds}{s - z}.$$

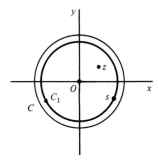

Figure 39

Now

$$\frac{1}{s - z} = \frac{1}{s}\left[\frac{1}{1 - (z/s)}\right];$$

and, by Exercise 18, Sec. 7,

$$\frac{1}{1 - c} = 1 + c + c^2 + \cdots + c^{N-1} + \frac{c^N}{1 - c} \qquad (N = 1, 2, \ldots)$$

when c is any complex number other than unity. Hence

$$\frac{1}{s - z} = \frac{1}{s}\left[1 + \left(\frac{z}{s}\right) + \left(\frac{z}{s}\right)^2 + \cdots + \left(\frac{z}{s}\right)^{N-1} + \frac{(z/s)^N}{1 - (z/s)}\right];$$

and consequently,

$$\frac{1}{s - z} = \frac{1}{s} + \frac{1}{s^2}z + \frac{1}{s^3}z^2 + \cdots + \frac{1}{s^N}z^{N-1} + z^N\frac{1}{(s - z)s^N}.$$

We next multiply through this equation by $f(s)/(2\pi i)$ and then integrate each of the terms with respect to s around C_1. In view of expression (2) and the fact (Sec. 39) that

$$\frac{1}{2\pi i} \int_{C_1} \frac{f(s)\,ds}{s^{n+1}} = \frac{f^{(n)}(0)}{n!} \qquad (n = 0, 1, 2, \ldots),$$

we can write the result as follows:

(3) $$f(z) = f(0) + \frac{f'(0)}{1!}z + \frac{f''(0)}{2!}z^2 + \cdots + \frac{f^{(N-1)}(0)}{(N-1)!}z^{N-1} + \rho_N(z)$$

where

(4)
$$\rho_N(z) = \frac{z^N}{2\pi i} \int_{C_1} \frac{f(s)\,ds}{(s-z)s^N}.$$

Recalling that $|z| = r$ and $|s| = R_1$, where $r < R_1$, we note that

$$|s - z| \geq ||s| - |z|| = R_1 - r.$$

Hence it follows from expression (4) that when M_1 denotes the maximum value of $|f(s)|$ on C_1,

$$|\rho_N(z)| \leq \frac{r^N}{2\pi} \frac{M_1}{(R_1-r)R_1^N} 2\pi R_1 = \frac{M_1 R_1}{R_1-r}\left(\frac{r}{R_1}\right)^N.$$

But $(r/R_1) < 1$, and therefore

$$\lim_{N\to\infty} \rho_N(z) = 0.$$

Thus, for each point z interior to C, the limit as N tends to infinity of the sum of the first N terms on the right in equation (3) is $f(z)$. That is, if f is analytic inside a circle centered at the origin with radius R, then $f(z)$ is represented by the series expansion

(5)
$$f(z) = f(0) + \frac{f'(0)}{1!}z + \frac{f''(0)}{2!}z^2 + \cdots + \frac{f^{(n)}(0)}{n!}z^n + \cdots$$

in the open disk $|z| < R$. This special case of series (1) when $z_0 = 0$ is called a *Maclaurin series*. If we agree, as we did in Sec. 39, that $f^{(0)}(z) = f(z)$ and $0! = 1$, we can write series (5) more compactly as

(6)
$$f(z) = \sum_{n=0}^{\infty} \frac{f^{(n)}(0)}{n!}z^n \qquad (|z| < R).$$

Suppose now that f is as in the statement of the theorem, where the center of the circle C of radius R is an arbitrary point z_0. Since $f(z)$ is analytic when $|z - z_0| < R$, the composite function $f(z + z_0)$ is analytic when $|(z + z_0) - z_0| < R$. But this last inequality is simply $|z| < R$; and if we write $g(z) = f(z + z_0)$, the analyticity of g inside the circle $|z| = R$ implies the existence of a Maclaurin series representation:

$$g(z) = \sum_{n=0}^{\infty} \frac{g^{(n)}(0)}{n!}z^n \qquad (|z| < R).$$

That is,

$$f(z + z_0) = \sum_{n=0}^{\infty} \frac{f^{(n)}(z_0)}{n!}z^n \qquad (|z| < R).$$

Replacing z by $z - z_0$ in this equation and its condition of validity, we arrive at the desired Taylor series representation for $f(z)$ about the point z_0:

(7)
$$f(z) = \sum_{n=0}^{\infty} \frac{f^{(n)}(z_0)}{n!}(z - z_0)^n \qquad (|z - z_0| < R).$$

45. Observations and Examples

When it is known that f is analytic at all points within a circle centered at z_0, convergence of the Taylor series about z_0 to $f(z)$ for each point z within that circle is ensured; no test for the convergence of the series is required. In fact, according to Taylor's theorem, the series converges to $f(z)$ within the circle about z_0 whose radius is the distance from z_0 to the nearest point z_1 where f fails to be analytic. In Sec. 49 we shall find that this is actually the largest circle centered at z_0 such that the series converges to $f(z)$ for all z interior to it.

Also, in Sec. 50 we shall see that if there are constants a_n ($n = 0, 1, 2, \ldots$) such that

$$(1) \qquad f(z) = \sum_{n=0}^{\infty} a_n (z - z_0)^n$$

for all points z interior to some circle centered at z_0, then the power series here must be *the* Taylor series for f about z_0. This observation will be useful in the following examples, where it is emphasized that the coefficients a_n in expansion (1) can often be found in more efficient ways than by appealing directly to the formula $a_n = f^{(n)}(z_0)/n!$ in Taylor's theorem.

Example 1. Since the function $f(z) = e^z$ is entire, it has a Maclaurin series representation which is valid for all z. Here $f^{(n)}(z) = e^z$; and because $f^{(n)}(0) = 1$, it follows that

$$(2) \qquad e^z = \sum_{n=0}^{\infty} \frac{z^n}{n!} \qquad\qquad (|z| < \infty).$$

Note that if $z = x + i0$, expansion (2) becomes

$$e^x = \sum_{n=0}^{\infty} \frac{x^n}{n!} \qquad\qquad (-\infty < x < \infty).$$

Example 2. If $f(z) = \sin z$, then $f^{(2n)}(0) = 0$ ($n = 0, 1, 2, \ldots$) and $f^{(2n+1)}(0) = (-1)^n$ ($n = 0, 1, 2, \ldots$). Hence

$$(3) \qquad \sin z = \sum_{n=0}^{\infty} (-1)^n \frac{z^{2n+1}}{(2n + 1)!} \qquad\qquad (|z| < \infty).$$

The condition $|z| < \infty$ once again follows from the fact that the function is entire.

Differentiating each side of equation (3) with respect to z and interchanging the symbols for differentiation and summation on the right-hand side, we have the expansion

$$(4) \qquad \cos z = \sum_{n=0}^{\infty} (-1)^n \frac{z^{2n}}{(2n)!} \qquad\qquad (|z| < \infty).$$

Such term by term differentiation of a Maclaurin or Taylor series is justified in Sec. 49.

Because $\sinh z = -i \sin(iz)$ (Sec. 24), we need only replace z by iz on each side of equation (3) and multiply through the resulting equation by $-i$ to see that

(5)
$$\sinh z = \sum_{n=0}^{\infty} \frac{z^{2n+1}}{(2n+1)!} \qquad (|z| < \infty).$$

Differentiation of each side of this equation then shows that

(6)
$$\cosh z = \sum_{n=0}^{\infty} \frac{z^{2n}}{(2n)!} \qquad (|z| < \infty).$$

Example 3. Another Maclaurin series representation is

(7)
$$\frac{1}{1-z} = \sum_{n=0}^{\infty} z^n \qquad (|z| < 1).$$

For the derivatives of the function $f(z) = 1/(1-z)$, which fails to be analytic at $z = 1$, are

$$f^{(n)}(z) = \frac{n!}{(1-z)^{n+1}} \qquad (n = 0, 1, 2, \ldots).$$

In particular, $f^{(n)}(0) = n!$. Note that expansion (7) gives us the sum of an infinite *geometric series*, where z is the common ratio of adjacent terms:

$$1 + z + z^2 + \cdots + z^n + \cdots = \frac{1}{1-z} \qquad (|z| < 1).$$

This is essentially the summation formula which was found in another way in Exercise 4, Sec. 43.

By substituting $-z$ for z in equation (7) and noting that $|z| < 1$ when $|-z| < 1$, we see that

(8)
$$\frac{1}{1+z} = \sum_{n=0}^{\infty} (-1)^n z^n \qquad (|z| < 1).$$

In a similar manner, such expansions as

$$\frac{1}{1-z^2} = \sum_{n=0}^{\infty} z^{2n} \qquad (|z| < 1)$$

are also obtained from equation (7). The condition on z here follows from the fact that $|z| < 1$ when $|z^2| < 1$.

Example 4. When $z \neq 0$ and $f(z) = 1/z$,

$$f^{(n)}(z) = \frac{(-1)^n n!}{z^{n+1}} \qquad (n = 0, 1, 2, \ldots);$$

so $f^{(n)}(1) = (-1)^n n!$. Hence the expansion of the function $1/z$ into a Taylor series about the point $z = 1$ is

(9)
$$\frac{1}{z} = \sum_{n=0}^{\infty} (-1)^n (z - 1)^n \qquad (|z - 1| < 1).$$

This expansion is valid in the indicated region since the function is analytic at all points except $z = 0$.

Observe that once expansion (7) is known, expansion (9) is actually obtained more easily by simply replacing z by $1 - z$ in that earlier representation and noting that the inequality $|1 - z| < 1$ is the same as $|z - 1| < 1$.

Example 5. For our final example, let us expand the function

$$f(z) = \frac{1 + 2z}{z^2 + z^3} = \frac{1}{z^2}\left(2 - \frac{1}{1 + z}\right)$$

into a series involving powers of z. We cannot find a Maclaurin series for $f(z)$ itself since it is not analytic at $z = 0$; but expansion (8) provides one for $1/(1 + z)$. Thus, when $0 < |z| < 1$, it follows that

$$\frac{1 + 2z}{z^2 + z^3} = \frac{1}{z^2}(2 - 1 + z - z^2 + z^3 - z^4 + z^5 - \cdots)$$

$$= \frac{1}{z^2} + \frac{1}{z} - 1 + z - z^2 + z^3 - \cdots.$$

We call such terms as $1/z^2$ and $1/z$ *negative* powers of z since they can be written z^{-2} and z^{-1}, respectively.

EXERCISES

1. Show that

$$e^z = e \sum_{n=0}^{\infty} \frac{(z - 1)^n}{n!} \qquad (|z| < \infty).$$

2. By using the definition (Sec. 23)

$$\sin z = \frac{e^{iz} - e^{-iz}}{2i},$$

   along with results in Exercises 8 and 9 of Sec. 43 to justify the steps, derive the Maclaurin series for $\sin z$ in Sec. 45 from the representation

$$e^z = \sum_{n=0}^{\infty} \frac{z^n}{n!} \qquad (|z| < \infty)$$

   for e^z there.

3. Expand $\cos z$ into a Taylor series about the point $z = \pi/2$.
4. Expand $\sinh z$ into a Taylor series about the point $z = \pi i$.
5. Derive the Maclaurin series in Sec. 45 for $\cosh z$ by
   (a) appealing directly to Taylor's theorem;
   (b) referring to the identity $\cosh z = \cos(iz)$ (Sec. 24) and using the Maclaurin series representation (Sec. 45)

$$\cos z = \sum_{n=0}^{\infty} (-1)^n \frac{z^{2n}}{(2n)!} \qquad (|z| < \infty).$$

6. What is the largest circle within which the Maclaurin series for the function tanh z converges to tanh z for all z? Write the first two nonzero terms of that series.

7. Show that

(a)
$$\frac{1}{z^2} = \sum_{n=0}^{\infty} (n + 1)(z + 1)^n \qquad (|z + 1| < 1);$$

(b)
$$\frac{1}{z^2} = \frac{1}{4} \sum_{n=0}^{\infty} (-1)^n (n + 1) \left(\frac{z - 2}{2}\right)^n \qquad (|z - 2| < 2).$$

8. Show that when $0 < |z| < 4$,

$$\frac{1}{4z - z^2} = \frac{1}{4z} + \sum_{n=0}^{\infty} \frac{z^n}{4^{n+2}}.$$

9. Substitute $1/z$ for z in expansion (8), Sec. 45, as well as in its condition of validity, to obtain a representation for $1/(1 + z)$ in negative powers of z that is valid when $1 < |z| < \infty$.

$$Ans. \quad \frac{1}{1 + z} = \sum_{n=1}^{\infty} \frac{(-1)^{n+1}}{z^n}.$$

10. Show that when $z \neq 0$,

$$\frac{\sin(z^2)}{z^4} = \frac{1}{z^2} - \frac{z^2}{3!} + \frac{z^6}{5!} - \frac{z^{10}}{7!} + \cdots.$$

11. Show that when $0 < |z - 1| < 2$,

$$\frac{z}{(z - 1)(z - 3)} = \frac{-1}{2(z - 1)} - 3 \sum_{n=0}^{\infty} \frac{(z - 1)^n}{2^{n+2}}.$$

46. Laurent Series

If a function f fails to be analytic at a point z_0, we cannot apply Taylor's theorem at that point. It is often possible, however, to find a series representation for $f(z)$ involving both positive and negative powers of $z - z_0$. This has been illustrated in Example 5 of Sec. 45, as well as in Exercises 8, 10, and 11 there. We now present the theory of such representations, and we begin with *Laurent's theorem*.

Theorem. *Let C_0 and C_1 denote two positively oriented circles centered at a point z_0, where C_0 is smaller than C_1 (Fig. 40). If a function f is analytic on C_0 and C_1*

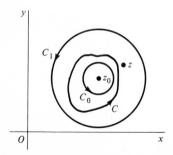

Figure 40

and throughout the annular domain between them, then at each point z in that domain f (z) is represented by the expansion

(1)
$$f(z) = \sum_{n=0}^{\infty} a_n(z - z_0)^n + \sum_{n=1}^{\infty} \frac{b_n}{(z - z_0)^n}$$

where

(2)
$$a_n = \frac{1}{2\pi i} \int_{C_1} \frac{f(z)\, dz}{(z - z_0)^{n+1}} \qquad (n = 0, 1, 2, \ldots)$$

and

(3)
$$b_n = \frac{1}{2\pi i} \int_{C_0} \frac{f(z)\, dz}{(z - z_0)^{-n+1}} \qquad (n = 1, 2, \ldots).$$

The series here is called a *Laurent series*.

We let R_0 and R_1 denote the radii of C_0 and C_1, respectively. Thus $R_0 < R_1$; and if f is analytic at every point inside and on C_1 except at the point z_0 itself, the radius R_0 may be taken arbitrarily small. Expansion (1) is then valid when

$$0 < |z - z_0| < R_1.$$

If f is analytic at *all* points inside and on C_1, we need only write the integrand in expression (3) as $f(z)(z - z_0)^{n-1}$ to see that it is analytic inside and on C_0. For $n - 1 \geqq 0$ when n is a positive integer. So all the coefficients b_n are zero; and because (Sec. 39)

$$\frac{1}{2\pi i} \int_{C_1} \frac{f(z)\, dz}{(z - z_0)^{n+1}} = \frac{f^{(n)}(z_0)}{n!} \qquad (n = 0, 1, 2, \ldots),$$

expansion (1) reduces to a Taylor series about z_0.

Since the two integrands $f(z)/(z - z_0)^{n+1}$ and $f(z)/(z - z_0)^{-n+1}$ in expressions (2) and (3) are analytic throughout the annular domain $R_0 < |z - z_0| < R_1$ and on its boundary, any simple closed contour C around that domain in the positive direction (Fig. 40) can be used as a path of integration in place of the circular paths C_0 and C_1. (See Exercise 3, Sec. 37.) Thus the Laurent series (1) can be written

(4)
$$f(z) = \sum_{n=-\infty}^{\infty} c_n(z - z_0)^n \qquad (R_0 < |z - z_0| < R_1)$$

where

(5)
$$c_n = \frac{1}{2\pi i} \int_C \frac{f(z)\, dz}{(z - z_0)^{n+1}} \qquad (n = 0, \pm 1, \pm 2, \ldots).$$

In particular cases, of course, some of the coefficients may be zero.

Example 1. The function

$$f(z) = \frac{1}{(z - 1)^2} \qquad (0 < |z - 1| < \infty)$$

already has the form (4), where $z_0 = 1$. Here $c_{-2} = 1$ and all the other coefficients are zero. This is in agreement with expression (5), where the path C can be a positively oriented circle $|z - 1| = R$. More precisely, expression (5) tells us that

$$c_n = \frac{1}{2\pi i} \int_C \frac{dz}{(z - 1)^{n+3}} \qquad (n = 0, \pm 1, \pm 2, \ldots)$$

when C is such a circle, and the above values of c_n can be obtained by evaluating these integrals (see Exercise 16, Sec. 32).

The coefficients in expansion (4) are usually found by other means than by using expression (5).

Example 2. The expansion

$$\frac{e^z}{z^2} = \frac{1}{z^2} + \frac{1}{z} + \frac{1}{2!} + \frac{z}{3!} + \frac{z^2}{4!} + \cdots \qquad (0 < |z| < \infty)$$

follows from the Maclaurin series representation

$$e^z = \sum_{n=0}^{\infty} \frac{z^n}{n!} = 1 + \frac{z}{1!} + \frac{z^2}{2!} + \frac{z^3}{3!} + \frac{z^4}{4!} + \cdots \qquad (|z| < \infty).$$

Example 3. Likewise,

$$e^{1/z} = 1 + \sum_{n=1}^{\infty} \frac{1}{n! z^n} \qquad (0 < |z| < \infty).$$

It will be shown in Sec. 50 that series representations of type (1) or (4) are unique. So the series found in Examples 2 and 3 must be Laurent series in powers of z for the domain $0 < |z| < \infty$.

We shall prove Laurent's theorem first when $z_0 = 0$, in which case the circles C_0 and C_1 are centered at the origin (Fig. 41). The verification of the theorem when z_0 is arbitrary will follow readily.

We begin the proof when $z_0 = 0$ by showing that if z is a fixed point in the annular domain between C_0 and C_1, then

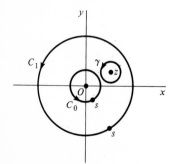

Figure 41

(6) $$f(z) = \frac{1}{2\pi i} \int_{C_1} \frac{f(s)\, ds}{s - z} - \frac{1}{2\pi i} \int_{C_0} \frac{f(s)\, ds}{s - z}.$$

This equation is valid in view of the remarks near the end of Sec. 39 regarding the Cauchy integral formula when the path of integration is the oriented boundary of a multiply connected domain. To give the details in this instance, we construct a positively oriented circle γ about z which is small enough to be completely contained in the annular domain, as shown in Fig. 41. It then follows from the adaptation of the Cauchy-Goursat theorem to functions analytic in a closed region whose interior is a multiply connected domain (Sec. 36) that

$$\int_{C_1} \frac{f(s)\, ds}{s - z} - \int_{C_0} \frac{f(s)\, ds}{s - z} - \int_{\gamma} \frac{f(s)\, ds}{s - z} = 0.$$

According to the Cauchy integral formula, the value of the third integral here is $2\pi i f(z)$; hence equation (6) is valid.

In the first integral in equation (6) we write, as we did in the proof of Taylor's theorem (Sec. 44),

(7) $$\frac{1}{s - z} = \frac{1}{s} + \frac{1}{s^2} z + \frac{1}{s^3} z^2 + \cdots + \frac{1}{s^N} z^{N-1} + z^N \frac{1}{(s - z)s^N}.$$

As for the second integral, we note that

$$-\frac{1}{s - z} = \frac{1}{z - s}.$$

Consequently, by simply interchanging z and s in equation (7), we can write

(8) $$-\frac{1}{s - z} = \frac{1}{z} + \frac{1}{s^{-1}} \frac{1}{z^2} + \cdots + \frac{1}{s^{-N+1}} \frac{1}{z^N} + \frac{1}{z^N} \left(\frac{s^N}{z - s} \right).$$

According to equation (6), then,

$$f(z) = a_0 + a_1 z + a_2 z^2 + \cdots + a_{N-1} z^{N-1} + \rho_N(z)$$

$$+ \frac{b_1}{z} + \frac{b_2}{z^2} + \cdots + \frac{b_N}{z^N} + \sigma_N(z)$$

where the numbers a_n $(n = 0, 1, 2, \ldots, N - 1)$ and b_n $(n = 1, 2, \ldots, N)$ are given by the equations

(9) $$a_n = \frac{1}{2\pi i} \int_{C_1} \frac{f(s)\, ds}{s^{n+1}}, \qquad b_n = \frac{1}{2\pi i} \int_{C_0} \frac{f(s)\, ds}{s^{-n+1}}$$

and where

$$\rho_N(z) = \frac{z^N}{2\pi i} \int_{C_1} \frac{f(s)\, ds}{(s - z)s^N}, \qquad \sigma_N(z) = \frac{1}{2\pi i z^N} \int_{C_0} \frac{s^N f(s)\, ds}{z - s}.$$

Now write $|z| = r$, so that $R_0 < r < R_1$. The demonstration that $\rho_N(z)$ approaches zero as N tends to infinity is the same as in the proof of Taylor's theorem in Sec. 44.

If M_0 is the maximum value of $|f(s)|$ on C_0, then

$$|\sigma_N(z)| \leq \frac{M_0 R_0}{r - R_0} \left(\frac{R_0}{r} \right)^N$$

where $(R_0/r) < 1$. Hence $\sigma_N(z)$ also approaches zero as N tends to infinity. This completes the proof of Laurent's theorem when $z_0 = 0$ since expressions (2) and (3) for a_n and b_n are the same as expressions (9) when $z_0 = 0$ and the variable of integration s is used.

To extend the proof so that the centers of the circles C_0 and C_1 may be any point z_0 in the finite plane, we let f be a function satisfying the conditions in the theorem; and, just as we did in the proof of Taylor's theorem, we write $g(z) = f(z + z_0)$. If R_0 and R_1 denote the radii of C_0 and C_1, respectively, then $f(z)$ is analytic in the annular region $R_0 \leq |z - z_0| \leq R_1$. Hence the function $f(z + z_0)$ is analytic when $R_0 \leq |(z + z_0) - z_0| \leq R_1$. That is, g is analytic in the region $R_0 \leq |z| \leq R_1$, centered at the origin; and this means that $g(z)$ has a Laurent series representation:

$$(10) \qquad g(z) = \sum_{n=0}^{\infty} a_n z^n + \sum_{n=1}^{\infty} \frac{b_n}{z^n} \qquad (R_0 < |z| < R_1).$$

Furthermore,

$$(11) \qquad a_n = \frac{1}{2\pi i} \int_{\Gamma_1} \frac{g(z) \, dz}{z^{n+1}} \qquad (n = 0, 1, 2, \ldots),$$

$$(12) \qquad b_n = \frac{1}{2\pi i} \int_{\Gamma_0} \frac{g(z) \, dz}{z^{-n+1}} \qquad (n = 1, 2, \ldots)$$

where Γ_1 and Γ_0 are the positively oriented circles $|z| = R_1$ and $|z| = R_0$, respectively.

Expansion (1), with coefficients (2) and (3), is finally obtained if we write $f(z + z_0)$ instead of $g(z)$ in equations (10) through (12) and then replace z by $z - z_0$ in the resulting equations, as well as in the condition of validity $R_0 < |z| < R_1$. [See Exercise 15(b), Sec. 32.] This finishes the proof of Laurent's theorem.

47. Further Properties of Series

If a series

$$(1) \qquad \sum_{n=1}^{\infty} z_n$$

of complex numbers $z_n = x_n + iy_n$ converges, we know from Theorem 2, Sec. 43, that the series of real numbers

$$(2) \qquad \sum_{n=1}^{\infty} x_n \qquad \text{and} \qquad \sum_{n=1}^{\infty} y_n$$

both converge. Now a necessary condition for the convergence of an infinite series of real numbers is that the nth term approach zero as n tends to infinity; so the numbers

x_n and y_n must approach zero if series (1) converges. Hence z_n approaches zero. This shows that *a necessary condition for the convergence of series* (1) *is that*

$$(3) \qquad\qquad \lim_{n\to\infty} z_n = 0.$$

The terms of a convergent series of complex numbers are therefore *bounded*. More precisely, there exists a positive constant M such that $|z_n| \leq M$ for each positive integer n. (See Exercise 10, Sec. 43.)

Suppose that series (1) is *absolutely convergent*; that is, the series

$$\sum_{n=1}^{\infty} |z_n| = \sum_{n=1}^{\infty} \sqrt{x_n^2 + y_n^2}$$

of real numbers converges. It follows from the comparison test for series of real numbers that the two series

$$\sum_{n=1}^{\infty} |x_n| \qquad \text{and} \qquad \sum_{n=1}^{\infty} |y_n|$$

converge. Both of the series (2) are then absolutely convergent, and they are therefore convergent because the absolute convergence of a series of real numbers implies the convergence of the series itself. But when the series (2) both converge, series (1) converges. Consequently, *absolute convergence of a series of complex numbers implies the convergence of that series.*

We now prove an important theorem on the convergence of power series. These and various results which follow apply as well to general power series of the form

$$\sum_{n=0}^{\infty} a_n(z - z_0)^n$$

but are given only for the special case in which $z_0 = 0$. Proofs in the general case are essentially the same, and many of our results are generalized by merely substituting $z - z_0$ for z.

Theorem. *If a power series*

$$(4) \qquad\qquad \sum_{n=0}^{\infty} a_n z^n$$

converges when $z = z_1$ $(z_1 \neq 0)$, *it is absolutely convergent for every value of z such that* $|z| < |z_1|$.

The proof here depends on a property of series pointed out at the beginning of the section. Assuming that the series

$$\sum_{n=0}^{\infty} a_n z_1^n \qquad\qquad\qquad (z_1 \neq 0)$$

converges, we note that the terms $a_n z_1^n$ are bounded; that is,

$$|a_n z_1^n| \leq M \qquad\qquad\qquad (n = 0, 1, 2, \ldots)$$

for some positive constant M. Write

$$\left|\frac{z}{z_1}\right| = \rho \qquad \text{where } |z| < |z_1|;$$

then

$$|a_n z^n| = |a_n z_1^n| \left|\frac{z}{z_1}\right|^n \leqq M\rho^n.$$

Now the series whose terms are the real numbers $M\rho^n$ is a geometric series which is convergent because $\rho < 1$. We therefore conclude from the comparison test for series of real numbers that the series

$$\sum_{n=0}^{\infty} |a_n z^n|$$

converges, and the theorem is proved.

The set of all points inside some circle about the origin is therefore a region of convergence for the power series (4). The greatest circle about the origin such that the series converges at each point inside is called the *circle of convergence* of the power series. *The series cannot converge at any point z_2 outside that circle,* according to the above theorem; for if it did, it would converge everywhere inside the circle centered at the origin and passing through z_2. The first circle could not then be the circle of convergence.

If in series (4) we replace z by $z - z_0$, we have the series

$$(5) \qquad \sum_{n=0}^{\infty} a_n (z - z_0)^n.$$

We see at once from the above theorem that if this series converges at a point z_1, where $z_1 \neq z_0$, then it is absolutely convergent when

$$|z - z_0| < |z_1 - z_0|.$$

That is, it is absolutely convergent at every point z interior to the circle centered at z_0 and passing through z_1. Likewise, if the series

$$\sum_{n=1}^{\infty} \frac{b_n}{(z - z_0)^n}$$

converges when $z = z_1$ ($z_1 \neq z_0$), then it is absolutely convergent at every point z *exterior to* the circle centered at z_0 and passing through z_1.

48. Uniform Convergence

Let C denote the circle of convergence $|z| = R$ of a power series about the point $z = 0$, and let the function $S(z)$ represent the sum of that series:

$$(1) \qquad S(z) = \sum_{n=0}^{\infty} a_n z^n \qquad (|z| < R).$$

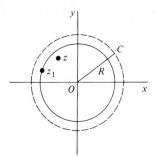

Figure 42

Then write the remainder function

(2)
$$\rho_N(z) = S(z) - \sum_{n=0}^{N-1} a_n z^n \qquad (|z| < R).$$

Since the power series converges for any fixed value of z where $|z| < R$, we know that the remainder $\rho_N(z)$ approaches zero for any such z as N tends to infinity. According to definition (1), Sec. 43, of the limit of a sequence, this means that corresponding to each positive number ε there is a positive integer N_ε such that

(3)
$$|\rho_N(z)| < \varepsilon \qquad \text{whenever} \qquad N > N_\varepsilon.$$

If we take any point z such that $|z| \leq |z_1|$ where $|z_1| < R$ (Fig. 42), condition (3) must, of course, hold; but we can also show that corresponding to each positive number ε, a single value of N_ε may be selected such that condition (3) holds regardless of what value of z is chosen in the closed disk $|z| \leq |z_1|$. When the choice of N_ε may thus depend on the value of ε but not on the choice of z in a given region, the convergence of a series is said to be *uniform* in that region.

To establish uniform convergence of the power series (1) in the region $|z| \leq |z_1|$, we let m and N denote any two positive integers where $m > N$ and write the remainder (2) in the form

(4)
$$\rho_N(z) = \lim_{m \to \infty} \sum_{n=N}^{m} a_n z^n.$$

Because there are points with modulus greater than $|z_1|$ for which series (1) converges, we know from the theorem in Sec. 47 that the series

$$\sum_{n=0}^{\infty} |a_n z_1^n|$$

converges. The remainder after N terms for this series is the nonnegative real number

(5)
$$\sigma_N = \lim_{m \to \infty} \sum_{n=N}^{m} |a_n z_1^n|.$$

Now, in view of Exercise 3, Sec. 43,

$$|\rho_N(z)| = \lim_{m \to \infty} \left| \sum_{n=N}^{m} a_n z^n \right|;$$

and when $|z| \leq |z_1|$,

$$\left| \sum_{n=N}^{m} a_n z^n \right| \leq \sum_{n=N}^{m} |a_n||z|^n \leq \sum_{n=N}^{m} |a_n||z_1|^n = \sum_{n=N}^{m} |a_n z_1^n|.$$

Hence

(6) $|\rho_N(z)| \leq \sigma_N$ when $|z| \leq |z_1|$.

Since σ_N are the remainders of a convergent series, they tend to zero as N tends to infinity. That is, for each positive number ε, an integer N_ε exists such that

(7) $\sigma_N < \varepsilon$ whenever $N > N_\varepsilon$.

Because of conditions (6) and (7), then, condition (3) holds for all points z in the region $|z| \leq |z_1|$, and the value of N_ε is independent of the choice of z. The convergence of the series (1) is therefore uniform in the region $|z| \leq |z_1|$. We state our result in the following theorem.

> **Theorem.** *The power series* (1) *is uniformly convergent for all points z inside and on any circle about the origin that is interior to the circle of convergence.*

For each N the partial sum

$$S_N(z) = \sum_{n=0}^{N-1} a_n z^n$$

of the series (1) is a polynomial and is therefore continuous at the point z_1, which was arbitrarily selected inside the circle C of radius R. We now show that the sum $S(z)$ of the series is also continuous at z_1. That is, for each positive number ε there is a positive number δ such that

(8) $|S(z) - S(z_1)| < \varepsilon$ whenever $|z - z_1| < \delta$,

the number δ being small enough that z lies in the domain of definition $|z| < R$ of $S(z)$.

To do this, we first note that when $|z| < R$, the equation

$$S(z) = S_N(z) + \rho_N(z)$$

implies that

$$|S(z) - S(z_1)| = |S_N(z) - S_N(z_1) + \rho_N(z) - \rho_N(z_1)|,$$

or

(9) $|S(z) - S(z_1)| \leq |S_N(z) - S_N(z_1)| + |\rho_N(z)| + |\rho_N(z_1)|.$

In view of the uniform convergence established above, a positive integer N_ε exists such that

(10) $|\rho_N(z)| < \dfrac{\varepsilon}{3}$ whenever $N > N_\varepsilon$

and where z is any point lying in some closed disk about the origin whose radius is larger than $|z_1|$ but less than the radius R of the circle C. In particular, condition (10) is satisfied for all z lying in a neighborhood $|z - z_1| < \delta$ of z_1 which is small enough to be contained in that closed disk.

Also, as already noted, the polynomial $S_N(z)$ is continuous at z_1 for each value of N. In particular, when $N = N_\varepsilon + 1$ we can choose our δ so small that

$$(11) \qquad |S_N(z) - S_N(z_1)| < \frac{\varepsilon}{3} \qquad \text{whenever} \qquad |z - z_1| < \delta.$$

By writing $N = N_\varepsilon + 1$ in inequality (9) and using the fact that statements (10) and (11) are true when $N = N_\varepsilon + 1$, we now find that

$$|S(z) - S(z_1)| < \frac{\varepsilon}{3} + \frac{\varepsilon}{3} + \frac{\varepsilon}{3} \qquad \text{whenever} \qquad |z - z_1| < \delta.$$

This is statement (8).

We have thus shown that *the power series* (1) *represents a continuous function of z at each point interior to its circle of convergence.*

By substituting $z - z_0$ or its reciprocal for z, we can extend the results in this section, with obvious modifications, to series of the types

$$\sum_{n=0}^{\infty} a_n(z - z_0)^n, \qquad \sum_{n=1}^{\infty} \frac{b_n}{(z - z_0)^n}.$$

Suppose, for instance, that the second series here converges at some point other than z_0. As pointed out at the end of Sec. 47, the series is absolutely convergent in a region exterior to a circle centered at z_0; and we now know that it represents a continuous function throughout the region.

Note, too, how it follows from the theorem in this section that if a Laurent series representation (Sec. 46)

$$f(z) = \sum_{n=0}^{\infty} a_n(z - z_0)^n + \sum_{n=1}^{\infty} \frac{b_n}{(z - z_0)^n}$$

is valid in an annulus $R_0 < |z - z_0| < R_1$, *both* of the series on the right converge uniformly in a closed annulus which is concentric to and interior to that region of validity.

49. Integration and Differentiation of Power Series

We have just seen that a power series

$$(1) \qquad S(z) = \sum_{n=0}^{\infty} a_n z^n$$

represents a continuous function at each point interior to its circle of convergence. We prove in this section that the sum $S(z)$ is actually analytic within that circle. Our proof depends on the following theorem, which is of interest in itself.

Theorem 1. *Let C denote any contour interior to the circle of convergence of the power series* (1) *and let g(z) be any function that is continuous on C. The series formed by multiplying each term of the power series by g(z) can be integrated term by term over C; that is,*

$$(2) \qquad \int_C g(z)S(z)\,dz = \sum_{n=0}^{\infty} a_n \int_C g(z)z^n\,dz.$$

To prove this theorem, we note that since both $g(z)$ and the sum $S(z)$ of the power series are continuous on C, the integral over C of the product

$$g(z)S(z) = \sum_{n=0}^{N-1} a_n g(z)z^n + g(z)\rho_N(z),$$

where $\rho_N(z)$ is the remainder of the given series after N terms, exists. The terms of the finite sum here are also continuous on the contour C, and so their integrals over C exist. Consequently, the integral of the quantity $g(z)\rho_N(z)$ must exist, and

$$(3) \qquad \int_C g(z)S(z)\,dz = \sum_{n=0}^{N-1} a_n \int_C g(z)z^n\,dz + \int_C g(z)\rho_N(z)\,dz.$$

Let M be the maximum value of $|g(z)|$ on C, and let L denote the length of C. In view of the uniform convergence of the given power series (Sec. 48), we know that for each positive number ε there exists a positive integer N_ε such that, for all points z on C,

$$|\rho_N(z)| < \varepsilon \qquad \text{whenever} \qquad N > N_\varepsilon.$$

Since N_ε is independent of z, we find that

$$\left| \int_C g(z)\rho_N(z)\,dz \right| < M\varepsilon L \qquad \text{whenever} \qquad N > N_\varepsilon;$$

that is,

$$\lim_{N\to\infty} \int_C g(z)\rho_N(z)\,dz = 0.$$

It follows, therefore, from equation (3) that

$$\int_C g(z)S(z)\,dz = \lim_{N\to\infty} \sum_{n=0}^{N-1} a_n \int_C g(z)z^n\,dz.$$

This is the same as equation (2), and Theorem 1 is proved.

If $g(z) = 1$ for each value of z in the open disk bounded by the circle of convergence of power series (1), then

$$\int_C g(z)z^n\,dz = \int_C z^n\,dz = 0 \qquad\qquad (n = 0, 1, 2, \ldots)$$

for every *closed* contour C lying in that domain. According to equation (2), then,

$$\int_C S(z)\,dz = 0$$

for every such contour; and, by Morera's theorem (Sec. 40), the function $S(z)$ is analytic throughout the domain. This result can be stated as follows.

> **Theorem 2.** *The power series* (1) *represents a function that is analytic at every point interior to its circle of convergence.*

Theorem 2 is often helpful in establishing the analyticity of functions and in evaluating limits.

Example 1. To illustrate, let us show that the function

$$f(z) = \begin{cases} \dfrac{\sin z}{z} & \text{when} \quad z \neq 0, \\[2mm] 1 & \text{when} \quad z = 0 \end{cases}$$

is entire. Since the Maclaurin series expansion

$$\sin z = \sum_{n=0}^{\infty} (-1)^n \frac{z^{2n+1}}{(2n+1)!}$$

represents $\sin z$ for every value of z, the series

$$(4) \qquad \sum_{n=0}^{\infty} (-1)^n \frac{z^{2n}}{(2n+1)!} = 1 - \frac{z^2}{3!} + \frac{z^4}{5!} - \cdots,$$

obtained by dividing each term of that Maclaurin series by z, converges to $f(z)$ when $z \neq 0$. But series (4) clearly converges to $f(0)$ when $z = 0$. Hence $f(z)$ is represented by the convergent power series (4) for all z, and f is therefore an entire function. Note that since f is continuous at $z = 0$ and since $(\sin z)/z = f(z)$ when $z \neq 0$,

$$(5) \qquad \lim_{z \to 0} \frac{\sin z}{z} = \lim_{z \to 0} f(z) = f(0) = 1.$$

This is a result known beforehand because the limit here is the definition of the derivative of $\sin z$ at $z = 0$.

We observed at the beginning of Sec. 45 that the Taylor series for a function f about a point z_0 converges to $f(z)$ at each point z interior to the circle centered at z_0 and passing through the nearest point z_1 where f fails to be analytic. In view of Theorem 2, we now know that *there is no larger circle about z_0 such that at each point z interior to it the Taylor series converges to $f(z)$.* For if there were such a circle, f would be analytic at z_1; but f is not analytic at z_1.

We now present a companion to Theorem 1. It was used to obtain the Maclaurin series for $\cos z$ and $\cosh z$ in Sec. 45.

Theorem 3. *The power series (1) can be differentiated term by term; that is, at each point z interior to the circle of convergence of that series,*

$$(6) \qquad\qquad S'(z) = \sum_{n=1}^{\infty} n a_n z^{n-1}.$$

To establish this result, let z denote any point interior to the circle of convergence; and let C be some positively oriented simple closed contour surrounding z and interior to that circle. Also, define the function

$$(7) \qquad\qquad g(s) = \frac{1}{2\pi i (s - z)^2}$$

at each point s on C. Since the function

$$S(z) = \sum_{n=0}^{\infty} a_n z^n$$

is analytic inside and on C, we can write

$$\int_C g(s) S(s)\, ds = \frac{1}{2\pi i} \int_C \frac{S(s)\, ds}{(s - z)^2} = S'(z)$$

with the aid of the integral representation for derivatives in Sec. 39. Furthermore,

$$\int_C g(s) s^n\, ds = \frac{1}{2\pi i} \int_C \frac{s^n\, ds}{(s - z)^2} = \frac{d}{dz} z^n \qquad (n = 0, 1, 2, \ldots).$$

So when the variable of integration z in equation (2) is replaced by the variable s and $g(s)$ is the function (7) along this closed contour C, equation (2) becomes

$$S'(z) = \sum_{n=0}^{\infty} a_n \frac{d}{dz} z^n,$$

which is the same as equation (6). This completes the proof.

All three theorems in this section are readily extended in the usual way to series involving positive or negative powers of $z - z_0$.

EXERCISES

1. By differentiating the Maclaurin series expansion (Sec. 45)

$$\frac{1}{1 - z} = \sum_{n=0}^{\infty} z^n \qquad\qquad (|z| < 1),$$

obtain the representations

$$\frac{1}{(1 - z)^2} = \sum_{n=0}^{\infty} (n + 1) z^n, \qquad \frac{2}{(1 - z)^3} = \sum_{n=0}^{\infty} (n + 1)(n + 2) z^n \qquad (|z| < 1).$$

2. Find the Taylor series for $1/z$ about the point $z = -3$. Then, by differentiating that series term by term, find a similar expansion for $1/z^2$. Give regions of validity.

3. In the w plane, integrate the Maclaurin series expansion (see Example 4, Sec. 45)

$$\frac{1}{w} = \sum_{n=0}^{\infty} (-1)^n (w - 1)^n \qquad\qquad (|w - 1| < 1)$$

along a contour interior to the circle of convergence from $w = 1$ to $w = z$ to obtain the representation

$$\text{Log } z = \sum_{n=1}^{\infty} \frac{(-1)^{n+1}}{n}(z - 1)^n \qquad\qquad (|z - 1| < 1).$$

4. With the aid of series, prove that if c is a complex constant and

$$f(z) = \begin{cases} \dfrac{e^{cz} - 1}{z} & \text{when } z \neq 0, \\[2mm] c & \text{when } z = 0, \end{cases}$$

then f is entire.

5. Expand $\sin z$ into powers of $z + \pi$ to prove that

$$\lim_{z \to -\pi} \frac{\sin z}{z + \pi} = -1.$$

6. Prove that if

$$f(z) = \begin{cases} \dfrac{\text{Log } z}{z - 1} & \text{when } z \neq 1, \\[2mm] 1 & \text{when } z = 1, \end{cases}$$

then f is analytic throughout the domain $|z| > 0$, $-\pi < \text{Arg } z < \pi$.
Suggestion: See Exercise 3.

7. Write

$$f(z) = \begin{cases} \dfrac{\cos z}{z^2 - (\pi/2)^2} & \text{when } z \neq \pm\pi/2, \\[2mm] -\dfrac{1}{\pi} & \text{when } z = \pm\pi/2. \end{cases}$$

Then prove that f is an entire function.

8. Suppose that a function f is analytic at z_0 and that $f(z_0) = 0$. Use series to show that

$$\lim_{z \to z_0} \frac{f(z)}{z - z_0} = f'(z_0).$$

Point out how this result also follows directly from the definition of $f'(z_0)$.

9. Suppose that f and g are analytic at z_0 and that $f(z_0) = g(z_0) = 0$, while $g'(z_0) \neq 0$. Prove that

$$\lim_{z \to z_0} \frac{f(z)}{g(z)} = \frac{f'(z_0)}{g'(z_0)}.$$

10. Prove that if f is analytic at z_0 and $f(z_0) = f'(z_0) = \cdots = f^{(m)}(z_0) = 0$, then the function

$$g(z) = \begin{cases} \dfrac{f(z)}{(z - z_0)^{m+1}} & \text{when } z \neq z_0, \\[3mm] \dfrac{f^{(m+1)}(z_0)}{(m + 1)!} & \text{when } z = z_0 \end{cases}$$

is analytic at z_0.

11. (a) Let z be any complex number, and let C denote the unit circle $w = e^{i\phi}$ ($-\pi \leq \phi \leq \pi$) in the w plane. Then use that contour in expression (5), Sec. 46, for the coefficients in a Laurent series, adapted to such series about the origin in the w plane, to show that

$$\exp\left[\frac{z}{2}\left(w - \frac{1}{w}\right)\right] = \sum_{n=-\infty}^{\infty} J_n(z)w^n \qquad (0 < |w| < \infty)$$

where

$$J_n(z) = \frac{1}{2\pi} \int_{-\pi}^{\pi} \exp[-i(n\phi - z \sin \phi)]\, d\phi \qquad (n = 0, \pm 1, \pm 2, \ldots).$$

(b) With the aid of Exercise 5, Sec. 30, regarding certain definite integrals of even and odd complex-valued functions of a real variable, show that the coefficients in part (a) can be written*

$$J_n(z) = \frac{1}{\pi} \int_0^{\pi} \cos(n\phi - z \sin \phi)\, d\phi \qquad (n = 0, \pm 1, \pm 2, \ldots).$$

12. (a) Let $f(z)$ denote a function which is analytic in some annular domain about the origin that includes the unit circle $z = e^{i\phi}$ ($-\pi \leq \phi \leq \pi$). By taking that circle as the path of integration in expressions (2) and (3), Sec. 46, for the coefficients a_n and b_n in a Laurent series in powers of z, show that

$$f(z) = \frac{1}{2\pi} \int_{-\pi}^{\pi} f(e^{i\phi})\, d\phi + \frac{1}{2\pi} \sum_{n=1}^{\infty} \int_{-\pi}^{\pi} f(e^{i\phi})\left[\left(\frac{z}{e^{i\phi}}\right)^n + \left(\frac{e^{i\phi}}{z}\right)^n\right] d\phi$$

when z is any point in the annular domain.

(b) Write $u(\theta) = \text{Re}[f(e^{i\theta})]$, and show how it follows from the expansion in part (a) that

$$u(\theta) = \frac{1}{2\pi} \int_{-\pi}^{\pi} u(\phi)\, d\phi + \frac{1}{\pi} \sum_{n=1}^{\infty} \int_{-\pi}^{\pi} u(\phi) \cos[n(\theta - \phi)]\, d\phi.$$

This is one form of the *Fourier series* expansion of the real-valued function $u(\theta)$ on the interval $-\pi \leq \theta \leq \pi$. The restriction on $u(\theta)$ is more severe than is necessary in order for it to be represented by a Fourier series.[†]

50. Uniqueness of Series Representations

According to Theorem 3 in Sec. 49,

$$S'(z) = \sum_{n=1}^{\infty} na_n z^{n-1}$$

---

*These coefficients $J_n(z)$ are called *Bessel functions* of the first kind. They play a prominent role in certain areas of applied mathematics. See, for example, the authors' "Fourier Series and Boundary Value Problems," 3d ed., chap. 8, 1978.

[†]For other sufficient conditions, see secs. 41 and 49 of the book cited in the footnote to Exercise 11.

everywhere within the circle of convergence C of the series

$$(1) \qquad\qquad S(z) = \sum_{n=0}^{\infty} a_n z^n.$$

Consequently, the series for $S'(z)$ can be differentiated term by term; that is,

$$S''(z) = \sum_{n=2}^{\infty} n(n-1)a_n z^{n-2}$$

for all z within C. In fact, the derivative of $S(z)$ of any order can be found by successively differentiating its series representation term by term. Evidently, then,

$$S(0) = a_0, \quad S'(0) = 1!a_1, \quad S''(0) = 2!a_2, \quad \ldots, \quad S^{(n)}(0) = n!a_n, \quad \ldots;$$

and the coefficients a_n are those in the Maclaurin series expansion for $S(z)$:

$$a_n = \frac{S^{(n)}(0)}{n!} \qquad\qquad (n = 0, 1, 2, \ldots).$$

The generalization to series involving positive powers of $z - z_0$ is immediate. We thus have the following theorem on the uniqueness of representations of functions in power series.

Theorem 1. *If a series*

$$(2) \qquad\qquad \sum_{n=0}^{\infty} a_n(z - z_0)^n$$

converges to $f(z)$ at all points interior to some circle $|z - z_0| = R$, then it is the Taylor series expansion for f in powers of $z - z_0$.

This result was anticipated in Sec. 45, where it has already been used. We give here another illustration.

Example. If we substitute z^2 for z in the equation (Sec. 45)

$$\sin z = \sum_{n=0}^{\infty} (-1)^n \frac{z^{2n+1}}{(2n+1)!} \qquad\qquad (|z| < \infty)$$

and its condition of validity, we see that

$$(3) \qquad\qquad \sin(z^2) = \sum_{n=0}^{\infty} (-1)^n \frac{z^{4n+2}}{(2n+1)!} \qquad\qquad (|z| < \infty).$$

This series must be identical to the series that would be found by expanding the function $\sin(z^2)$ directly into a Maclaurin series.

Note how it follows from Theorem 1 that if series (2) converges to zero throughout some neighborhood of z_0, then the coefficients a_n must all be zero.

Theorem 2. *If a series*

$$(4) \qquad \sum_{n=-\infty}^{\infty} c_n(z - z_0)^n = \sum_{n=0}^{\infty} a_n(z - z_0)^n + \sum_{n=1}^{\infty} \frac{b_n}{(z - z_0)^n}$$

*converges to $f(z)$ at all points in some annular domain about z_0, then it is the
Laurent series expansion for f in powers of $z - z_0$ for that domain.*

We prove this theorem with the aid of Theorem 1 in Sec. 49, extended to series
involving both positive and negative powers of $z - z_0$. Using the index of summation
m in the given series, we write

$$(5) \qquad \int_C g(z) f(z) \, dz = \sum_{m=-\infty}^{\infty} c_m \int_C g(z) (z - z_0)^m \, dz$$

where $g(z)$ is any one of the functions

$$g(z) = \frac{1}{2\pi i (z - z_0)^{n+1}} \qquad (n = 0, \pm 1, \pm 2, \ldots)$$

and C is a circle around the given annulus, centered at z_0 and taken in the positive sense.
Since (see Exercise 16, Sec. 32)

$$\frac{1}{2\pi i} \int_C \frac{dz}{(z - z_0)^{n-m+1}} = \begin{cases} 0 & \text{when } m \neq n, \\ 1 & \text{when } m = n, \end{cases}$$

equation (5) reduces to

$$\frac{1}{2\pi i} \int_C \frac{f(z) \, dz}{(z - z_0)^{n+1}} = c_n \,,$$

which is an expression in Sec. 46 for the coefficients in the Laurent series for f in
the annulus.

51. Multiplication and Division of Series

Suppose that each of the power series

$$(1) \qquad \sum_{n=0}^{\infty} a_n z^n \qquad \text{and} \qquad \sum_{n=0}^{\infty} b_n z^n$$

converges within some circle $|z| = R$. The sums $f(z)$ and $g(z)$ are then analytic functions
in the disk $|z| < R$ (Sec. 49), and the product of those sums has a Maclaurin series
expansion which is valid there:

$$(2) \qquad f(z)g(z) = \sum_{n=0}^{\infty} c_n z^n \qquad\qquad (|z| < R).$$

Since the series (1) are the Maclaurin series for $f(z)$ and $g(z)$, the first three
coefficients in expansion (2) are given by the equations

$$c_0 = f(0)g(0) = a_0b_0 \,,$$

$$c_1 = \frac{f(0)g'(0) + f'(0)g(0)}{1!} = a_0b_1 + a_1b_0 \,,$$

and

$$c_2 = \frac{f(0)g''(0) + 2f'(0)g'(0) + f''(0)g(0)}{2!} = a_0b_2 + a_1b_1 + a_2b_0 \,.$$

The general expression for any coefficient c_n is easily obtained by referring to the formula (Exercise 12, Sec. 53)

$$(3) \qquad [f(z)g(z)]^{(n)} = \sum_{k=0}^{n} \binom{n}{k} f^{(k)}(z)g^{(n-k)}(z),$$

where

$$\binom{n}{k} = \frac{n!}{k!(n-k)!} \qquad (k = 0, 1, 2, \ldots, n),$$

for the nth derivative of the product of two differentiable functions. As usual, $f^{(0)}(z) = f(z)$ and $0! = 1$. Evidently,

$$c_n = \sum_{k=0}^{n} \frac{f^{(k)}(0)}{k!} \frac{g^{(n-k)}(0)}{(n-k)!} = \sum_{k=0}^{n} a_k b_{n-k};$$

and so expansion (2) can be written

$$(4) \qquad f(z)g(z) = a_0b_0 + (a_0b_1 + a_1b_0)z + (a_0b_2 + a_1b_1 + a_2b_0)z^2 + \cdots$$

$$+ \left(\sum_{k=0}^{n} a_k b_{n-k} \right) z^n + \cdots \qquad (|z| < R).$$

Series (4) is the same as the series obtained by formally multiplying the two series (1) together term by term and collecting the resulting terms in like powers of z; it is called the *Cauchy product* of the two given series.

We have just proved the following theorem.

Theorem. *The Cauchy product of the two power series* (1) *converges to the product of their sums at all points interior to both of their circles of convergence.*

Continuing to let $f(z)$ and $g(z)$ denote the sums of series (1), suppose now that $g(z) \neq 0$ throughout the open disk $|z| < R$ in which both series converge. The quotient $h(z) = f(z)/g(z)$ is analytic there and thus has a Maclaurin series representation:

$$(5) \qquad h(z) = \sum_{n=0}^{\infty} d_n z^n \qquad (|z| < R).$$

Since $d_0 = h(0)$, $d_1 = h'(0)/1!$, $d_2 = h''(0)/2!$, etc., the first few of these coefficients can be found by differentiating the quotient $f(z)/g(z)$ successively. The results are the same as those found by formally carrying out the division of the first of series (1) by the second. Such a method identifies the first few terms of the quotient of the two given power series with those of the Maclaurin series for $f(z)/g(z)$. This is generally the result that is needed, although it can be shown that the two methods give rise to series that are entirely identical.

52. Examples

Our first three examples involve the function

(1) $$f(z) = \frac{-1}{(z-1)(z-2)} = \frac{1}{z-1} - \frac{1}{z-2},$$

which is analytic everywhere except at the two points $z = 1$ and $z = 2$.

Example 1. First we obtain the Maclaurin series for $f(z)$, which represents $f(z)$ in the open disk $|z| < 1$. Observe that $|z/2| < 1$ at each point in that disk. Consequently, by recalling that (Example 3, Sec. 45)

$$\frac{1}{1-z} = \sum_{n=0}^{\infty} z^n \qquad (|z| < 1),$$

we can write

$$f(z) = \frac{1}{2}\left[\frac{1}{1-(z/2)}\right] - \frac{1}{1-z} = \sum_{n=0}^{\infty}\left[\frac{1}{2}\left(\frac{z}{2}\right)^n - z^n\right] \qquad (|z| < 1).$$

That is,

(2) $$f(z) = \sum_{n=0}^{\infty} (2^{-n-1} - 1)z^n \qquad (|z| < 1).$$

Since this series in nonnegative powers of z converges to $f(z)$ when $|z| < 1$, it is the Maclaurin series for $f(z)$.

Note how it follows from expansion (2) that $f^{(n)}(0) = n!(2^{-n-1} - 1)$, since the coefficient of z^n there must have the value $f^{(n)}(0)/n!$.

Example 2. Let us now find the Laurent series that represents $f(z)$ throughout the annulus $1 < |z| < 2$. Each point z in that annular domain satisfies the inequalities $|1/z| < 1$ and $|z/2| < 1$. Hence

$$f(z) = \frac{1}{z}\left[\frac{1}{1-(1/z)}\right] + \frac{1}{2}\left[\frac{1}{1-(z/2)}\right] = \sum_{n=0}^{\infty}\frac{1}{z^{n+1}} + \sum_{n=0}^{\infty}\frac{z^n}{2^{n+1}} \qquad (1 < |z| < 2).$$

If we replace the index of summation n in the first of these series by $n - 1$ and then interchange the two series, we arrive at an expansion having the same form as the one in the statement of Laurent's theorem (Sec. 46):

(3) $$f(z) = \sum_{n=0}^{\infty}\frac{z^n}{2^{n+1}} + \sum_{n=1}^{\infty}\frac{1}{z^n} \qquad (1 < |z| < 2).$$

Since there is only one such representation for $f(z)$ in the annulus $1 < |z| < 2$, expansion (3) is, in fact, the Laurent series for $f(z)$ there.

Observe that if C is any simple closed contour around the annulus in the positive direction, we can, according to Sec. 46, write the coefficients $b_n = 1$ of $1/z^n$ in expansion (3) as

$$(4) \qquad\qquad b_n = \frac{1}{2\pi i} \int_C \frac{f(z)\, dz}{z^{-n+1}} \qquad\qquad (n = 1, 2, \ldots).$$

In particular, when $n = 1$,

$$\int_C f(z)\, dz = 2\pi i b_1 = 2\pi i.$$

Example 3. To obtain the Laurent series for $f(z)$ in the unbounded domain $2 < |z| < \infty$, we note that $|1/z| < 1$ and $|2/z| < 1$ for each point z in that domain. Thus

$$(5) \qquad f(z) = \frac{1}{z}\left[\frac{1}{1 - (1/z)} - \frac{1}{1 - (2/z)}\right] = \sum_{n=0}^{\infty} \frac{1 - 2^n}{z^{n+1}} = \sum_{n=1}^{\infty} \frac{1 - 2^{n-1}}{z^n}$$

$$(2 < |z| < \infty).$$

This is the required series. Here the coefficient b_1 of $1/z$ is zero; and if the contour C in expression (4) now denotes any simple closed contour about the origin and exterior to the circle $|z| = 2$, the value of the integral of $f(z)$ around C is seen to be zero.

Example 4. The Maclaurin series for $e^z/(1 + z)$ is valid in the disk $|z| < 1$. The first three nonzero terms are easily found by writing

$$\frac{e^z}{1 + z} = \left(1 + z + \frac{1}{2}z^2 + \frac{1}{6}z^3 + \cdots\right)(1 - z + z^2 - z^3 + \cdots)$$

and multiplying these two series term by term. The desired result is

$$\frac{e^z}{1 + z} = 1 + \frac{1}{2}z^2 - \frac{1}{3}z^3 + \cdots \qquad\qquad (|z| < 1).$$

Example 5. As pointed out in Sec. 24, the zeros of $\sinh z$ are the numbers $z = n\pi i$ $(n = 0, \pm 1, \pm 2, \ldots)$. So the function

$$\frac{1}{z^2 \sinh z} = \frac{1}{z^3}\left(\frac{1}{1 + z^2/3! + z^4/5! + \cdots}\right)$$

has a Laurent series representation in the domain $0 < |z| < \pi$. The denominator of the fraction in parentheses here is a power series that converges to $(\sinh z)/z$ when $z \neq 0$ and to unity when $z = 0$. Thus the sum of that series is not zero anywhere in the domain $|z| < \pi$; and the power series representation of the fraction is found by division to be

$$\frac{1}{1 + z^2/3! + z^4/5! + \cdots} = 1 - \frac{1}{3!}z^2 + \left[\frac{1}{(3!)^2} - \frac{1}{5!}\right]z^4 + \cdots,$$

or

(6) $$\frac{1}{1 + z^2/3! + z^4/5! + \cdots} = 1 - \frac{1}{6}z^2 + \frac{7}{360}z^4 + \cdots \qquad (|z| < \pi).$$

Hence

(7) $$\frac{1}{z^2 \sinh z} = \frac{1}{z^3} - \frac{1}{6}\frac{1}{z} + \frac{7}{360}z + \cdots \qquad (0 < |z| < \pi).$$

Although we have given only the first three nonzero terms of this Laurent series, any number of terms can, of course, be found by continuing the division.

To illustrate an alternative to straightforward division, let us obtain identity (6) by writing

(8) $$\frac{1}{1 + z^2/3! + z^4/5! + \cdots} = d_0 + d_1 z + d_2 z^2 + d_3 z^3 + d_4 z^4 + \cdots,$$

where the coefficients in the Maclaurin series on the right are to be determined. If equation (8) is to hold, then

$$1 = \left(1 + \frac{1}{3!}z^2 + \frac{1}{5!}z^4 + \cdots\right)(d_0 + d_1 z + d_2 z^2 + d_3 z^3 + d_4 z^4 + \cdots);$$

that is,

$$(d_0 - 1) + d_1 z + \left(d_2 + \frac{1}{3!}d_0\right)z^2 + \left(d_3 + \frac{1}{3!}d_1\right)z^3$$

$$+ \left(d_4 + \frac{1}{3!}d_2 + \frac{1}{5!}d_0\right)z^4 + \cdots = 0$$

when $|z| < \pi$. Since all the coefficients of powers of z in this last power series must be zero, then $d_0 - 1 = 0$, $d_1 = 0$, etc. After successively computing the values of d_0, d_1, d_2, d_3, and d_4, we find that with those values, equation (8) is the same as equation (6).

53. Zeros of Analytic Functions

If a function f is analytic at z_0, there is a circle of radius R about z_0 interior to which f is represented by a Taylor series. That is,

(1) $$f(z) = a_0 + \sum_{n=1}^{\infty} a_n(z - z_0)^n \qquad (|z - z_0| < R)$$

where $a_0 = f(z_0)$ and $a_n = f^{(n)}(z_0)/n!$ $(n = 1, 2, \ldots)$. If $f(z_0) = 0$, then $a_0 = 0$. If, in addition,

(2) $$f'(z_0) = f''(z_0) = \cdots = f^{(m-1)}(z_0) = 0$$

but $f^{(m)}(z_0) \neq 0$, one can write

$$f(z) = (z - z_0)^m \sum_{n=0}^{\infty} a_{m+n}(z - z_0)^n \qquad (|z - z_0| < R),$$

where $a_m \neq 0$. Thus

(3) $$f(z) = (z - z_0)^m g(z) \qquad\qquad (|z - z_0| < R),$$

where g is analytic and $g(z_0) = a_m \neq 0$. When a function $f(z)$ that is analytic at z_0 can be written in the form (3), with the stated conditions on $g(z)$, z_0 is called a *zero of order m* of f.

Since g is continuous at z_0, we know (Sec. 13) that for each positive number ε there exists a positive number δ such that

$$|g(z) - a_m| < \varepsilon \qquad \text{whenever} \qquad |z - z_0| < \delta.$$

If $\varepsilon = |a_m|/2$ and δ_0 is a corresponding value of δ, then

(4) $$|g(z) - a_m| < \frac{|a_m|}{2} \qquad \text{whenever} \qquad |z - z_0| < \delta_0 .$$

It follows that $g(z) \neq 0$ at any point in the neighborhood $|z - z_0| < \delta_0$ because the first of inequalities (4) becomes $|a_m| < |a_m|/2$ if $g(z) = 0$ at such a point.

We have now established the following theorem.

Theorem. *Let a function f be analytic at a point z_0 which is a zero of f. There is a neighborhood of z_0 throughout which f has no other zeros, unless f is identically zero. That is, the zeros of an analytic function which is not identically zero are isolated.*

EXERCISES

1. Write $f(z) = \sin(z^2)$. With the aid of the Maclaurin series expansion (Sec. 50)

$$\sin(z^2) = \sum_{n=0}^{\infty} (-1)^n \frac{z^{4n+2}}{(2n+1)!} \qquad\qquad (|z| < \infty),$$

show that $f^{(2n+1)}(0) = 0 \ (n = 0, 1, 2, \dots)$ and $f^{(4n)}(0) = 0 \ (n = 1, 2, \dots)$.

2. Use the expansion

$$\frac{1}{z^2 \sinh z} = \frac{1}{z^3} - \frac{1}{6}\frac{1}{z} + \frac{7}{360}z + \cdots \qquad\qquad (0 < |z| < \pi)$$

in Example 5, Sec. 52, to show that if C is the circle $|z| = 1$, taken counterclockwise, then

$$\int_C \frac{dz}{z^2 \sinh z} = -\frac{\pi i}{3}.$$

3. Obtain the Maclaurin series representation

$$z \cosh(z^2) = \sum_{n=0}^{\infty} \frac{z^{4n+1}}{(2n)!} \qquad\qquad (|z| < \infty).$$

4. Represent the function $(z + 1)/(z - 1)$ by

   (a) its Maclaurin series, and give the region of validity for that representation;

(b) its Laurent series for the domain $1 < |z| < \infty$.

$$\text{Ans.} \quad (a) \; -1 - 2 \sum_{n=1}^{\infty} z^n \quad (|z| < 1); \quad (b) \; 1 + 2 \sum_{n=1}^{\infty} \frac{1}{z^n}.$$

5. Obtain the expansion of the function $(z - 1)/z^2$ into
   (a) its Taylor series in powers of $z - 1$, and give the region of validity;
   (b) its Laurent series for the domain $1 < |z - 1| < \infty$.

$$\text{Ans.} \quad (a) \; \sum_{n=1}^{\infty} (-1)^{n+1} n(z - 1)^n \quad (|z - 1| < 1); \quad (b) \; \sum_{n=1}^{\infty} \frac{(-1)^{n+1} n}{(z - 1)^n}.$$

6. Show that

$$\frac{\sinh z}{z^2} = \frac{1}{z} + \sum_{n=0}^{\infty} \frac{z^{2n+1}}{(2n + 3)!} \qquad (0 < |z| < \infty).$$

7. Give two Laurent series expansions in powers of z for the function

$$f(z) = \frac{1}{z^2(1 - z)},$$

and specify the regions in which those expansions are valid.

$$\text{Ans.} \quad \frac{1}{z^2} + \frac{1}{z} + \sum_{n=0}^{\infty} z^n \;\; (0 < |z| < 1); \quad -\sum_{n=3}^{\infty} \frac{1}{z^n} \;\; (1 < |z| < \infty).$$

8. Write the two Laurent series in powers of z that represent the function

$$f(z) = \frac{1}{z(1 + z^2)}$$

in certain domains, and specify those domains.

9. Show that

$$\frac{e^z}{z(z^2 + 1)} = \frac{1}{z} + 1 - \frac{1}{2}z - \frac{5}{6}z^2 + \cdots \qquad (0 < |z| < 1).$$

10. Show that

$$(a) \; \csc z = \frac{1}{z} + \frac{1}{3!}z + \left[\frac{1}{(3!)^2} - \frac{1}{5!}\right]z^3 + \cdots \quad (0 < |z| < \pi);$$

$$(b) \; \frac{1}{e^z - 1} = \frac{1}{z} - \frac{1}{2} + \frac{1}{12}z - \frac{1}{720}z^3 + \cdots \quad (0 < |z| < 2\pi).$$

11. Write the Laurent series expansion of the function $1/(z - a)$ for the domain $|a| < |z| < \infty$, where a is real and $-1 < a < 1$. Then write $z = e^{i\theta}$ to obtain the summation formulas

$$\sum_{n=1}^{\infty} a^n \cos n\theta = \frac{a \cos \theta - a^2}{1 - 2a \cos \theta + a^2}, \qquad \sum_{n=1}^{\infty} a^n \sin n\theta = \frac{a \sin \theta}{1 - 2a \cos \theta + a^2}.$$

(Compare Exercise 5, Sec. 43.)

12. Use mathematical induction to verify formula (3), Sec. 51, for the nth derivative of the product of two differentiable functions.

13. Let $f(z)$ be an entire function which is represented by a series of the form

$$f(z) = z + a_2 z^2 + a_3 z^3 + \cdots \qquad (|z| < \infty).$$

(a) By differentiating the composite function $g(z) = f[f(z)]$ successively, find the first three nonzero terms in the Maclaurin series for $g(z)$ and thus show that

$$f[f(z)] = z + 2a_2 z^2 + 2(a_2^2 + a_3)z^3 + \cdots \qquad (|z| < \infty).$$

(b) Obtain the result in part (a) in a *formal* manner by writing

$$f[f(z)] = f(z) + a_2[f(z)]^2 + a_3[f(z)]^3 + \cdots,$$

replacing $f(z)$ on the right-hand side here by its series representation, and then collecting terms in like powers of z.

(c) By applying the result in part (a) to the function $f(z) = \sin z$, show that

$$\sin(\sin z) = z - \frac{1}{3}z^3 + \cdots \qquad (|z| < \infty).$$

SIX

RESIDUES AND POLES

The Cauchy-Goursat theorem (Sec. 33) tells us that if a function is analytic at all points interior to and on a simple closed contour C, the value of the integral of the function around that contour is zero. If, however, the function fails to be analytic at a finite number of points interior to C, there is, as we shall see in this chapter, a specific number, called a residue, which each of those points contributes to the value of the integral.

We develop here the theory of residues and illustrate that theory by using it to evaluate certain types of definite and improper real integrals occurring in applied mathematics.

54. Residues

Recall (Sec. 19) that a point z_0 is called a singular point of a function f if f fails to be analytic at z_0 but is analytic at some point in every neighborhood of z_0. A singular point z_0 is said to be *isolated* if, in addition, there is some neighborhood of z_0 throughout which f is analytic except at the point itself.

Example 1. Since $1/z$ is analytic everywhere except at $z = 0$, the origin is an isolated singular point of that function.

Example 2. The function

$$\frac{z + 1}{z^3(z^2 + 1)}$$

has the three isolated singular points $z = 0$ and $z = \pm i$.

Example 3. The origin is a singular point of Log z, but it is *not* an isolated singular point since every neighborhood of the origin contains points on the negative real axis and Log z fails to be analytic at each of those points (see Sec. 25).

Example 4. The function

$$\frac{1}{\sin(\pi/z)}$$

has the singular points $z = 0$ and $z = 1/n$ $(n = \pm 1, \pm 2, \ldots)$, all lying on the segment of the real axis from $z = -1$ to $z = 1$. Each singular point except $z = 0$ is isolated. The singular point $z = 0$ is not isolated because every neighborhood of the origin contains other singular points of the function.

When z_0 is an isolated singular point of a function f, there is a positive number R_1 such that f is analytic at each point z for which $0 < |z - z_0| < R_1$. Consequently, the function is represented by a Laurent series

$$(1) \quad f(z) = \sum_{n=0}^{\infty} a_n(z - z_0)^n + \frac{b_1}{z - z_0} + \frac{b_2}{(z - z_0)^2} + \cdots + \frac{b_n}{(z - z_0)^n} + \cdots$$

$$(0 < |z - z_0| < R_1),$$

where the coefficients a_n and b_n have certain integral representations (Sec. 46). In particular,

$$b_n = \frac{1}{2\pi i} \int_C \frac{f(z)\, dz}{(z - z_0)^{-n+1}} \qquad (n = 1, 2, \ldots)$$

where C is any positively oriented simple closed contour around z_0 and lying in the domain $0 < |z - z_0| < R_1$. When $n = 1$, this expression for b_n can be written

$$(2) \qquad \int_C f(z)\, dz = 2\pi i b_1.$$

The complex number b_1, which is the coefficient of $1/(z - z_0)$ in expansion (1), is called the *residue* of f at the isolated singular point z_0.

Equation (2) provides a powerful method for evaluating certain integrals around simple closed contours.

Example 5. Consider the integral

$$(3) \qquad \int_C \frac{e^{-z}}{(z - 1)^2}\, dz$$

where C is the circle $|z| = 2$, described in the positive sense. The integrand

$$f(z) = \frac{e^{-z}}{(z - 1)^2}$$

is analytic within and on C except at the isolated singular point $z = 1$. Thus, according to equation (2), the value of integral (3) is $2\pi i$ times the residue of f at $z = 1$. To determine this residue, we recall (Sec. 45) the Maclaurin series expansion

(4)
$$e^z = \sum_{n=0}^{\infty} \frac{z^n}{n!} \qquad (|z| < \infty),$$

from which it follows that

$$\frac{e^{-z}}{(z-1)^2} = \frac{e^{-1}e^{-(z-1)}}{(z-1)^2} = \sum_{n=0}^{\infty} \frac{(-1)^n}{n!e}(z-1)^{n-2} \qquad (0 < |z-1| < \infty).$$

In this Laurent series expansion, which can be written in the form (1), the coefficient of $1/(z-1)$ is $-1/e$. That is, the residue of f at $z = 1$ is $-1/e$. Hence

(5)
$$\int_C \frac{e^{-z}}{(z-1)^2} \, dz = -\frac{2\pi i}{e}.$$

Example 6. Let us show that

(6)
$$\int_C \exp\left(\frac{1}{z^2}\right) dz = 0$$

where C is the same contour as in Example 5. Since $1/z^2$ is analytic everywhere except at the origin, so is the integrand. The isolated singular point $z = 0$ is interior to C; and, with the aid of the Maclaurin series (4), we can write the Laurent expansion

$$\exp\left(\frac{1}{z^2}\right) = 1 + \frac{1}{1!}\frac{1}{z^2} + \frac{1}{2!}\frac{1}{z^4} + \frac{1}{3!}\frac{1}{z^6} + \cdots \qquad (0 < |z| < \infty).$$

The residue of the integrand at its isolated singular point $z = 0$ is therefore zero ($b_1 = 0$), and the value of integral (6) is established.

We are reminded by this example that although the analyticity of a function within and on a simple closed contour C is a sufficient condition for the value of the integral around C to be zero, it is not a necessary condition.

55. Residue Theorem

If a function f has only a finite number of singular points interior to a given simple closed contour C, they must be isolated. The following theorem is a precise statement of the fact that if, moreover, f is analytic on C and C is described in the positive sense, the value of the integral of f around C is $2\pi i$ times the sum of the residues at those singular points.

> **Theorem.** *Let C be a positively oriented simple closed contour within and on which a function f is analytic except for a finite number of singular points z_1, $z_2, \ldots, z_n$ interior to C. If $B_1, B_2, \ldots, B_n$ denote the residues of f at those respective points, then*

(1)
$$\int_C f(z) \, dz = 2\pi i (B_1 + B_2 + \cdots + B_n).$$

To prove the theorem, let the singular points z_j $(j = 1, 2, \ldots, n)$ be centers of

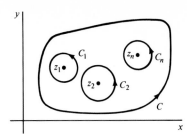

Figure 43

positively oriented circles C_j which are interior to C and are so small that no two of the circles have points in common (Fig. 43). The circles C_j together with the simple closed contour C form the boundary of a closed region throughout which f is analytic and whose interior is a multiply connected domain. Hence, according to the extension of the Cauchy-Goursat theorem to such regions (Sec. 36),

$$\int_C f(z)\,dz - \int_{C_1} f(z)\,dz - \int_{C_2} f(z)\,dz - \cdots - \int_{C_n} f(z)\,dz = 0.$$

This reduces to equation (1) because (Sec. 54)

$$\int_{C_j} f(z)\,dz = 2\pi i B_j \qquad\qquad (j = 1, 2, \ldots, n),$$

and the proof is complete.

Example. Let us use the theorem to evaluate the integral

(2)
$$\int_C \frac{5z - 2}{z(z - 1)}\,dz$$

where C is the circle $|z| = 2$, described counterclockwise. The integrand has the two singularities $z = 0$ and $z = 1$, both of which are interior to C. We can find the residues B_1 at $z = 0$ and B_2 at $z = 1$ with the aid of the Maclaurin series (Sec. 45)

$$\frac{1}{1 - z} = 1 + z + z^2 + \cdots \qquad\qquad (|z| < 1).$$

We first write the Laurent expansion

$$\frac{5z - 2}{z(z - 1)} = \left(\frac{5z - 2}{z}\right)\left(\frac{-1}{1 - z}\right) = \left(5 - \frac{2}{z}\right)(-1 - z - z^2 - \cdots)$$

$$= \frac{2}{z} - 3 - 3z - \cdots \qquad\qquad (0 < |z| < 1)$$

of the integrand and conclude that $B_1 = 2$. Next, we observe that

$$\frac{5z - 2}{z(z - 1)} = \left[\frac{5(z - 1) + 3}{z - 1}\right]\left[\frac{1}{1 + (z - 1)}\right]$$

$$= \left(5 + \frac{3}{z - 1}\right)[1 - (z - 1) + (z - 1)^2 - \cdots]$$

when $0 < |z - 1| < 1$. The coefficient of $1/(z - 1)$ in the Laurent expansion which is valid for $0 < |z - 1| < 1$ is therefore 3. Thus $B_2 = 3$, and

$$\int_C \frac{5z - 2}{z(z - 1)} \, dz = 2\pi i (B_1 + B_2) = 10\pi i.$$

In this example it is, of course, simpler to write the integrand as the sum of its partial fractions. Then

$$\int_C \frac{5z - 2}{z(z - 1)} \, dz = \int_C \frac{2}{z} \, dz + \int_C \frac{3}{z - 1} \, dz = 4\pi i + 6\pi i = 10\pi i.$$

[Also, see Exercise 6(b), Sec. 56.]

56. Principal Part of a Function

We have seen that if a function f has an isolated singular point z_0, then f can be represented by a Laurent series

$$(1) \quad f(z) = \sum_{n=0}^{\infty} a_n(z - z_0)^n + \frac{b_1}{z - z_0} + \frac{b_2}{(z - z_0)^2} + \cdots + \frac{b_n}{(z - z_0)^n} + \cdots$$

in a domain $0 < |z - z_0| < R_1$, centered at that point. The portion of the series involving negative powers of $z - z_0$ is called the *principal part* of f at z_0. We now use the principal part to distinguish between three types of isolated singular points. The behavior of f near z_0 is fundamentally different in each case.

If the principal part of f at z_0 contains at least one nonzero term but the number of such terms is finite, there exists a positive integer m such that $b_m \neq 0$ and $b_{m+1} = b_{m+2} = \cdots = 0$. That is, expansion (1) takes the form

$$(2) \quad f(z) = \sum_{n=0}^{\infty} a_n(z - z_0)^n + \frac{b_1}{z - z_0} + \frac{b_2}{(z - z_0)^2} + \cdots + \frac{b_m}{(z - z_0)^m}$$

$$(0 < |z - z_0| < R_1)$$

where $b_m \neq 0$. In this case the isolated singular point z_0 is called a *pole of order m*. A pole of order $m = 1$ is called a *simple pole*.

Example 1. Observe that the function

$$\frac{z^2 - 2z + 3}{z - 2} = z + \frac{3}{z - 2} = 2 + (z - 2) + \frac{3}{z - 2} \qquad (0 < |z - 2| < \infty)$$

has a simple pole at $z = 2$. Its residue there is 3.

Example 2. The function

$$\frac{\sinh z}{z^4} = \frac{1}{z^4}\left(z + \frac{z^3}{3!} + \frac{z^5}{5!} + \frac{z^7}{7!} + \cdots\right) = \frac{1}{z^3} + \frac{1}{3!}\frac{1}{z} + \frac{1}{5!}z + \frac{1}{7!}z^3 + \cdots$$

$$(0 < |z| < \infty)$$

has a pole of order 3 at $z = 0$, with residue $1/6$.

It can be shown (Exercise 12, Sec. 58) that $f(z)$ always tends to ∞ as z approaches a pole.

When the principal part of f at z_0 has an infinite number of nonzero terms, the point is called an *essential singular point*.

Example 3. The function

$$\exp\left(\frac{1}{z}\right) = 1 + \sum_{n=1}^{\infty} \frac{1}{n!} \frac{1}{z^n} \qquad (0 < |z| < \infty)$$

has an essential singular point at $z = 0$. Its residue there is unity.

An important result concerning the behavior of a function near an essential singular point is due to Picard. It states that *in each neighborhood of an essential singular point a function assumes every finite value, with one possible exception, an infinite number of times*. We will not prove Picard's theorem but will prove a closely related result in Chap. 12 (Sec. 107).*

Example 4. For an illustration of Picard's theorem, let us show that the function $\exp(1/z)$ in Example 3 assumes the value -1 an infinite number of times in each neighborhood of the origin. To do this, we recall from Sec. 22 that $\exp z = -1$ when $z = (2n + 1)\pi i$ $(n = 0, \pm 1, \pm 2, \ldots)$. This means that $\exp(1/z) = -1$ at the points

$$z = \frac{1}{(2n + 1)\pi i} \qquad (n = 0, \pm 1, \pm 2, \ldots),$$

and an infinite number of these points lie in any given neighborhood of the origin. Since $\exp(1/z) \neq 0$ for any value of z, zero is the exceptional value in Picard's theorem which may not be assumed by the function.

When all the coefficients b_n in the principal part of f at an isolated singular point z_0 are zero, the point z_0 is called a *removable singular point* of f. In this case the Laurent series (1) contains only nonnegative powers of $z - z_0$, and the series is, in fact, a power series. Note that the residue at a removable singular point is always zero. If we define $f(z)$ as a_0 at z_0, the function becomes analytic at z_0 (see Theorem 2, Sec. 49). Thus a function f with a removable singular point can be made analytic at that point by assigning a suitable value to the function there.

Example 5. Consider, for instance, the function

$$f(z) = \frac{e^z - 1}{z} = \frac{1}{z}\left(1 + \frac{z}{1!} + \frac{z^2}{2!} + \frac{z^3}{3!} + \cdots - 1\right)$$

$$= 1 + \frac{z}{2!} + \frac{z^2}{3!} + \cdots \qquad (0 < |z| < \infty).$$

---

*For a proof of Picard's theorem, see sec. 51 in vol. III of the book by Markushevich which is cited in Appendix 1.

The point $z = 0$ is a removable singular point; and if we write $f(0) = 1$, the function becomes entire.

EXERCISES

1. In each case, write the principal part of the function at its isolated singular point and determine whether that point is a pole, an essential singular point, or a removable singular point.

   (a) $z \exp\left(\dfrac{1}{z}\right)$;   (b) $\dfrac{z^2}{1+z}$;   (c) $\dfrac{\sin z}{z}$;   (d) $\dfrac{\cos z}{z}$;   (e) $\dfrac{1}{(2-z)^3}$.

2. Show that the singular point of each of the following functions is a pole. Determine the order m of that pole and the corresponding residue B.

   (a) $\dfrac{1 - \cosh z}{z^3}$;   (b) $\dfrac{1 - \exp(2z)}{z^4}$;   (c) $\dfrac{\exp(2z)}{(z-1)^2}$.

   Ans.   (a) $m = 1$, $B = -1/2$;   (b) $m = 3$, $B = -4/3$;   (c) $m = 2$, $B = 2e^2$.

3. Find the residue at $z = 0$ of the function

   (a) $\dfrac{1}{z + z^2}$;   (b) $z \cos\left(\dfrac{1}{z}\right)$;   (c) $\dfrac{z - \sin z}{z}$;   (d) $\dfrac{\cot z}{z^4}$;   (e) $\dfrac{\sinh z}{z^4(1 - z^2)}$.

   Ans.   (a) 1;   (b) $-1/2$;   (c) 0;   (d) $-1/45$;   (e) $7/6$.

4. Use residues to evaluate the integrals of the following functions around the circle $|z| = 3$ in the positive sense:

   (a) $\dfrac{\exp(-z)}{z^2}$;   (b) $z^2 \exp\left(\dfrac{1}{z}\right)$;   (c) $\dfrac{z+1}{z^2 - 2z}$.

   Ans.   (a) $-2\pi i$;   (b) $\pi i/3$;   (c) $2\pi i$.

5. Let f be a function which is analytic at a point z_0. Show that
   (a) if $f(z_0) = 0$, then z_0 is a removable singular point of the function

   $$g(z) = \frac{f(z)}{z - z_0};$$

   (b) if $f(z_0) \neq 0$, the point z_0 is a simple pole of the function g in part (a), with residue $f(z_0)$.

6. Suppose that a function $f(z)$ is analytic everywhere in the finite plane except for a finite number of isolated singular points which are *all* interior to a positively oriented circle C centered at the origin. According to Sec. 46, there is a Laurent series expansion

   $$f(z) = \sum_{n=0}^{\infty} a_n z^n + \sum_{n=1}^{\infty} \frac{b_n}{z^n}$$

   which is valid for all points z exterior to C and whose coefficient b_1 is given by the equation

   $$b_1 = \frac{1}{2\pi i} \int_C f(z)\, dz.$$

(a) After noting that this number b_1 is *not,* in general, the residue of f at the point $z = 0$, which may not even be a singular point, state why it is actually the sum of all the residues of f at its singular points in the finite plane.

(b) Use the above integral representation for b_1 to evaluate the integral of the function

$$f(z) = \frac{5z - 2}{z(z - 1)} = \left(\frac{5}{z} - \frac{2}{z^2}\right)\frac{1}{1 - (1/z)}$$

around the circle $|z| = 2$ in the positive direction. (This integral was evaluated in Sec. 55 by adding two residues that were found separately.)

7. By the method used in Exercise 6(b), evaluate the integral of $f(z)$ around the positively oriented circle $|z| = 2$ when $f(z)$ is

(a) $\dfrac{z^5}{1 - z^3}$; (b) $\dfrac{1}{1 + z^2}$; (c) $\dfrac{1}{z}$.

Ans. (a) $-2\pi i$; (b) 0; (c) $2\pi i$.

8. Let C denote the circle $|z| = 1$, taken counterclockwise. Follow the steps indicated below to show that

$$\int_C \exp\left(z + \frac{1}{z}\right) dz = 2\pi i \sum_{n=0}^{\infty} \frac{1}{n!(n + 1)!}.$$

(a) Using the Maclaurin series expansion for e^z and referring to Theorem 1, Sec. 49, which justifies the term by term integration to be used, write the above integral as

$$\sum_{n=0}^{\infty} \frac{1}{n!} \int_C z^n \exp\left(\frac{1}{z}\right) dz.$$

(b) With the aid of the residue theorem in Sec. 55, evaluate the integrals appearing in part (a) to arrive at the desired result.

57. Residues at Poles

The basic method for finding the residue of a function at an isolated singular point z_0 is that of appealing directly to the appropriate Laurent series and noting the coefficient of $1/(z - z_0)$. When z_0 is an essential singular point, we offer no other method. But for residues at poles there are some useful formulas which we now derive.

Suppose that a function $f(z)$ can be written in the form

(1) $$f(z) = \frac{\phi(z)}{z - z_0}$$

where $\phi(z)$ is analytic at z_0 and $\phi(z_0) \neq 0$. From the Taylor series

(2) $$\phi(z) = \sum_{n=0}^{\infty} \frac{\phi^{(n)}(z_0)}{n!}(z - z_0)^n$$

$$= \phi(z_0) + \frac{\phi'(z_0)}{1!}(z - z_0) + \frac{\phi''(z_0)}{2!}(z - z_0)^2 + \frac{\phi'''(z_0)}{3!}(z - z_0)^3 + \cdots,$$

which is valid in an open disk $|z - z_0| < R$, it follows that

$$f(z) = \frac{\phi(z_0)}{z - z_0} + \frac{\phi'(z_0)}{1!} + \frac{\phi''(z_0)}{2!}(z - z_0) + \frac{\phi'''(z_0)}{3!}(z - z_0)^2 + \cdots$$

$$(0 < |z - z_0| < R).$$

Since this is the unique Laurent series representation for $f(z)$ in the domain $0 < |z - z_0| < R$ and since $\phi(z_0) \neq 0$, *f has a simple pole at z_0 and the residue b_1 there is given by the equation*

(3) $$b_1 = \phi(z_0).$$

Example 1. The function $f(z) = (z + 1)/(z^2 + 9)$ has an isolated singular point at $z = 3i$ and can be written as

$$f(z) = \frac{\phi(z)}{z - 3i} \qquad \text{where} \qquad \phi(z) = \frac{z + 1}{z + 3i}.$$

Since $\phi(z)$ is analytic at $z = 3i$ and $\phi(3i) = (3 - i)/6 \neq 0$, that point is a simple pole of the function f; the residue there is $(3 - i)/6$. The point $z = -3i$ is also a simple pole of f, with residue $(3 + i)/6$.

The above discussion can be extended so as to apply to functions of the type

(4) $$f(z) = \frac{\phi(z)}{(z - z_0)^m} \qquad (m = 2, 3, \ldots)$$

where $\phi(z)$ is analytic at z_0 and $\phi(z_0) \neq 0$. For series (2) enables us to write

$$f(z) = \frac{\phi(z_0)}{(z - z_0)^m} + \frac{\phi'(z_0)/1!}{(z - z_0)^{m-1}} + \frac{\phi''(z_0)/2!}{(z - z_0)^{m-2}} + \cdots + \frac{\phi^{(m-1)}(z_0)/(m - 1)!}{z - z_0}$$

$$+ \sum_{n=m}^{\infty} \frac{\phi^{(n)}(z_0)}{n!}(z - z_0)^{n-m} \qquad (0 < |z - z_0| < R);$$

and this shows that *f has a pole of order m at z_0, with residue*

(5) $$b_1 = \frac{\phi^{(m-1)}(z_0)}{(m - 1)!}.$$

Example 2. If $f(z) = (z^3 + 2z)/(z - i)^3$, then

$$f(z) = \frac{\phi(z)}{(z - i)^3} \qquad \text{where} \qquad \phi(z) = z^3 + 2z.$$

The function $\phi(z)$ is entire, and $\phi(i) = i \neq 0$. Hence f has a pole of order 3 at $z = i$. The residue is

$$b_1 = \frac{\phi''(i)}{2!} = 3i.$$

It is straightforward to show (Exercise 11, Sec. 58) that if a function f has a pole at a point $z = z_0$, then $f(z)$ can always be expressed in the form (1) or (4). In spite of that fact, however, the identification of an isolated singular point as a pole of a certain order is often done most efficiently by the basic method of appealing directly to a Laurent series.

Example 3. If, for instance, the residue of the function $f(z) = (\sinh z)/z^4$ is needed at the singularity $z = 0$, it would be incorrect to write $f(z) = \phi(z)/z^4$, where $\phi(z) = \sinh z$, and to attempt an application of formula (5) with $m = 4$. For it is necessary that $\phi(z_0) \neq 0$ if formula (5) is to be used. The simplest way to find the residue at $z = 0$ is to write out a few terms of the Laurent series for $f(z)$, as was done in Example 2 of Sec. 56. There it was shown that $z = 0$ is a pole of the *third* order and that the residue is $1/6$.

Sometimes the series approach can be effectively combined with formula (3) or (5).

Example 4. Since $z(e^z - 1)$ is entire and its zeros are therefore isolated (Sec. 53), the point $z = 0$ is clearly an isolated singular point of the function

$$f(z) = \frac{1}{z(e^z - 1)}.$$

From the Maclaurin series

$$e^z = 1 + \frac{z}{1!} + \frac{z^2}{2!} + \frac{z^3}{3!} + \cdots \qquad (|z| < \infty),$$

we see that

$$z(e^z - 1) = z\left(\frac{z}{1!} + \frac{z^2}{2!} + \frac{z^3}{3!} + \cdots\right) = z^2\left(1 + \frac{z}{2!} + \frac{z^2}{3!} + \cdots\right) \qquad (|z| < \infty).$$

Thus

$$(6) \qquad f(z) = \frac{\phi(z)}{z^2} \qquad \text{where} \qquad \phi(z) = \frac{1}{1 + z/2! + z^2/3! + \cdots}.$$

Since $\phi(z)$ is analytic at $z = 0$ and $\phi(0) = 1 \neq 0$, the point $z = 0$ is a pole of the *second* order; and, according to formula (5), the residue is $b_1 = \phi'(0)$. Because

$$\phi'(z) = \frac{-(1/2 + z/3 + \cdots)}{(1 + z/2! + z^2/3! + \cdots)^2}$$

in a neighborhood of the origin, then, $b_1 = 1/2$.

The residue can also be found by dividing the series representation for $z(e^z - 1)$ into 1, or by multiplying the Laurent series for $1/(e^z - 1)$ in Exercise 10(b), Sec. 53, by $1/z$.

58. Quotients of Analytic Functions

Suppose that a function $f(z)$ can be written as a quotient

$$(1) \qquad f(z) = \frac{p(z)}{q(z)}$$

where p and q are both analytic at z_0. It is easy to show that z_0 *is an isolated singular point of f if and only if $q(z_0) = 0$.* First of all, if $q(z_0) = 0$, then $q(z) \neq 0$ at any other point in some neighborhood of z_0; this is because the zeros of an analytic function which is not identically zero are isolated (Sec. 53). It follows that f is analytic everywhere in that neighborhood except at the point z_0 itself, and z_0 is therefore an isolated singular point of f. Conversely, if z_0 is an isolated singular point of f, then $q(z_0) = 0$. For when $q(z_0) \neq 0$, it follows by continuity that $q(z) \neq 0$ throughout some neighborhood of z_0 (see Exercise 10, Sec. 15). Hence f would be analytic at z_0, and this contradicts the fact that z_0 is a singular point of f.

Assume now that in addition to the condition that the functions p and q in equation (1) be analytic at z_0, $p(z_0) \neq 0$ *and $q(z)$ has a zero of order m at z_0.* That is, $p(z_0) \neq 0$ and

$$(2) \qquad q(z) = (z - z_0)^m r(z)$$

where m is some positive integer $(m = 1, 2, \ldots)$, r is analytic at z_0, and $r(z_0) \neq 0$ (see Sec. 53). In view of equations (1) and (2), $f(z)$ can be written

$$(3) \qquad f(z) = \frac{\phi(z)}{(z - z_0)^m} \qquad \text{where} \qquad \phi(z) = \frac{p(z)}{r(z)}.$$

Observe that since p and r are both analytic at z_0 and since $r(z_0) \neq 0$, the quotient $\phi(z)$ is also analytic there; furthermore, $\phi(z_0) = p(z_0)/r(z_0) \neq 0$. Consequently, f has a pole of order m at z_0; and formula (3) or (5) in Sec. 57 can be used to find the residue.

In particular, let us show how it follows that *if a function f is of the type* (1) *where $p(z_0) \neq 0$, $q(z_0) = 0$, and $q'(z_0) \neq 0$, the point z_0 is a simple pole of f and the residue there is*

$$(4) \qquad b_1 = \frac{p(z_0)}{q'(z_0)}.$$

According to Sec. 53, z_0 is a zero of order $m = 1$ of the function q; and we now know that this means z_0 is a simple pole of f. Referring to expression (3), we see that the corresponding residue is the number

$$b_1 = \phi(z_0) = \frac{p(z_0)}{r(z_0)}$$

where $r(z)$ is the analytic function related to $q(z)$ by equation (2) when $m = 1$:

$$(5) \qquad q(z) = (z - z_0)r(z).$$

Differentiation of each side of equation (5) reveals that $q'(z_0) = r(z_0)$, and the desired expression for b_1 follows.

Example 1. Consider the function

$$f(z) = \cot z = \frac{\cos z}{\sin z},$$

which is a quotient of the entire functions $p(z) = \cos z$ and $q(z) = \sin z$. The singularities of that quotient occur at the zeros of q, or at the points $z = n\pi$ $(n = 0, \pm 1, \pm 2, \ldots)$. Since $p(n\pi) = (-1)^n \neq 0$, $q(n\pi) = 0$, and $q'(n\pi) = (-1)^n \neq 0$, each singular point $z = n\pi$ is a simple pole, with residue

$$b_1 = \frac{p(n\pi)}{q'(n\pi)} = 1.$$

Example 2. Let us find the residue of the function $f(z) = z/(z^4 + 4)$ at the isolated singular point $z_0 = \sqrt{2}\, e^{i\pi/4} = 1 + i$ by writing $p(z) = z$ and $q(z) = z^4 + 4$. Since z_0 is a zero of $q(z)$ and since $p(z_0) = z_0 \neq 0$ and $q'(z_0) = 4z_0^3 \neq 0$, f has a simple pole at z_0. The residue is the number

$$b_1 = \frac{1}{4z_0^2} = \frac{1}{8i} = -\frac{i}{8}.$$

Although this residue can also be found by the method in Sec. 57, the computation is somewhat more involved.

There are formulas similar to formula (4) for residues at poles of higher order, but they are lengthier and, in general, not practical.

EXERCISES

1. In each case show that the singular points of the function are poles. Determine the order m of each pole, and find the corresponding residue B.

(a) $\dfrac{z^2 + 2}{z - 1}$;   (b) $\left(\dfrac{z}{2z + 1}\right)^3$;   (c) $\tanh z$;   (d) $\dfrac{\exp z}{z^2 + \pi^2}$;   (e) $\dfrac{z}{\cos z}$;

(f) $\dfrac{z^{1/4}}{z + 1}$ $(|z| > 0, 0 < \arg z < 2\pi)$.

Ans.   (a) $m = 1, B = 3$;   (b) $m = 3, B = -3/16$;
     (c) $m = 1, B = 1$;   (f) $m = 1, B = (1 + i)/\sqrt{2}$.

2. Show that the point $z = 0$ is a simple pole of the function $f(z) = \csc z$ and that the residue there is 1 by (a) appealing to the Laurent series for $\csc z$ in Exercise 10(a), Sec. 53; (b) applying results in Sec. 58.

3. Find the value of the integral

$$\int_c \frac{3z^3 + 2}{(z - 1)(z^2 + 9)}\, dz$$

taken counterclockwise around the circle   (a) $|z - 2| = 2$;   (b) $|z| = 4$.

Ans.   (a) πi;   (b) $6\pi i$.

4. Find the value of the integral

$$\int_C \frac{dz}{z^3(z+4)}$$

taken counterclockwise around the circle   (a) $|z| = 2$;   (b) $|z + 2| = 3$.

Ans.   (a) $\pi i/32$;   (b) 0.

5. Let C be the circle $|z| = 2$, described in the positive sense, and evaluate the integral

(a) $\int_C \tan z\, dz$;   (b) $\int_C \frac{dz}{\sinh 2z}$;   (c) $\int_C \frac{\cosh \pi z\, dz}{z(z^2 + 1)}$.

Ans.   (a) $-4\pi i$;   (b) $-\pi i$.

6. Show that if C is the positively oriented circle $|z| = 8$, then

$$\frac{1}{2\pi i} \int_C \frac{e^{zt}}{\sinh z}\, dz = 1 - 2\cos \pi t + 2\cos 2\pi t.$$

7. Show that

$$\int_C \frac{dz}{(z^2 - 1)^2 + 3} = \frac{\pi}{2\sqrt{2}}$$

where C is the positively oriented boundary of the rectangle whose sides lie along the lines $x = \pm 2$, $y = 0$, and $y = 1$.

Suggestion: By observing that the four zeros of the polynomial $q(z) = (z^2 - 1)^2 + 3$ are the square roots of the numbers $1 \pm \sqrt{3}\,i$, show that the function $1/q(z)$ is analytic within and on C except at the points $z_0 = (\sqrt{3} + i)/\sqrt{2}$ and $-\bar{z}_0 = (-\sqrt{3} + i)/\sqrt{2}$. Then apply the method in Sec. 58 for identifying a simple pole and finding the corresponding residue.

8. Suppose that

$$f(z) = \frac{1}{[q(z)]^2}$$

where the function q is analytic at z_0, $q(z_0) = 0$, and $q'(z_0) \neq 0$. Show that z_0 is a pole of order 2 of the function f, with residue

$$b_1 = -\frac{q''(z_0)}{[q'(z_0)]^3}.$$

Suggestion: Note that z_0 is a zero of order $m = 1$ of the function q, so that equation (5), Sec. 58, holds. Then write

$$f(z) = \frac{\phi(z)}{(z - z_0)^2} \quad \text{where} \quad \phi(z) = \frac{1}{[r(z)]^2}.$$

The desired form of the residue $b_1 = \phi'(z_0)$ can be obtained by showing that $q'(z_0) = r(z_0)$ and $q''(z_0) = 2r'(z_0)$.

9. Apply the result in Exercise 8 to find the residue at $z = 0$ of the function

(a) $\csc^2 z$;   (b) $\dfrac{1}{(z + z^2)^2}$.

Ans.   (a) 0;   (b) -2.

10. Let C_N denote the positively oriented boundary of the square whose edges lie along the lines

$$x = \pm(N + \tfrac{1}{2})\pi \quad \text{and} \quad y = \pm(N + \tfrac{1}{2})\pi,$$

where N is a positive integer. Show that

$$\int_{C_N} \frac{dz}{z^2 \sin z} = 2\pi i \left[\frac{1}{6} + 2 \sum_{n=1}^{N} \frac{(-1)^n}{n^2 \pi^2} \right].$$

Then, using the fact that the value of this integral tends to zero as N tends to infinity (Exercise 19, Sec. 32), point out how it follows that

$$\sum_{n=1}^{\infty} \frac{(-1)^{n+1}}{n^2} = \frac{\pi^2}{12}.$$

11. Suppose that $f(z)$ has a pole of order m at $z = z_0$, so that there is a Laurent series expansion of the type

$$f(z) = \sum_{n=0}^{\infty} a_n(z - z_0)^n + \frac{b_1}{z - z_0} + \frac{b_2}{(z - z_0)^2} + \cdots + \frac{b_m}{(z - z_0)^m} \quad (0 < |z - z_0| < R_1)$$

where $b_m \neq 0$ (Sec. 56). Let $\phi(z)$ be a function defined on the open disk $|z - z_0| < R_1$ by means of the equations $\phi(z) = (z - z_0)^m f(z) \ (0 < |z - z_0| < R_1)$ and $\phi(z_0) = b_m$. Show that $\phi(z)$ can be represented by a power series in powers of $z - z_0$ and hence that it is analytic at z_0. Thus prove that *if a function f has a pole of order m at a point z_0, it can be written*

$$f(z) = \frac{\phi(z)}{(z - z_0)^m}$$

where $\phi(z)$ is analytic at z_0 and $\phi(z_0) \neq 0$. (The converse of this statement was established in Sec. 57.)

12. Let f denote a function which has a pole of order m at a point $z = z_0$. With the aid of the expression for $f(z)$ obtained in Exercise 11, show that $1/f(z)$ tends to zero as z tends to z_0 and hence that

$$\lim_{z \to z_0} f(z) = \infty.$$

[See Exercise 11(b), Sec. 12.]

13. Let a function $f(z)$ be analytic throughout a simply connected domain D, and let z_0 be the only zero of $f(z)$ in D. Prove that if C is a positively oriented simple closed contour in D that encloses z_0, then

$$\frac{1}{2\pi i} \int_C \frac{f'(z)}{f(z)} \, dz = m$$

where m is the order of the zero. The quotient $f'(z)/f(z)$ is the derivative of any branch of $\log f(z)$ and is known as the *logarithmic derivative* of $f(z)$.

14. Using the result in Exercise 13, prove the following property of the logarithmic derivative defined there. Let D be a simply connected domain throughout which a function f is analytic and $f'(z) \neq 0$. Let C denote a simple closed contour in D, described in the positive sense, such that $f(z) \neq 0$ at any point on C. Then, if f has N zeros interior to C, that number is given by the equation

$$N = \frac{1}{2\pi i} \int_C \frac{f'(z)}{f(z)} \, dz.$$

59. Evaluation of Improper Real Integrals

An important application of the theory of residues is the evaluation of certain types of definite and improper integrals occurring in *real* analysis. The examples treated here and in the remainder of the chapter illustrate this application of the theory.

In calculus the improper integral of a continuous function $f(x)$ over the semi-infinite interval $x \geq 0$ is defined by means of the equation

$$(1) \qquad \int_0^\infty f(x)\, dx = \lim_{R \to \infty} \int_0^R f(x)\, dx .$$

When the limit on the right exists, the improper integral is said to *converge* and its value is the value of that limit.

If $f(x)$ is continuous for all x, the improper integral

$$(2) \qquad \int_{-\infty}^\infty f(x)\, dx$$

is defined by writing

$$(3) \qquad \int_{-\infty}^\infty f(x)\, dx = \lim_{R_1 \to \infty} \int_{-R_1}^0 f(x)\, dx + \lim_{R_2 \to \infty} \int_0^{R_2} f(x)\, dx .$$

When the individual limits here both exist, integral (2) converges, its value being the sum of the values of these two limits. Another value assigned to integral (2) is also useful. Namely, the *Cauchy principal value* (P.V.) of integral (2) is the number

$$(4) \qquad \text{P.V.} \int_{-\infty}^\infty f(x)\, dx = \lim_{R \to \infty} \int_{-R}^R f(x)\, dx ,$$

provided this single limit exists.

If integral (2) converges, the value obtained is the same as the Cauchy principal value. On the other hand, when $f(x) = x$, for example, the Cauchy principal value of integral (2) is 0, whereas that integral does not converge according to definition (3). But suppose that f is an *even* function; that is, $f(-x) = f(x)$ for all real numbers x. Then if the Cauchy principal value of integral (2) exists, integral (2) actually converges. Also, note that when f is even and either one of the integrals (1) or (2) converges, then so does the other; in fact,

$$(5) \qquad \int_0^\infty f(x)\, dx = \frac{1}{2} \int_{-\infty}^\infty f(x)\, dx .$$

Suppose now that the integrand $f(x)$ in integral (2) can be written

$$f(x) = \frac{p(x)}{q(x)}$$

where $p(x)$ and $q(x)$ are real polynomials with no factors in common and where $q(x)$ has no real zeros. When the integral converges, its value can often be found quite easily by determining its Cauchy principal value with the aid of residues.

Example. Let us show that the integral

(6)
$$\int_0^\infty \frac{2x^2 - 1}{x^4 + 5x^2 + 4}\, dx = \frac{1}{2}\int_{-\infty}^\infty \frac{2x^2 - 1}{x^4 + 5x^2 + 4}\, dx$$

converges and find its value. Note that the integral on the right represents an integration of the function

$$f(z) = \frac{2z^2 - 1}{z^4 + 5z^2 + 4} = \frac{2z^2 - 1}{(z^2 + 1)(z^2 + 4)}$$

along the entire real axis in the complex plane. This function has isolated singularities at the points $z = \pm i$, $z = \pm 2i$ and is analytic everywhere else.

When $R > 2$, the singular points of f in the upper half plane lie in the interior of the semicircular region bounded by the segment $z = x\ (-R \le x \le R)$ of the real axis and the upper half C_R of the circle $|z| = R$ from $z = R$ to $z = -R$ (Fig. 44).

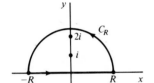

Figure 44

Integrating f counterclockwise around the boundary of this semicircular region, we see that

(7)
$$\int_{-R}^R f(x)\, dx + \int_{C_R} f(z)\, dz = 2\pi i\,(B_1 + B_2)$$

where B_1 is the residue of f at the point $z = i$ and B_2 is the residue of f at the point $z = 2i$. According to Sec. 57, we need only write

$$f(z) = \frac{\phi(z)}{z - i} \qquad \text{where} \qquad \phi(z) = \frac{2z^2 - 1}{(z + i)(z^2 + 4)}$$

to see that the point $z = i$ is a simple pole of f and that $B_1 = \phi(i) = -1/(2i)$. The point $z = 2i$ is also a simple pole, and $B_2 = 3/(4i)$. Hence $2\pi i(B_1 + B_2) = \pi/2$, and equation (7) can be put in the form

(8)
$$\int_{-R}^R f(x)\, dx = \frac{\pi}{2} - \int_{C_R} f(z)\, dz,$$

which is valid for all values of R greater than 2.

We now show that the value of the integral on the right in equation (8) approaches 0 as R tends to ∞. To do this, we observe that when z is a point on C_R,

$$|2z^2 - 1| \le 2|z|^2 + 1 = 2R^2 + 1$$

and

$$|z^4 + 5z^2 + 4| = |z^2 + 1|\,|z^2 + 4| \ge \big||z|^2 - 1\big|\,\big||z|^2 - 4\big| = (R^2 - 1)(R^2 - 4).$$

Consequently,

$$\left| \int_{C_R} \frac{2z^2 - 1}{z^4 + 5z^2 + 4} \, dz \right| \leq \frac{2R^2 + 1}{(R^2 - 1)(R^2 - 4)} \pi R,$$

the length of C_R being πR. The desired limit is now evident; that is,

$$\lim_{R \to \infty} \int_{C_R} f(z) \, dz = 0.$$

It thus follows from equation (8) that

$$\lim_{R \to \infty} \int_{-R}^{R} \frac{2x^2 - 1}{x^4 + 5x^2 + 4} \, dx = \frac{\pi}{2},$$

or

$$\text{P.V.} \int_{-\infty}^{\infty} \frac{2x^2 - 1}{x^4 + 5x^2 + 4} \, dx = \frac{\pi}{2}.$$

Since the integrand here is even, we know from remarks at the beginning of this section that the integral converges to its Cauchy principal value; and, according to statement (5),

$$(9) \qquad \int_{0}^{\infty} \frac{2x^2 - 1}{x^4 + 5x^2 + 4} \, dx = \frac{\pi}{4}.$$

60. Improper Integrals Involving Sines and Cosines

Residue theory can be useful in evaluating convergent improper integrals of the form

$$(1) \qquad \int_{-\infty}^{\infty} \frac{p(x)}{q(x)} \sin x \, dx \qquad \text{or} \qquad \int_{-\infty}^{\infty} \frac{p(x)}{q(x)} \cos x \, dx,$$

where $p(x)$ and $q(x)$ are real polynomials and $q(x)$ has no real zeros. As in Sec. 59, it is assumed that $p(x)$ and $q(x)$ have no factors in common. The method of that section cannot be applied directly here since $|\sin z|$ and $|\cos z|$ increase like $\sinh y$, or e^y, as y tends to infinity (Sec. 23). The modification illustrated below is suggested by the fact that

$$\int_{-R}^{R} \frac{p(x)}{q(x)} \cos x \, dx + i \int_{-R}^{R} \frac{p(x)}{q(x)} \sin x \, dx = \int_{-R}^{R} \frac{p(x)}{q(x)} e^{ix} \, dx,$$

together with the observation that the modulus e^{-y} of e^{iz} is bounded in the upper half plane.

Example 1. Let us show that

$$(2) \qquad \int_{-\infty}^{\infty} \frac{\cos x}{(x^2 + 1)^2} \, dx = \frac{\pi}{e}.$$

Because the integrand is even, it is sufficient to show that the Cauchy principal value of the integral exists and to find that value.

We introduce the function

$$(3) \qquad f(z) = \frac{e^{iz}}{(z^2 + 1)^2},$$

which is analytic everywhere on and above the real axis except at the point $z = i$. The singularity $z = i$ lies in the interior of the semicircular region whose boundary consists of the segment $-R \leq x \leq R$ of the real axis and the upper half C_R of the circle $|z| = R$ $(R > 1)$ from $z = R$ to $z = -R$. Integration of f counterclockwise around that boundary yields the equation

$$(4) \qquad \int_{-R}^{R} \frac{e^{ix}}{(x^2 + 1)^2} \, dx = 2\pi i B_1 - \int_{C_R} \frac{e^{iz}}{(z^2 + 1)^2} \, dz$$

where B_1 is the residue of f at $z = i$.

Since

$$f(z) = \frac{\phi(z)}{(z - i)^2} \qquad \text{where} \qquad \phi(z) = \frac{e^{iz}}{(z + i)^2},$$

the point $z = i$ is evidently a pole of order 2 of the function f, and

$$B_1 = \phi'(i) = -\frac{i}{2e}.$$

By equating the real parts of each side of equation (4), then, we find that

$$(5) \qquad \int_{-R}^{R} \frac{\cos x}{(x^2 + 1)^2} \, dx = \frac{\pi}{e} - \text{Re} \int_{C_R} \frac{e^{iz}}{(z^2 + 1)^2} \, dz.$$

Finally,

$$|e^{iz}| = e^{-y} \leq 1$$

when z is on C_R, since $y \geq 0$ there; and it follows that

$$\left| \text{Re} \int_{C_R} \frac{e^{iz}}{(z^2 + 1)^2} \, dz \right| \leq \left| \int_{C_R} \frac{e^{iz}}{(z^2 + 1)^2} \, dz \right| \leq \frac{\pi R}{(R^2 - 1)^2}.$$

Since this last quotient tends to 0 as R tends to ∞ and because of equation (5), we arrive at the desired result (2).

The method of Sec. 59 needs to be modified still further in order to evaluate certain other improper integrals of the type (1). Specifically, to show that the real or imaginary part of the value of the integral along the semicircle C_R tends to 0 as R tends to ∞, it is sometimes necessary to use the inequality

$$(6) \qquad \int_{0}^{\pi/2} e^{-R \sin \theta} \, d\theta < \frac{\pi}{2R} \qquad\qquad (R > 0),$$

which is known as *Jordan's inequality*. To verify it, we first note from the graphs of the functions $y = \sin\theta$ and $y = 2\theta/\pi$ when $0 \leq \theta \leq \pi/2$ that $\sin\theta \geq 2\theta/\pi$ for all values of θ in that interval. Consequently, if $R > 0$,

$$e^{-R\sin\theta} \leq e^{-2R\theta/\pi} \qquad \text{when} \qquad 0 \leq \theta \leq \frac{\pi}{2};$$

and so

$$\int_0^{\pi/2} e^{-R\sin\theta}\, d\theta \leq \int_0^{\pi/2} e^{-2R\theta/\pi}\, d\theta = \frac{\pi}{2R}(1 - e^{-R}) < \frac{\pi}{2R}.$$

Example 2. For an illustration of how Jordan's inequality is to be used, we now find the Cauchy principal value of the integral

$$\int_{-\infty}^{\infty} \frac{x\sin x\, dx}{x^2 + 2x + 2}.$$

As before, the existence of that value will be established by our actually finding it.
    We write

$$f(z) = \frac{ze^{iz}}{z^2 + 2z + 2} = \frac{ze^{iz}}{(z - z_1)(z - \bar{z}_1)},$$

where $z_1 = -1 + i$. The point $z = z_1$, which lies above the x axis, is a simple pole of f, with residue

$$(7) \qquad\qquad B_1 = \frac{z_1 e^{iz_1}}{z_1 - \bar{z}_1}.$$

Hence, when $R > \sqrt{2}$ and C_R denotes the upper half of the positively oriented circle $|z| = R$,

$$\int_{-R}^{R} \frac{xe^{ix}\, dx}{x^2 + 2x + 2} = 2\pi i B_1 - \int_{C_R} f(z)\, dz;$$

and it follows that

$$(8) \qquad \int_{-R}^{R} \frac{x\sin x\, dx}{x^2 + 2x + 2} = \text{Im}(2\pi i B_1) - \text{Im}\int_{C_R} f(z)\, dz.$$

    Now

$$(9) \qquad\qquad |f(z)| \leq \frac{R}{(R - \sqrt{2})^2}|e^{iz}|$$

when z is on C_R. Noting that $|\exp(iz)| \leq 1$ for such a point z and proceeding as we did in Example 1, we *cannot* conclude that the value of the integral of $f(z)$ along C_R tends to 0 as R tends to ∞. For the quotient $\pi R^2/(R - \sqrt{2})^2$ does not tend to 0 as R tends to ∞. Observe, however, that when z is a point on C_R, it can be written $z = Re^{i\theta}$ where $0 \leq \theta \leq \pi$. When z is on C_R, then,

$$|\exp(iz)| = |\exp(iRe^{i\theta})| = \exp(-R\sin\theta);$$

and, according to inequality (9),

(10)
$$|f(Re^{i\theta})| \le \frac{R}{(R - \sqrt{2})^2} e^{-R \sin \theta}$$

when $R > \sqrt{2}$ and $0 \le \theta \le \pi$.

Since the semicircle C_R has the parametric representation $z = Re^{i\theta}$ ($0 \le \theta \le \pi$),

$$\int_{C_R} f(z)\, dz = iR \int_0^\pi f(Re^{i\theta}) e^{i\theta}\, d\theta;$$

and inequality (10) now shows that

$$\left| \int_{C_R} f(z)\, dz \right| \le R \int_0^\pi |f(Re^{i\theta})|\, d\theta \le \frac{R^2}{(R - \sqrt{2})^2} \int_0^\pi e^{-R \sin \theta}\, d\theta.$$

Since the graph of $y = \sin \theta$ is symmetric with respect to the vertical line $\theta = \pi/2$ on the interval $0 \le \theta \le \pi$, so that $\sin(\pi - \theta) = \sin \theta$, and because of Jordan's inequality, we can therefore write

$$\left| \int_{C_R} f(z)\, dz \right| \le \frac{2R^2}{(R - \sqrt{2})^2} \int_0^{\pi/2} e^{-R \sin \theta}\, d\theta < \frac{\pi R}{(R - \sqrt{2})^2}.$$

We may now conclude that the value of the integral of $f(z)$ along C_R tends to 0 as R tends to ∞; and it follows from the inequality

$$\left| \mathrm{Im} \int_{C_R} f(z)\, dz \right| \le \left| \int_{C_R} f(z)\, dz \right|$$

and equation (8), as well as expression (7) for B_1, that

(11) $$\mathrm{P.V.} \int_{-\infty}^\infty \frac{x \sin x\, dx}{x^2 + 2x + 2} = \mathrm{Im}(2\pi i B_1) = \frac{\pi}{e} (\sin 1 + \cos 1).$$

EXERCISES

Establish the following integration formulas with the aid of residues:

1. (a) $\displaystyle\int_0^\infty \frac{dx}{x^2 + 1} = \frac{\pi}{2}$; (b) $\displaystyle\int_0^\infty \frac{dx}{(x^2 + 1)^2} = \frac{\pi}{4}$.

2. (a) $\displaystyle\int_0^\infty \frac{x^2\, dx}{(x^2 + 1)(x^2 + 4)} = \frac{\pi}{6}$; (b) $\displaystyle\int_0^\infty \frac{x^2\, dx}{(x^2 + 9)(x^2 + 4)^2} = \frac{\pi}{200}$.

3. (a) $\displaystyle\int_0^\infty \frac{dx}{x^4 + 1} = \frac{\pi}{2\sqrt{2}}$; (b) $\displaystyle\int_0^\infty \frac{x^2\, dx}{x^6 + 1} = \frac{\pi}{6}$.

4. $\displaystyle\int_{-\infty}^\infty \frac{\cos x\, dx}{(x^2 + a^2)(x^2 + b^2)} = \frac{\pi}{a^2 - b^2}\left(\frac{e^{-b}}{b} - \frac{e^{-a}}{a}\right)$ $(a > b > 0)$.

5. $\displaystyle\int_0^\infty \frac{\cos ax}{x^2 + 1}\, dx = \frac{\pi}{2} e^{-a}$ $(a \ge 0)$.

6. $\displaystyle\int_0^\infty \frac{\cos ax}{(x^2 + b^2)^2}\, dx = \frac{\pi}{4b^3}(1 + ab)e^{-ab}$ $(a > 0, b > 0)$.

7. $\displaystyle\int_0^\infty \frac{x \sin 2x}{x^2 + 3}\, dx = \frac{\pi}{2} \exp(-2\sqrt{3})$.

8. $\displaystyle\int_{-\infty}^\infty \frac{x \sin ax}{x^4 + 4}\, dx = \frac{\pi}{2}e^{-a} \sin a$ $(a > 0)$.

9. $\displaystyle\int_{-\infty}^\infty \frac{x^3 \sin ax}{x^4 + 4}\, dx = \pi e^{-a} \cos a$ $(a > 0)$.

Use methods in Secs. 59 and 60 to establish the convergence of the following integrals and to find their values:

10. $\displaystyle\int_{-\infty}^\infty \frac{x^2\, dx}{(x^2 + 1)^2}$.

11. $\displaystyle\int_{-\infty}^\infty \frac{x \sin x\, dx}{(x^2 + 1)(x^2 + 4)}$.

12. $\displaystyle\int_0^\infty \frac{x^3 \sin x\, dx}{(x^2 + 1)(x^2 + 9)}$.

Use methods in Secs. 59 and 60 to show that the Cauchy principal values of the following integrals exist and to find those values:

13. $\displaystyle\int_{-\infty}^\infty \frac{dx}{x^2 + 2x + 2}$.

14. $\displaystyle\int_{-\infty}^\infty \frac{x\, dx}{(x^2 + 1)(x^2 + 2x + 2)}$.   *Ans.* $-\pi/5$.

15. $\displaystyle\int_{-\infty}^\infty \frac{\sin x\, dx}{x^2 + 4x + 5}$.   *Ans.* $-(\pi/e) \sin 2$.

16. $\displaystyle\int_{-\infty}^\infty \frac{(x + 1) \cos x}{x^2 + 4x + 5}\, dx$.   *Ans.* $\dfrac{\pi}{e}(\sin 2 - \cos 2)$.

17. $\displaystyle\int_{-\infty}^\infty \frac{\cos x\, dx}{(x + a)^2 + b^2}$ $(b > 0)$.

18. Use residues and the contour shown in Fig. 45 to establish the integration formula

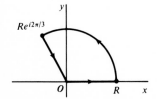

Figure 45

$$\int_0^\infty \frac{dx}{x^3 + 1} = \frac{2\pi}{3\sqrt{3}}.$$

19. The integration formula

$$\int_0^\infty \frac{dx}{[(x^2 - a)^2 + 1]^2} = \frac{\pi}{8\sqrt{2}\,A^3}[(2a^2 + 3)\sqrt{A + a} + a\sqrt{A - a}],$$

where a is any real number and $A = \sqrt{a^2 + 1}$, arises in the theory of case-hardening of steel by means of radio-frequency heating.* Follow the steps below to derive it.

(a) Point out why the four zeros of the polynomial

$$q(z) = (z^2 - a)^2 + 1$$

are the square roots of the numbers $a \pm i$. Then, using the fact that the numbers

$$z_0 = \frac{1}{\sqrt{2}}(\sqrt{A + a} + i\sqrt{A - a})$$

and $-z_0$ are the square roots of $a + i$ (Exercise 10, Sec. 7), verify that $\pm\bar{z}_0$ are the square roots of $a - i$ and hence that z_0 and $-\bar{z}_0$ are the only zeros of $q(z)$ in the upper half plane Im $z \geq 0$.

(b) Using the method derived in Exercise 8, Sec. 58, and keeping in mind that $z_0^2 = a + i$ for purposes of simplification, show that the point z_0 in part (a) is a pole of order 2 of the function $f(z) = 1/[q(z)]^2$ and that the residue B_1 at z_0 can be written

$$B_1 = -\frac{q''(z_0)}{[q'(z_0)]^3} = \frac{a - i(2a^2 + 3)}{16A^2 z_0}.$$

After observing that $q'(-\bar{z}) = -\overline{q'(z)}$ and $q''(-\bar{z}) = \overline{q''(z)}$, use the same method to show that the point $-\bar{z}_0$ in part (a) is also a pole of order 2 of the function $f(z)$, with residue

$$B_2 = \left\{\overline{\frac{q''(z_0)}{[q'(z_0)]^3}}\right\} = -\bar{B}_1.$$

Then obtain the expression

$$B_1 + B_2 = \frac{1}{8A^2 i} \operatorname{Im}\left[\frac{-a + i(2a^2 + 3)}{z_0}\right]$$

for the sum of these residues.

(c) With the aid of the results in parts (a) and (b) and using the fact that there is a positive constant R_0 such that $|q(z)| > |z|^4/2$ whenever $|z| > R_0$ (see Exercise 8, Sec. 42), complete the derivation of the integration formula.

61. Definite Integrals Involving Sines and Cosines

The method of residues is also useful in evaluating certain definite integrals of the type

(1) $$\int_0^{2\pi} F(\sin \theta, \cos \theta)\, d\theta.$$

*See pp. 359–364 of the book by Brown, Hoyler, and Bierwirth that is listed in Appendix 1.

The fact that θ varies from 0 to 2π suggests that we consider θ as an argument of a point z on the unit circle C centered at the origin; hence we write $z = e^{i\theta}\,(0 \leq \theta \leq 2\pi)$. When we make this substitution using the equations

$$(2) \qquad \sin\theta = \frac{z - z^{-1}}{2i}, \qquad \cos\theta = \frac{z + z^{-1}}{2}, \qquad d\theta = \frac{dz}{iz},$$

integral (1) becomes the contour integral

$$(3) \qquad \int_C F\left(\frac{z - z^{-1}}{2i}, \frac{z + z^{-1}}{2}\right)\frac{dz}{iz}$$

of a function of z around the circle C in the positive direction. The original integral (1) is, of course, simply a parametric form of integral (3), in accordance with expression (2), Sec. 31. When the integrand of integral (3) is a rational function of z, we can evaluate that integral by means of the residue theorem once the zeros of the polynomial in the denominator have been located and provided none lie on C.

Example. Let us show that

$$(4) \qquad \int_0^{2\pi} \frac{d\theta}{1 + a\sin\theta} = \frac{2\pi}{\sqrt{1 - a^2}} \qquad (-1 < a < 1).$$

This integration formula is clearly valid when $a = 0$, and we exclude that case in our derivation. With substitutions (2), the integral takes the form

$$(5) \qquad \int_C \frac{2/a}{z^2 + (2i/a)z - 1}\,dz,$$

where C is the positively oriented circle $|z| = 1$. The denominator of the integrand here has the pure imaginary zeros

$$z_1 = \left(\frac{-1 + \sqrt{1 - a^2}}{a}\right)i, \qquad z_2 = \left(\frac{-1 - \sqrt{1 - a^2}}{a}\right)i.$$

So if $f(z)$ denotes the integrand, then

$$f(z) = \frac{2/a}{(z - z_1)(z - z_2)}.$$

Note that because $|a| < 1$,

$$|z_2| = \frac{1 + \sqrt{1 - a^2}}{|a|} > 1.$$

Also, since $|z_1 z_2| = 1$, it follows that $|z_1| < 1$. Hence there are no singular points on C, and the only one interior to it is the point z_1. The corresponding residue B_1 is found by writing

$$f(z) = \frac{\phi(z)}{z - z_1} \qquad \text{where} \qquad \phi(z) = \frac{2/a}{z - z_2}.$$

This shows that z_1 is a simple pole and that

$$B_1 = \phi(z_1) = \frac{2/a}{z_1 - z_2} = \frac{1}{i\sqrt{1 - a^2}}.$$

Consequently,

$$\int_C \frac{2/a}{z^2 + (2i/a)z - 1} \, dz = 2\pi i B_1 = \frac{2\pi}{\sqrt{1 - a^2}};$$

and the integration formula (4) follows.

The method just illustrated applies equally well when the arguments of the sine and cosine are integral multiples of θ. One can write, for example, $\cos 2\theta = (z^2 + z^{-2})/2$ when $z = e^{i\theta}$.

62. Integration through a Branch Cut

The residue theorem can often be used to evaluate real integrals when the function $f(z)$ that is introduced has a branch cut.

Example. Let x^{-a}, where $x > 0$ and $0 < a < 1$, denote the principal value of the indicated power of x; that is, x^{-a} is the positive real number $\exp(-a \log x)$. We shall evaluate here the improper real integral

(1)
$$\int_0^\infty \frac{x^{-a}}{x + 1} \, dx \qquad (0 < a < 1),$$

which is important in the study of the gamma function.* Note that integral (1) is improper not only because of its upper limit of integration but also because its integrand has an infinite discontinuity at $x = 0$. The integral converges when $0 < a < 1$ since the integrand behaves like x^{-a} near $x = 0$ and like x^{-a-1} as x tends to infinity. We do not, however, need to establish convergence separately; for that will be contained in our evaluation of the integral.

We begin by letting C_ρ and C_R denote the circles $|z| = \rho$ and $|z| = R$, respectively, where $\rho < 1 < R$; and we assign them the orientations shown in Fig. 46. Since the branch

*See, for example, p. 4 of the book by Lebedev cited in Appendix 1.

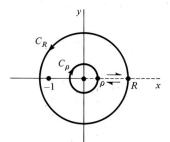

Figure 46

(2)
$$f(z) = \frac{z^{-a}}{z+1} \qquad (|z| > 0, 0 < \arg z < 2\pi)$$

of the multiple-valued function $z^{-a}/(z+1)$, with branch cut $\arg z = 0$, is piecewise continuous on C_ρ and C_R, the integrals

(3)
$$\int_{C_\rho} f(z)\,dz, \qquad \int_{C_R} f(z)\,dz$$

exist. Our technique for evaluating integral (1) is based on an expression for the sum of these two contour integrals.

To derive that expression, we introduce the two simple closed contours shown in Fig. 47 and obtained by dividing the annulus $\rho \le |z| \le R$ in Fig. 46 into two pieces. The legs L and $-L$ of those contours are directed line segments along any ray $\arg z = \theta_0$ where $\pi < \theta_0 < 3\pi/2$. Also, Γ_ρ and γ_ρ are the indicated portions of C_ρ, while Γ_R and γ_R make up C_R.

Now the branch

$$f_1(z) = \frac{z^{-a}}{z+1} \qquad \left(|z| > 0, -\frac{\pi}{2} < \arg z < \frac{3\pi}{2}\right)$$

of $z^{-a}/(z+1)$ is, except for a simple pole at $z = -1$, analytic within and on the closed contour on the left in Fig. 47, the branch cut being the lower half of the imaginary axis. By writing

$$\phi(z) = z^{-a} = \exp[-a \log z] \qquad \left(|z| > 0, -\frac{\pi}{2} < \arg z < \frac{3\pi}{2}\right),$$

we find that the residue of f_1 at $z = -1$ is

$$\phi(-1) = \exp[-a(\operatorname{Log} 1 + i\pi)] = e^{-ia\pi}.$$

Consequently,

(4)
$$\int_\rho^R \frac{\exp[-a(\operatorname{Log} r + i0)]}{r+1}\,dr$$

$$+ \int_{\Gamma_R} f_1(z)\,dz + \int_L f_1(z)\,dz + \int_{\Gamma_\rho} f_1(z)\,dz = 2\pi i e^{-ia\pi}.$$

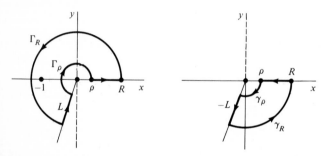

Figure 47

The first integral here is the integral of $f_1(z)$ along the horizontal leg from $z = \rho$ to $z = R$ when the parametric representation $z = re^{i0}$ $(\rho \leq r \leq R)$ is used.

The branch

$$f_2(z) = \frac{z^{-a}}{z+1} \qquad \left(|z| > 0, \frac{\pi}{2} < \arg z < \frac{5\pi}{2} \right)$$

is, on the other hand, analytic everywhere within and on the closed contour on the right in Fig. 47. Its branch cut is the upper half of the imaginary axis. Keeping in mind that $\arg z$ is 2π, rather than 0, when z is a point on the line segment joining the points ρ and R on the real axis, we can write

$$
(5) \qquad -\int_{\rho}^{R} \frac{\exp[-a(\operatorname{Log} r + i2\pi)]}{r+1}\, dr
$$

$$
+ \int_{\gamma_\rho} f_2(z)\, dz - \int_{L} f_2(z)\, dz + \int_{\gamma_R} f_2(z)\, dz = 0.
$$

The desired expression for the sum of integrals (3) follows if we equate the sum of all the terms on the left-hand sides of equations (4) and (5) to the sum of the right-hand sides. In doing this, we observe that except for the legs along the real axis, $f_1(z) = f(z)$ on the first closed contour in Fig. 47 and $f_2(z) = f(z)$ on the second. Also,

$$
\int_{\Gamma_\rho} f(z)\, dz + \int_{\gamma_\rho} f(z)\, dz = \int_{C_\rho} f(z)\, dz
$$

and

$$
\int_{\Gamma_R} f(z)\, dz + \int_{\gamma_R} f(z)\, dz = \int_{C_R} f(z)\, dz .
$$

The result can be written

$$
(6) \qquad \int_{C_\rho} f(z)\, dz + \int_{C_R} f(z)\, dz = 2\pi i e^{-ia\pi} + (e^{-i2a\pi} - 1)\int_{\rho}^{R} \frac{r^{-a}}{r+1}\, dr .
$$

Note that this equation can be obtained *formally* by applying the residue theorem in the following way which, of course, requires a justification such as the one given above. Consider the closed contour, indicated in Fig. 46, traced out by a point moving from ρ to R along the branch cut for $f(z)$, next around C_R back to R, then along the cut to ρ, and finally around C_ρ back to ρ. If one uses equation (2) with $\arg z = 0$ and $\arg z = 2\pi$ to define $f(z)$ on the upper and lower "edges," respectively, of the cut annulus, the residue theorem suggests that equation (6) *may* be valid for that region.

Referring now to definition (2) of $f(z)$, we see that

$$
\left| \int_{C_\rho} f(z)\, dz \right| \leq \frac{\rho^{-a}}{1-\rho} 2\pi\rho = \frac{2\pi}{1-\rho}\rho^{1-a}
$$

and

$$
\left| \int_{C_R} f(z)\, dz \right| \leq \frac{R^{-a}}{R-1} 2\pi R = \frac{2\pi R}{R-1}\frac{1}{R^a} .
$$

Since $0 < a < 1$, the values of these two integrals tend to 0 as ρ and R tend to 0 and ∞, respectively. Rewriting equation (6) as an expression for the last integral there and then letting ρ tend to 0, we thus find that

$$\int_0^R \frac{r^{-a}}{r+1} \, dr = \frac{1}{e^{-i2a\pi} - 1} \left[\int_{C_R} f(z) \, dz - 2\pi i e^{-ia\pi} \right].$$

Letting R tend to ∞ in this last equation, we arrive at the result

$$\int_0^\infty \frac{r^{-a}}{r+1} \, dr = 2\pi i \frac{e^{-ia\pi}}{1 - e^{-i2a\pi}},$$

which is the same as

(7)
$$\int_0^\infty \frac{x^{-a}}{x+1} \, dx = \frac{\pi}{\sin a\pi} \qquad (0 < a < 1).$$

EXERCISES

Use residues to establish the following integration formulas:

1. $\displaystyle\int_0^{2\pi} \frac{d\theta}{5 + 4\sin\theta} = \frac{2\pi}{3}.$ **2.** $\displaystyle\int_{-\pi}^{\pi} \frac{d\theta}{1 + \sin^2\theta} = \sqrt{2}\,\pi.$

3. $\displaystyle\int_0^{2\pi} \frac{\cos^2 3\theta \, d\theta}{5 - 4\cos 2\theta} = \frac{3\pi}{8}.$

4. $\displaystyle\int_0^{2\pi} \frac{d\theta}{1 + a\cos\theta} = \frac{2\pi}{\sqrt{1 - a^2}} \quad (-1 < a < 1).$

5. $\displaystyle\int_0^{\pi} \frac{\cos 2\theta \, d\theta}{1 - 2a\cos\theta + a^2} = \frac{a^2\pi}{1 - a^2} \quad (-1 < a < 1).$

6. $\displaystyle\int_0^{\pi} \frac{d\theta}{(a + \cos\theta)^2} = \frac{a\pi}{(\sqrt{a^2 - 1})^3} \quad (a > 1).$

7. $\displaystyle\int_0^{\pi} \sin^{2n}\theta \, d\theta = \frac{(2n)!}{2^{2n}(n!)^2}\pi \quad (n = 1, 2, \ldots).$

8. Derive the integration formula

$$\int_0^\infty \exp(-x^2) \cos(2bx) \, dx = \frac{\sqrt{\pi}}{2} \exp(-b^2) \qquad (b > 0)$$

by integrating the function $\exp(-z^2)$ around the rectangular path shown in Fig. 48 and

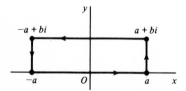

Figure 48

then letting a tend to infinity. Use the known integration formula*

$$\int_0^\infty \exp(-x^2)\,dx = \frac{\sqrt{\pi}}{2}.$$

9. Show that

(a) $\displaystyle\int_0^\infty \frac{\text{Log }x}{x^2 + 1}\,dx = 0;$ (b) $\displaystyle\int_0^\infty \frac{\text{Log }x}{(x^2 + 1)^2}\,dx = -\frac{\pi}{4}.$

Suggestion: Using the branch

$$\log z = \text{Log } r + i\theta \qquad \left(r > 0, -\frac{\pi}{2} < \theta < \frac{3\pi}{2}\right)$$

of the logarithmic function, integrate the functions

$$\frac{\log z}{z^2 + 1} \quad \text{and} \quad \frac{\log z}{(z^2 + 1)^2}$$

around the closed contour in Fig. 49. Also, use the integration formulas derived in Exercise 1, Sec. 60.

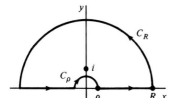

Figure 49

10. The *beta function* is this function of two real variables:

$$B(p, q) = \int_0^1 t^{p-1}(1 - t)^{q-1}\,dt \qquad (p > 0, q > 0).$$

Make the substitution $t = 1/(x + 1)$, and use the result obtained in Sec. 62 to show that

$$B(p, 1 - p) = \frac{\pi}{\sin p\pi} \qquad (0 < p < 1).$$

11. With the aid of the contours shown in Figs. 46 and 47, derive the following integration formulas:

(a) $\displaystyle\int_0^\infty \frac{x^{-1/2}}{x^2 + 1}\,dx = \frac{\pi}{\sqrt{2}}$ where $x^{-1/2} = \exp\left(-\frac{1}{2}\text{Log }x\right);$

(b) $\displaystyle\int_0^\infty \frac{x^a}{(x^2 + 1)^2}\,dx = \frac{(1 - a)\pi}{4\cos(a\pi/2)}$ $(-1 < a < 3)$ where $x^a = \exp(a\,\text{Log }x).$

12. The Fresnel integrals

$$\int_0^\infty \cos(x^2)\,dx = \int_0^\infty \sin(x^2)\,dx = \frac{\sqrt{\pi}}{2\sqrt{2}}$$

*The usual way to derive this particular integration formula is by means of a certain double integral. See, for example, A. E. Taylor and W. R. Mann, "Advanced Calculus," 3d ed., pp. 680–681, 1983.

are important in diffraction theory. Given that (see the footnote to Exercise 8)

$$\int_0^\infty \exp(-x^2)\, dx = \frac{\sqrt{\pi}}{2},$$

evaluate these integrals by integrating $\exp(iz^2)$ around the boundary of the sector $0 \leq r \leq R$, $0 \leq \theta \leq \pi/4$ and then letting R tend to ∞. Use Jordan's inequality (Sec. 60) to verify that the integral of $\exp(iz^2)$ along the circular arc $r = R$, $0 \leq \theta \leq \pi/4$ approaches zero as R tends to ∞.

13. Let the point $z = x_0$ on the x axis be a simple pole of a function f, and let B_0 denote the residue of f at that pole. It follows that there is a positive number ρ_0 such that

$$f(z) = \frac{B_0}{z - x_0} + g(z) \qquad (0 < |z - x_0| \leq \rho_0)$$

where g is continuous throughout the closed disk $|z - x_0| \leq \rho_0$. Let C_ρ denote the upper half of the circle $|z - x_0| = \rho$, where $\rho < \rho_0$ and where the orientation is in the positive sense (Fig. 50). Show that

$$\lim_{\rho \to 0} \int_{C_\rho} f(z)\, dz = -B_0 \pi i.$$

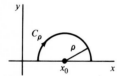

Figure 50

14. The integration formula

$$\int_0^\infty \frac{\sin x}{x}\, dx = \frac{\pi}{2}$$

is important in the theory of the Fourier integral.* Derive this formula by integrating the function $f(z) = e^{iz}/z$ around the simple closed contour shown in Fig. 49 and then letting ρ and R tend to 0 and ∞, respectively.

   Suggestion: Use the result in Exercise 13 to verify that the integral of $f(z)$ along the smaller semicircle in Fig. 49 approaches $-\pi i$ as ρ tends to 0. Use Jordan's inequality (Sec. 60) to show that the value of the integral along the larger semicircle approaches 0 as R tends to ∞.

15. Derive the integration formula

$$\int_0^\infty \frac{\sin^2 x}{x^2}\, dx = \frac{\pi}{2}.$$

   Suggestion: Note that $2 \sin^2 x = \mathrm{Re}(1 - e^{i2x})$, and integrate the function $f(z) = (1 - e^{i2z})/z^2$ around the closed contour shown in Fig. 49.

*See the authors' "Fourier Series and Boundary Value Problems," 3d ed., pp. 151–152, 1978.

16. The integral of the function

$$f(x) = \frac{1}{x(x^2 - 4x + 5)}$$

over an interval that includes the origin does not exist. Show that the *principal value* of the integral of that function along the entire x axis,

$$\text{P.V.} \int_{-\infty}^{\infty} f(x)\,dx = \lim_{\rho \to 0}\left[\int_{-\infty}^{-\rho} f(x)\,dx + \int_{\rho}^{\infty} f(x)\,dx\right] \qquad (\rho > 0),$$

does exist by finding that value with the aid of the contour in Fig. 49 and the result found in Exercise 13.

Ans. $2\pi/5$.

SEVEN

MAPPING BY ELEMENTARY FUNCTIONS

The geometric interpretation of a function of a complex variable as a mapping, or transformation, was introduced in Sec. 10. We saw there how the nature of such a function can be displayed graphically, to some extent, by the manner in which it maps certain curves and regions.

In this chapter we see further examples of how various curves and regions are mapped by elementary analytic functions. Applications of such results to physical problems are illustrated in Chaps. 9 and 10.

63. Linear Functions

The mapping of the z plane onto the w plane by means of the equation

$$(1) \qquad\qquad w = z + C,$$

where C is a complex constant, is a translation by means of the vector representing C. That is, if

$$w = u + iv, \qquad z = x + iy, \qquad \text{and} \qquad C = C_1 + iC_2,$$

then the image of any point (x, y) in the z plane is the point

$$(u, v) = (x + C_1, y + C_2)$$

in the w plane. Since each point in any given region of the z plane is mapped into the w plane in this manner, the image region is geometrically congruent to the original one.

To study the mapping

$$(2) \qquad\qquad w = Bz,$$

where B is a nonzero complex constant, we write B and z ($z \neq 0$) in exponential form as $B = be^{i\beta}$ and $z = re^{i\theta}$. Then

$$w = br\, e^{i(\beta + \theta)},$$

and transformation (2) maps the nonzero point z with polar coordinates (r, θ) onto the nonzero point in the w plane whose polar coordinates are $(\rho, \phi) = (br, \beta + \theta)$. This mapping consists of an expansion or contraction of the radius vector for z by the factor $b = |B|$, together with a rotation through an angle $\beta = \arg B$ about the origin. The image of a given region in the z plane is therefore geometrically similar to that region.

A general *linear transformation*

(3) $$w = Bz + C \qquad (B \neq 0),$$

which is a composition of the transformations

$$Z = Bz \; (B \neq 0) \qquad \text{and} \qquad w = Z + C,$$

is evidently an expansion, or contraction, and a rotation, followed by a translation.

Example. The transformation

(4) $$w = (1 + i)z + (2 - i)$$

maps the rectangular region shown in the z plane of Fig. 51 onto the rectangular region shown in the w plane there. This is seen by writing transformation (4) as a composition of the transformations $Z = (1 + i)z$ and $w = Z + (2 - i)$. Since $1 + i = \sqrt{2} \exp(i\pi/4)$, the first of these transformations is an expansion by the factor $\sqrt{2}$ and a rotation through the angle $\pi/4$. The second is a translation two units to the right and one unit downward.

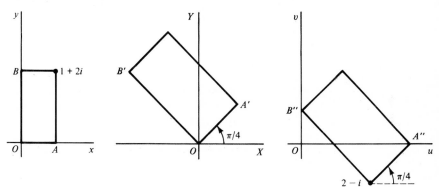

Figure 51. $w = (1 + i)z + (2 - i)$.

64. The Function $1/z$

The equation

(1) $$w = \frac{1}{z}$$

establishes a one to one correspondence between the nonzero points of the z and the w planes. Since $z\bar{z} = |z|^2$, the mapping can be described by means of the successive transformations

(2) $$Z = \frac{1}{|z|^2} z, \qquad w = \bar{Z}.$$

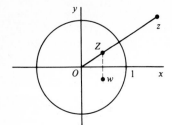

Figure 52. $w = 1/z$.

The first of these transformations is an inversion with respect to the unit circle $|z| = 1$. That is, the image of a nonzero point z is the point Z with the properties

$$|Z| = \frac{1}{|z|} \qquad \text{and} \qquad \arg Z = \arg z.$$

Thus the points exterior to the circle $|z| = 1$ are mapped onto the nonzero points interior to it (Fig. 52), and conversely. Any point on the circle is mapped onto itself. The second of transformations (2) is simply a reflection in the real axis.

Because

$$\lim_{z \to \infty} \frac{1}{z} = 0 \qquad \text{and} \qquad \lim_{z \to 0} \frac{1}{z} = \infty$$

(see Exercise 11, Sec. 12), it is natural to define a one to one transformation $w = T(z)$ from the extended z plane onto the extended w plane by writing $T(\infty) = 0$, $T(0) = \infty$, and $T(z) = 1/z$ for the remaining values of z. Observe that if we allow definition (3), Sec. 13, of continuity to include the point at infinity, then T is continuous throughout the extended z plane. For this reason, when the point at infinity is involved in a discussion of the function $1/z$, it is tacitly assumed that $T(z)$ is intended.

When a point $w = u + iv$ is the image of a nonzero point $z = x + iy$ under the transformation $w = 1/z$, writing $w = \bar{z}/|z|^2$ reveals that

$$(3) \qquad u = \frac{x}{x^2 + y^2}, \qquad v = \frac{-y}{x^2 + y^2}.$$

Also, since $z = 1/w = \bar{w}/|w|^2$,

$$(4) \qquad x = \frac{u}{u^2 + v^2}, \qquad y = \frac{-v}{u^2 + v^2}.$$

The following argument, based on these relations between coordinates, shows that *the mapping $w = 1/z$ transforms circles and lines into circles and lines.* When a, b, c, and d are all real numbers satisfying the condition $b^2 + c^2 > 4ad$, the equation

$$(5) \qquad a(x^2 + y^2) + bx + cy + d = 0$$

represents an arbitrary circle or line, where $a \neq 0$ for a circle and $a = 0$ for a line. If x and y satisfy equation (5), we can use relations (4) to substitute for those variables. After some simplifications, we find that u and v satisfy the equation

(6) $$d(u^2 + v^2) + bu - cv + a = 0,$$

which also represents a circle or line. Conversely, if u and v satisfy equation (6), it follows from relations (3) that x and y satisfy equation (5).

It is now clear from equations (5) and (6) that a circle ($a \neq 0$) not passing through the origin ($d \neq 0$) in the z plane is transformed into a circle not passing through the origin in the w plane. A circle through the origin in the z plane is transformed into a line which does not pass through the origin in the w plane. A line not passing through the origin in the z plane is transformed into a circle through the origin in the w plane, and a line through the origin in the z plane is transformed into a line through the origin in the w plane.

Example 1. According to equations (5) and (6), a vertical line $x = c_1$ ($c_1 \neq 0$) is transformed by $w = 1/z$ into the circle

(7) $$\left(u - \frac{1}{2c_1}\right)^2 + v^2 = \left(\frac{1}{2c_1}\right)^2,$$

centered on the u axis and tangent to the v axis. The image of a typical point (c_1, y) on the line is, by equations (3),

$$(u, v) = \left(\frac{c_1}{c_1^2 + y^2}, \frac{-y}{c_1^2 + y^2}\right).$$

If $c_1 > 0$, the circle (7) is evidently to the right of the v axis. As the point (c_1, y) moves up the entire line, its image traverses the circle once in the clockwise direction, the point at infinity in the extended z plane corresponding to the origin in the w plane. For if $y < 0$, then $v > 0$; and as y increases through negative values to 0, u increases from 0 to $1/c_1$. Then as y increases through positive values, v is negative and u decreases to 0.

If, on the other hand, $c_1 < 0$, the circle lies to the left of the v axis. As the point (c_1, y) moves upward, its image still makes one cycle, but in the counterclockwise direction. See Fig. 53, where the cases $c_1 = 1/3$ and $c_1 = -1/2$ are illustrated.

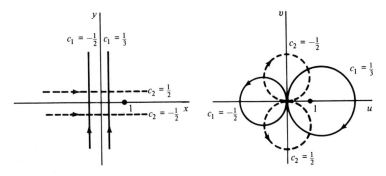

Figure 53. $w = 1/z$.

Example 2. The transformation $w = 1/z$ maps a horizontal line $y = c_2$ ($c_2 \neq 0$) onto the circle

(8)
$$u^2 + \left(v + \frac{1}{2c_2}\right)^2 = \left(\frac{1}{2c_2}\right)^2,$$

which is centered on the v axis and tangent to the u axis. Two special cases are shown in Fig. 53, where the corresponding orientations of the lines and circles are also indicated.

Example 3. We see from the first of equations (4) that when $w = 1/z$, the images of points in a half plane $x > c_1$ $(c_1 > 0)$ lie in the region

(9)
$$\frac{u}{u^2 + v^2} > c_1,$$

or

$$\left(u - \frac{1}{2c_1}\right)^2 + v^2 < \left(\frac{1}{2c_1}\right)^2.$$

That is, the image of any point in the half plane lies inside the circle

(10)
$$\left(u - \frac{1}{2c_1}\right)^2 + v^2 = \left(\frac{1}{2c_1}\right)^2.$$

Conversely, every point inside the circle satisfies inequality (9) and is therefore the image of a point in the half plane. Hence the image of the half plane is the entire circular domain.

Another way to obtain this result is to refer to Example 1 and note that the image of any line $x = c$, where $c > c_1$, is the circle

(11)
$$\left(u - \frac{1}{2c}\right)^2 + v^2 = \left(\frac{1}{2c}\right)^2.$$

As c increases through all values greater than c_1, the lines $x = c$ move to the right and the image circles (11) are interior to the circle (10) and shrink in size. Furthermore, the lines $x = c$ pass through all points in the half plane $x > c_1$; and the circles (11) pass through all the points interior to the circle (10).

EXERCISES

1. State why the transformation $w = iz$ is a rotation of the z plane through the angle $\pi/2$. Find the image of the infinite strip $0 < x < 1$.

Ans. $0 < v < 1$.

2. Show that the transformation $w = iz + i$ maps the half plane $x > 0$ onto the half plane $v > 1$.

3. Find the region onto which the half plane $y > 0$ is mapped by the transformation $w = (1 + i)z$ by using (*a*) polar coordinates; (*b*) cartesian coordinates. Sketch this region.

Ans. $v > u$.

4. Find the image of the half plane $y > 1$ under the transformation $w = (1 - i)z$.

5. Find the image of the semi-infinite strip $x > 0$, $0 < y < 2$ under the transformation $w = iz + 1$. Sketch the strip and its image.

Ans. $-1 < u < 1, v > 0$.

6. Give a geometric description of the transformation $w = B(z + C)$, where B and C are complex constants and $B \neq 0$.

7. Show that if $c_1 < 0$, the image of the half plane $x < c_1$ under the transformation $w = 1/z$ is the interior of a circle. What is the image when $c_1 = 0$?

8. Show that the image of the half plane $y > c_2$ under the transformation $w = 1/z$ is the interior of a circle, provided $c_2 > 0$. Find the image when $c_2 < 0$; also find it when $c_2 = 0$.

9. Find the image of the infinite strip $0 < y < 1/(2c)$ under the transformation $w = 1/z$. Sketch the strip and its image.

Ans. $u^2 + (v + c)^2 > c^2, v < 0$.

10. Find the image of the quadrant $x > 1$, $y > 0$ under the transformation $w = 1/z$.

Ans. $(u - \frac{1}{2})^2 + v^2 < (\frac{1}{2})^2, v < 0$.

11. Verify the mapping, where $w = 1/z$, of the regions and parts of the boundaries indicated in (a) Fig. 4, Appendix 2; (b) Fig. 5, Appendix 2.

12. Describe geometrically the transformation $w = 1/(z - 1)$.

13. Describe geometrically the transformation $w = i/z$. State why it transforms circles and lines into circles and lines.

14. Find the image of the semi-infinite strip $x > 0$, $0 < y < 1$ under the transformation $w = i/z$. Sketch the strip and its image.

Ans. $(u - \frac{1}{2})^2 + v^2 > (\frac{1}{2})^2, u > 0, v > 0$.

15. By writing $w = \rho \exp(i\phi)$, show that the mapping $w = 1/z$ transforms the hyperbola $x^2 - y^2 = 1$ into the lemniscate $\rho^2 = \cos 2\phi$. (See Exercise 19, Sec. 4.)

16. Let the circle $|z| = 1$ have a positive, or counterclockwise, orientation. Determine the orientation of its image under the transformation $w = 1/z$.

17. Show that when a circle is transformed into a circle under the transformation $w = 1/z$, the center of the original circle is never mapped onto the center of the image circle.

65. Linear Fractional Transformations

The transformation

(1)
$$w = \frac{az + b}{cz + d} \qquad (ad - bc \neq 0),$$

where $a, b, c,$ and d are complex constants, is called a *linear fractional transformation*, or *Möbius transformation*. When $c = 0$, it is simply a linear transformation (Sec. 63). When $c \neq 0$, equation (1) can be written

(2)
$$w = \frac{a}{c} + \frac{bc - ad}{c} \frac{1}{cz + d};$$

and it is clear how the condition $ad - bc \neq 0$ ensures that a linear fractional transformation is never a constant function. Note that the function $w = 1/z$, treated in Sec. 64, is a special case of transformation (1).

When cleared of fractions, equation (1) takes the form

(3) $$Azw + Bz + Cw + D = 0 \qquad (AD - BC \neq 0),$$

which is linear in z and linear in w, or bilinear in z and w. Hence another name for a linear fractional transformation is *bilinear transformation*.

Solving equation (1) for z, we find that

(4) $$z = \frac{-dw + b}{cw - a}.$$

Thus if $c = 0$, each point in the w plane is the image of one and only one point in the z plane; the same is true if $c \neq 0$, except for the point $w = a/c$. We can enlarge the domain of definition of transformation (1) in order to define a linear fractional transformation T on the *extended* z plane. We first write

(5) $$T(z) = \frac{az + b}{cz + d} \qquad (ad - bc \neq 0).$$

We then write $T(\infty) = \infty$ if $c = 0$; and we write $T(\infty) = a/c$ and $T(-d/c) = \infty$ if $c \neq 0$. This agrees with the way in which we enlarged the domain of definition of the function $1/z$ in Sec. 64 (see Exercise 17, Sec. 66).

When its domain of definition is enlarged in this way, the linear fractional transformation (5) is a *one to one* mapping of the extended z plane *onto* the extended w plane. That is, $T(z_1) \neq T(z_2)$ whenever $z_1 \neq z_2$, and for each point w in the second plane there is a point z in the first one such that $T(z) = w$. Hence, associated with the transformation T, there is an *inverse transformation* T^{-1} which is defined on the extended w plane as follows:

$$T^{-1}(w) = z \qquad \text{if and only if} \qquad T(z) = w.$$

From equation (4) we see that

$$T^{-1}(w) = \frac{-dw + b}{cw - a} \qquad (ad - bc \neq 0).$$

Evidently, T^{-1} is itself a linear fractional transformation, where $T^{-1}(\infty) = \infty$ if $c = 0$ and where $T^{-1}(a/c) = \infty$ and $T^{-1}(\infty) = -d/c$ if $c \neq 0$.

If T and S are two linear fractional transformations, then so is the composition $S[T(z)]$. This can be verified by combining expressions of the type (5).

We have already observed that if $c = 0$, the linear fractional transformation (1) is of the special type $w = Bz + C$ $(B \neq 0)$. If, on the other hand, $c \neq 0$, the form (2) of equation (1) reveals that the linear fractional transformation is a composition of the mappings

$$Z = cz + d, \qquad W = \frac{1}{Z}, \qquad w = \frac{a}{c} + \frac{bc - ad}{c} W.$$

It thus follows that *a linear fractional transformation always transforms circles and lines into circles and lines* because these special linear fractional transformations do this. (See Secs. 63 and 64.)

There is always a linear fractional transformation that maps three given distinct points z_1, z_2, and z_3 onto three specified distinct points w_1, w_2, and w_3, respectively. Indeed, the equation

(6) $$\frac{(w - w_1)(w_2 - w_3)}{(w - w_3)(w_2 - w_1)} = \frac{(z - z_1)(z_2 - z_3)}{(z - z_3)(z_2 - z_1)}$$

defines such a transformation when none of those points is the point at infinity. To verify this, let us write equation (6) as

(7) $$(z - z_3)(w - w_1)(z_2 - z_1)(w_2 - w_3)$$
$$= (z - z_1)(w - w_3)(z_2 - z_3)(w_2 - w_1).$$

If $z = z_1$, the right-hand side of equation (7) is zero and it follows that $w = w_1$. Similarly, if $z = z_3$, the left-hand side is zero and consequently $w = w_3$. If $z = z_2$, we have the linear equation

$$(w - w_1)(w_2 - w_3) = (w - w_3)(w_2 - w_1),$$

whose unique solution is $w = w_2$. The transformation defined by equation (6) is actually a linear fractional transformation, as is seen by expanding the products in equation (7) and writing the result in the form (3). The condition $AD - BC \neq 0$ with equation (3) is clearly satisfied since, as just demonstrated, equation (6) does not define a constant function. It is left to the reader (Exercise 16, Sec. 66) to show that equation (6) defines the *only* linear fractional transformation mapping the points z_1, z_2, and z_3 onto w_1, w_2, and w_3, respectively.

Example 1. To find the linear fractional transformation which maps the points $z_1 = -1$, $z_2 = 0$, and $z_3 = 1$ onto the points $w_1 = -i$, $w_2 = 1$, and $w_3 = i$, respectively, we need only substitute these numbers into equation (6) and then solve for w in terms of z. The result is

$$w = \frac{i - z}{i + z}.$$

In equation (6) the point at infinity can be introduced as one of the prescribed points in either the z or the w plane.

Example 2. If, for instance, w_2 is to be ∞, we replace w_2 by $1/w_2$ in equation (6) and then, after clearing fractions in the numerator and denominator on the left-hand side, we write $w_2 = 0$. This yields the equation

(8) $$\frac{w - w_1}{w - w_3} = \frac{(z - z_1)(z_2 - z_3)}{(z - z_3)(z_2 - z_1)}.$$

To be specific, suppose that $z_1 = 1$, $z_2 = 0$, $z_3 = -1$ and $w_1 = i$, $w_2 = \infty$, $w_3 = 1$.

Substituting the given values of z_1, z_2, z_3 and w_1, w_3 into equation (8), we find that

$$w = \frac{(1 + i)z + (i - 1)}{2z}.$$

66. Special Linear Fractional Transformations

Let us determine all linear fractional transformations that map the upper half plane Im $z > 0$ onto the open disk $|w| < 1$ and the boundary Im $z = 0$ onto the boundary $|w| = 1$ (Fig. 54).

Keeping in mind that points on the line Im $z = 0$ are to be transformed into points on the circle $|w| = 1$, we start by selecting the points $z = 0$, $z = 1$, and $z = \infty$ on the line and determining conditions on a linear fractional transformation

$$(1) \qquad\qquad w = \frac{az + b}{cz + d} \qquad\qquad (ad - bc \neq 0)$$

which are necessary in order for the images of those points to have unit modulus.

Observe that $c \neq 0$ since, according to Sec. 65, the image of the point $z = \infty$ is $w = \infty$ when $c = 0$. We also saw in Sec. 65 that the image of the point $z = \infty$ is $w = a/c$ when $c \neq 0$. So the requirement that $|w| = 1$ when $z = \infty$ means that

$$(2) \qquad\qquad |a| = |c| \neq 0.$$

In addition, since $|w| = 1$ when $z = 0$, we find from equation (1) that

$$(3) \qquad\qquad |b| = |d| \neq 0.$$

Hence

$$w = \left(\frac{a}{c}\right)\frac{z + (b/a)}{z + (d/c)},$$

or, since $|a/c| = 1$,

$$(4) \qquad\qquad w = e^{i\alpha}\frac{z - z_0}{z - z_1}$$

where α is a real constant and z_0, z_1 are nonzero complex constants. Since $|d/c| = |b/a|$, according to equations (2) and (3), we know that $|z_1| = |z_0|$.

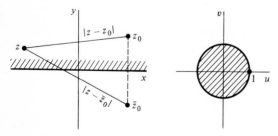

Figure 54. $w = e^{i\alpha}\dfrac{z - z_0}{z - \bar{z}_0}$ (Im $z_0 > 0$).

Next, we impose on transformation (4) the condition that $|w| = 1$ when $z = 1$. This shows that

$$|1 - z_1| = |1 - z_0|,$$

or

$$(1 - z_1)(1 - \bar{z}_1) = (1 - z_0)(1 - \bar{z}_0).$$

But $z_1 \bar{z}_1 = z_0 \bar{z}_0$ since $|z_1| = |z_0|$, and the above relation reduces to

$$z_1 + \bar{z}_1 = z_0 + \bar{z}_0;$$

that is, Re $z_1 = $ Re z_0. It follows that either $z_1 = z_0$ or $z_1 = \bar{z}_0$, again since $|z_1| = |z_0|$. If $z_1 = z_0$, transformation (4) becomes the constant function $w = \exp(i\alpha)$; hence $z_1 = \bar{z}_0$.

Transformation (4), with $z_1 = \bar{z}_0$, maps the point z_0 onto the origin $w = 0$; and since points interior to the circle $|w| = 1$ are to be the images of points *above* the real axis in the z plane, we may conclude that Im $z_0 > 0$. Any linear fractional transformation having the mapping property stated in the first paragraph of this section must therefore be of the form

(5)
$$w = e^{i\alpha} \frac{z - z_0}{z - \bar{z}_0} \qquad (\text{Im } z_0 > 0),$$

where α is real.

It remains to show that, conversely, any linear fractional transformation of the form (5) has the desired mapping property. This is easily done by taking absolute values of each side of equation (5) and interpreting the resulting equation,

$$|w| = \frac{|z - z_0|}{|z - \bar{z}_0|},$$

geometrically. If a point z lies above the real axis, both it and the point z_0 lie on the same side of that axis, which is the perpendicular bisector of the line segment joining z_0 and $\bar{z}_0$. It follows that the distance $|z - z_0|$ is less than the distance $|z - \bar{z}_0|$ (Fig. 54); that is, $|w| < 1$. Likewise, if z lies below the real axis, the distance $|z - z_0|$ is greater than the distance $|z - \bar{z}_0|$; and so $|w| > 1$. Finally, if z is on the real axis, $|w| = 1$ because then $|z - z_0| = |z - \bar{z}_0|$. Since any linear fractional transformation is a one to one mapping of the extended z plane onto the extended w plane, this shows that *transformation* (5) *maps the half plane* Im $z > 0$ *onto the disk* $|w| < 1$ *and the boundary of that half plane onto the boundary of that disk.*

Example. The transformation $w = (i - z)/(i + z)$ in Example 1 of Sec. 65 can be written

$$w = e^{i\pi} \frac{z - i}{z - \bar{i}}.$$

Hence it has the above mapping property.

EXERCISES

1. Find the linear fractional transformation that maps the points $z_1 = 2$, $z_2 = i$, $z_3 = -2$ onto the points $w_1 = 1$, $w_2 = i$, $w_3 = -1$.

   Ans. $w = (3z + 2i)/(iz + 6)$.

2. Find the linear fractional transformation that maps the points $z_1 = -i$, $z_2 = 0$, $z_3 = i$ onto the points $w_1 = -1$, $w_2 = i$, $w_3 = 1$. Into what curve is the imaginary axis $x = 0$ transformed?

3. Find the bilinear transformation that maps the points $z_1 = \infty$, $z_2 = i$, $z_3 = 0$ onto the points $w_1 = 0$, $w_2 = i$, $w_3 = \infty$.

   Ans. $w = -1/z$.

4. Find the bilinear transformation that maps the points z_1, z_2, z_3 onto the points $w_1 = 0$, $w_2 = 1$, $w_3 = \infty$.

   $$Ans. \quad w = \frac{(z - z_1)(z_2 - z_3)}{(z - z_3)(z_2 - z_1)}.$$

5. Show that a composition of two linear fractional transformations is again a linear fractional transformation, as stated in Sec. 65.

6. A *fixed point* of a transformation $w = f(z)$ is a point z_0 such that $f(z_0) = z_0$. Show that every linear fractional transformation, with the exception of the identity transformation $w = z$, has at most two fixed points in the extended plane.

7. Find the fixed points (Exercise 6) of the transformation
   (a) $w = (z - 1)/(z + 1)$;   (b) $w = (6z - 9)/z$.

   Ans.   (a) $z = \pm i$;   (b) $z = 3$.

8. Modify equation (6), Sec. 65, for the case in which both z_2 and w_2 are the point at infinity. Then show that any linear fractional transformation must be of the form $w = az$ $(a \neq 0)$ when its fixed points (Exercise 6) are 0 and ∞.

9. Prove that if the origin is a fixed point (Exercise 6) of a linear fractional transformation, the transformation can be written in the form $w = z/(cz + d)$, where $d \neq 0$.

10. Recall (Sec. 66) that the transformation $w = (i - z)/(i + z)$ maps the half plane $\text{Im } z > 0$ onto the disk $|w| < 1$ and the boundary of the half plane onto the boundary of the disk. Show that a point $z = x$ is mapped onto the point

    $$w = \frac{1 - x^2}{1 + x^2} + i\frac{2x}{1 + x^2},$$

    and then complete the verification of the mapping illustrated in Fig. 13, Appendix 2, by showing that segments of the x axis are mapped as indicated there.

11. Verify the mapping shown in Fig. 12, Appendix 2, where $w = (z - 1)/(z + 1)$.
    Suggestion: Write the given transformation as a composition of the mappings

    $$Z = iz, \quad W = \frac{i - Z}{i + Z}, \quad w = -W.$$

    Then refer to the mapping verified in Exercise 10.

12. (a) By finding the inverse of the transformation $w = (i - z)/(i + z)$ and appealing to Fig. 13, Appendix 2, which was verified in Exercise 10, show that the transformation $w = i(1 - z)/(1 + z)$ maps the disk $|z| \leq 1$ onto the upper half plane $\text{Im } w \geq 0$.

(b) Show that the linear fractional transformation $w = (z - 2)/z$ can be written

$$Z = z - 1, \qquad W = i\frac{1 - Z}{1 + Z}, \qquad w = iW.$$

Then, with the aid of the result in part (a), verify that the transformation $w = (z - 2)/z$ maps the disk $|z - 1| \leq 1$ onto the left half plane Re $w \leq 0$.

13. Transformation (5), Sec. 66, maps the point $z = \infty$ onto the point $w = \exp(i\alpha)$, which lies on the boundary of the disk $|w| \leq 1$. Show that if $0 < \alpha < 2\pi$ and the points $z = 0$ and $z = 1$ are to be mapped onto the points $w = 1$ and $w = \exp(i\alpha/2)$, respectively, the transformation can be written

$$w = e^{i\alpha}\frac{z + \exp(-i\alpha/2)}{z + \exp(i\alpha/2)}.$$

14. Note that when $\alpha = \pi/2$, the transformation in Exercise 13 becomes

$$w = \frac{iz + \exp(i\pi/4)}{z + \exp(i\pi/4)}.$$

Verify that this special case maps points on the x axis in the manner indicated in Fig. 55.

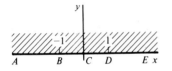

Figure 55. $w = \dfrac{iz + \exp(i\pi/4)}{z + \exp(i\pi/4)}$.

15. Show that when Im $z_0 < 0$, transformation (5), Sec. 66, maps the lower half plane Im $z \leq 0$ onto the unit disk $|w| \leq 1$.

16. Show that there is only one linear fractional transformation that maps three given distinct points z_1, z_2, and z_3 in the extended z plane onto three specified distinct points w_1, w_2, and w_3 in the extended w plane.

Suggestion: Let T and S be two such linear fractional transformations. Then, after pointing out why $S^{-1}[T(z_k)] = z_k$ ($k = 1, 2, 3$), use the results in Exercises 5 and 6 to show that $S^{-1}[T(z)] = z$ for all z. Thus show that $T(z) = S(z)$ for all z.

17. With the aid of the results in Exercises 11 and 12 of Sec. 12, show that when

$$T(z) = \frac{az + b}{cz + d} \qquad\qquad (ad - bc \neq 0),$$

(a) $\lim\limits_{z \to \infty} T(z) = \infty$ if $c = 0$;

(b) $\lim\limits_{z \to \infty} T(z) = \dfrac{a}{c}$ and $\lim\limits_{z \to -d/c} T(z) = \infty$ if $c \neq 0$.

18. With the aid of equation (6), Sec. 65, prove that if a linear fractional transformation maps each point of the x axis onto the u axis, then the coefficients in the transformation are all real, except possibly for a common complex factor. The converse statement is evident.

67. The Function z^2

In this section we consider the transformation

(1) $$w = z^2,$$

which is easily described in terms of polar coordinates. If $z = re^{i\theta}$ and $w = \rho e^{i\phi}$, then

$$\rho e^{i\phi} = r^2 e^{i2\theta}.$$

Hence the image of any nonzero point z is found by squaring the modulus of z and doubling a value of arg z; that is, $|w| = |z|^2$ and arg $w = 2$ arg z.

Example 1. Transformation (1) is a one to one mapping of the first quadrant $r \geq 0, 0 \leq \theta \leq \pi/2$ in the z plane onto the upper half $\rho \geq 0, 0 \leq \phi \leq \pi$ of the w plane (Fig. 56). It is also a mapping of the upper half plane $r \geq 0, 0 \leq \theta \leq \pi$ onto the entire w plane. However, in this case the transformation is not one to one since both the positive and negative real axes in the z plane are mapped onto the positive real axis in the w plane.

A circle $r = r_0$ is transformed into the circle $\rho = r_0^2$, and the sector $r \leq r_0$, $0 \leq \theta \leq \pi/2$ is mapped in a one to one manner onto the semicircular region $\rho \leq r_0^2$, $0 \leq \phi \leq \pi$ (see Fig. 56).

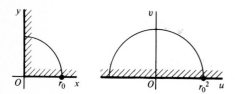

Figure 56. $w = z^2$.

When $z = x + iy$ and $w = u + iv$, transformation (1) becomes

$$u + iv = x^2 - y^2 + i2xy,$$

or

(2) $$u = x^2 - y^2, \qquad v = 2xy.$$

This form involving cartesian coordinates is often useful.

Example 2. When $w = z^2$, each branch of a hyperbola $x^2 - y^2 = c_1$ $(c_1 > 0)$ is mapped in a one to one manner onto the vertical line $u = c_1$. To see this, we note from the first of equations (2) that $u = c_1$ when (x, y) is a point lying on either branch. When, in particular, it lies on the right-hand branch, the second of equations (2) tells us that $v = 2y\sqrt{y^2 + c_1}$. Thus the image of the right-hand branch can be expressed parametrically as

$$u = c_1, \qquad v = 2y\sqrt{y^2 + c_1} \qquad (-\infty < y < \infty);$$

and it is evident that the image of a point (x, y) on that branch moves upward along the

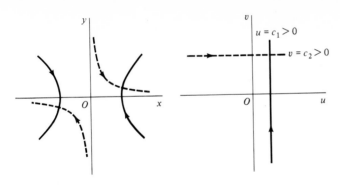

Figure 57. $w = z^2$.

entire line as (x, y) traces out the branch in the upward direction (Fig. 57). Likewise, since the pair of equations

$$u = c_1, \qquad v = -2y\sqrt{y^2 + c_1} \qquad (-\infty < y < \infty)$$

furnishes a parametric representation for the image of the left-hand branch of the hyperbola, the image of a point going *downward* along the entire left-hand branch is seen to move up the entire line $u = c_1$.

It is left as an exercise to show that each branch of a hyperbola $2xy = c_2$ $(c_2 > 0)$ is transformed into the line $v = c_2$, as indicated in Fig. 57. The cases in which c_1 and c_2 are negative are also treated in the exercises. Images of regions whose boundaries involve such hyperbolas are easily obtained.

Example 3. The domain $x > 0, y > 0, xy < 1$ consists of all points lying on the upper branches of hyperbolas from the family $xy = c$ where $0 < c < 1$. The image of the domain when $w = z^2$ therefore consists of all points lying on the lines $v = 2c$ $(0 < c < 1)$. That is, the image of the domain is the horizontal strip $0 < v < 2$.

When n is any positive integer, various mapping properties of the transformation

(3) $$w = z^n, \qquad \text{or} \qquad \rho e^{i\phi} = r^n e^{in\theta},$$

are similar to those of $w = z^2$.

Example 4. Transformation (3) maps the region $r \geq 0, 0 \leq \theta \leq \pi/n$ onto the upper half plane $\rho \geq 0, 0 \leq \phi \leq \pi$. It maps the entire z plane onto the entire w plane, where each nonzero point in the w plane is the image of n distinct points in the z plane. The circle $r = r_0$ is mapped onto the circle $\rho = r_0^n$; and the sector $r \leq r_0, 0 \leq \theta \leq 2\pi/n$ is mapped onto the disk $\rho \leq r_0^n$, but not in a one to one manner.

68. The Function $z^{1/2}$

As in Sec. 7, the values of $z^{1/2}$ are the two square roots of z when $z \neq 0$. According to that section, if polar coordinates are used and $z = r \exp(i\Theta)$ $(-\pi < \Theta \leq \pi)$, then

(1)
$$z^{1/2} = \sqrt{r} \, \exp \frac{i(\Theta + 2k\pi)}{2} \qquad (k = 0, 1),$$

the principal root occurring when $k = 0$.

In Sec. 26 we saw that $z^{1/2}$ can also be written

(2)
$$z^{1/2} = \exp\left(\frac{1}{2} \log z\right) \qquad (z \neq 0).$$

The *principal branch* $F_0(z)$ of the double-valued function $z^{1/2}$ is then obtained by taking the principal branch of $\log z$ and writing (see Sec. 27)

$$F_0(z) = \exp\left(\frac{1}{2} \operatorname{Log} z\right) \qquad (|z| > 0, -\pi < \operatorname{Arg} z < \pi),$$

or

(3)
$$F_0(z) = \sqrt{r} \, \exp \frac{i\Theta}{2} \qquad (r > 0, -\pi < \Theta < \pi).$$

The right-hand side of this equation is, of course, the same as the right-hand side of equation (1) when $k = 0$ and $-\pi < \Theta < \pi$ there. The ray $\Theta = \pi$ is the branch cut for F_0, and the point $z = 0$ is the branch point. Although the right-hand side of equation (3) is defined for points on the branch cut of F_0, the function obtained by that extension is not even continuous there. For there are values of Θ close to π as well as values close to $-\pi$ in any neighborhood of a point on the negative real axis.

Images of curves and regions under the transformation $w = F_0(z)$ may be obtained by writing $w = \rho \exp(i\phi)$ where $\rho = \sqrt{r}$ and $\phi = \Theta/2$.

Example. The transformation $w = F_0(z)$ maps the domain $r > 0, -\pi < \Theta < \pi$ onto the right-hand half $\rho > 0, -\pi/2 < \phi < \pi/2$ of the w plane. In particular, it maps the region $0 < r \leq r_0, -\pi < \Theta < \pi$ onto the half disk $0 < \rho \leq \sqrt{r_0}, -\pi/2 < \phi < \pi/2$ (Fig. 58).

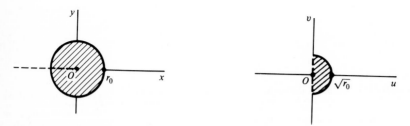

Figure 58. $w = F_0(z) = \sqrt{r} \, \exp \dfrac{i\Theta}{2} \quad (r > 0, -\pi < \Theta < \pi)$.

The choice $k = 1$ in equation (1) corresponds to the branch

4)
$$F_1(z) = \sqrt{r} \, \exp \frac{i(\Theta + 2\pi)}{2} \qquad (r > 0, -\pi < \Theta < \pi)$$

of the double-valued function (2). Since $\exp(i\pi) = -1$, it follows that $F_1(z) = -F_0(z)$. The values $\pm F_0(z)$ thus represent the totality of values of $z^{1/2}$ at all points in the domain $r > 0$, $-\pi < \Theta < \pi$. If, by means of expression (3), we extend the domain of definition of F_0 to include the ray $\Theta = \pi$ and if we write $F_0(0) = 0$, then the values $\pm F_0(z)$ represent the totality of values of $z^{1/2}$ in the entire z plane.

Other branches of $z^{1/2}$ are obtained by using other branches of $\log z$ in expression (2). A branch where the ray $\theta = \alpha$ is the branch cut is given by the equation

$$(5) \qquad f_\alpha(z) = \sqrt{r}\, \exp\frac{i\theta}{2} \qquad (r > 0,\ \alpha < \theta < \alpha + 2\pi).$$

Observe that when $\alpha = -\pi$ we have the branch $F_0(z)$ and that when $\alpha = \pi$ we have the branch $F_1(z)$. Just as in the case F_0, the domain of definition of f_α can be extended to the entire complex plane by using expression (5) to define f_α at the nonzero points on the branch cut and by writing $f_\alpha(0) = 0$. Such extensions are, however, never continuous in the entire complex plane.

It is sometimes useful to note that if $w = f_\alpha(z)$, then $z = w^2$. For instance, we know (Example 2, Sec. 67) that the function $w = z^2$ is a one to one mapping of the branch of the hyperbola $2xy = 1$ which lies in the first quadrant of the z plane onto the line $v = 1$ in the w plane. Hence, after interchanging the z and the w planes, we find that the branch f_0, with branch cut $\theta = 0$, maps the line $y = 1$ in the z plane in a one to one manner onto the branch of the hyperbola $2uv = 1$ which lies in the first quadrant of the w plane.

69. Related Functions

Let n be any positive integer greater than 1. The values of $z^{1/n}$ are the nth roots of z when $z \neq 0$; and, according to Secs. 7 and 26, the multiple-valued function $z^{1/n}$ can be written

$$(1) \qquad z^{1/n} = \exp\!\left(\frac{1}{n}\log z\right) = \sqrt[n]{r}\, \exp\frac{i(\Theta + 2k\pi)}{n} \qquad (k = 0, 1, 2, \dots, n-1),$$

where $r = |z|$ and $\Theta = \operatorname{Arg} z$.

The case $n = 2$ was considered in Sec. 68. In the general case, each of the n functions

$$(2) \qquad F_k(z) = \sqrt[n]{r}\, \exp\frac{i(\Theta + 2k\pi)}{n} \qquad (k = 0, 1, 2, \dots, n-1)$$

is a branch of $z^{1/n}$, defined on the domain $r > 0$, $-\pi < \Theta < \pi$. When $w = \rho e^{i\phi}$, the transformation $w = F_k(z)$ is a one to one mapping of that domain onto the domain $\rho > 0$, $(2k-1)\pi/n < \phi < (2k+1)\pi/n$. These n branches of $z^{1/n}$ yield the n distinct nth roots of z at any point z in the domain $r > 0$, $-\pi < \Theta < \pi$. The principal branch occurs when $k = 0$, and further branches of the type (5) in Sec. 68 are readily constructed.

We now consider some mappings which are compositions of polynomials and square roots of z.

Example 1. Branches of the double-valued function $(z - z_0)^{1/2}$ can be obtained by noting that it is a composition of the translation $Z = z - z_0$ with the double-valued function $Z^{1/2}$. Each branch of $Z^{1/2}$ yields a branch of $(z - z_0)^{1/2}$. When $Z = Re^{i\theta}$, branches of $Z^{1/2}$ are

$$Z^{1/2} = \sqrt{R} \exp \frac{i\theta}{2} \qquad (R > 0, \alpha < \theta < \alpha + 2\pi).$$

Hence, if we write $R = |z - z_0|$, $\theta = \arg(z - z_0)$, and $\Theta = \text{Arg}(z - z_0)$, two branches of $(z - z_0)^{1/2}$ are

(3)
$$G_0(z) = \sqrt{R} \exp \frac{i\Theta}{2} \qquad (R > 0, -\pi < \Theta < \pi)$$

and

(4)
$$g_0(z) = \sqrt{R} \exp \frac{i\theta}{2} \qquad (R > 0, 0 < \theta < 2\pi).$$

The branch of $Z^{1/2}$ used in writing $G_0(z)$ is defined at all points in the Z plane except for the origin and points on the ray $\text{Arg } Z = \pi$. The transformation $w = G_0(z)$ is therefore a one to one mapping of the domain $|z - z_0| > 0, -\pi < \text{Arg}(z - z_0) < \pi$ onto the right half $\text{Re } w > 0$ of the w plane (Fig. 59). The transformation $w = g_0(z)$ maps the domain $|z - z_0| > 0, 0 < \arg(z - z_0) < 2\pi$ in a one to one manner onto the upper half plane $\text{Im } w > 0$.

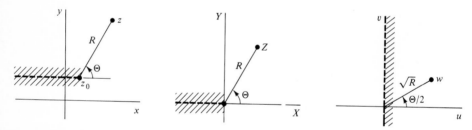

Figure 59. $w = G_0(z)$.

Example 2. For an instructive but less elementary example, we now consider the double-valued function $(z^2 - 1)^{1/2}$. Using established properties of logarithms, we can write

$$(z^2 - 1)^{1/2} = \exp\left[\frac{1}{2} \log(z^2 - 1)\right] = \exp\left[\frac{1}{2} \log(z - 1) + \frac{1}{2} \log(z + 1)\right],$$

or

(5)
$$(z^2 - 1)^{1/2} = (z - 1)^{1/2}(z + 1)^{1/2} \qquad (z \neq \pm 1).$$

Thus, if $f_1(z)$ is a branch of $(z - 1)^{1/2}$ defined on a domain D_1 and $f_2(z)$ is a branch of $(z + 1)^{1/2}$ defined on a domain D_2, the product $f(z) = f_1(z)f_2(z)$ is a branch of $(z^2 - 1)^{1/2}$ defined at all points lying in both D_1 and D_2.

To obtain a specific branch of $(z^2 - 1)^{1/2}$, we use the branch of $(z - 1)^{1/2}$ and the branch of $(z + 1)^{1/2}$ given by equation (4). If we write $r_1 = |z - 1|$ and $\theta_1 = \arg(z - 1)$, that branch of $(z - 1)^{1/2}$ is

$$f_1(z) = \sqrt{r_1} \exp \frac{i\theta_1}{2} \qquad (r_1 > 0, 0 < \theta_1 < 2\pi).$$

The branch of $(z + 1)^{1/2}$ given by equation (4) is

$$f_2(z) = \sqrt{r_2} \exp \frac{i\theta_2}{2} \qquad (r_2 > 0, 0 < \theta_2 < 2\pi)$$

where $r_2 = |z + 1|$ and $\theta_2 = \arg(z + 1)$. The product of these two branches is therefore the branch f of $(z^2 - 1)^{1/2}$ defined by the equation

(6) $$f(z) = \sqrt{r_1 r_2} \exp \frac{i(\theta_1 + \theta_2)}{2}$$

where

$$r_k > 0, \qquad 0 < \theta_k < 2\pi \qquad\qquad (k = 1, 2).$$

As illustrated in Fig. 60, the branch f is defined everywhere in the z plane except on the ray $r_2 \geq 0$, $\theta_2 = 0$, which is the portion $x \geq -1$ of the x axis.

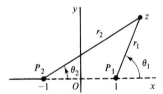

Figure 60

The branch f of $(z^2 - 1)^{1/2}$ given in equation (6) can be extended to a function

(7) $$F(z) = \sqrt{r_1 r_2} \exp \frac{i(\theta_1 + \theta_2)}{2}$$

where

$$r_k > 0, \qquad 0 \leq \theta_k < 2\pi \quad (k = 1, 2), \qquad \text{and} \qquad r_1 + r_2 > 2.$$

As we shall now see, this function is analytic everywhere in its domain of definition, which is the entire z plane except for the segment $-1 \leq x \leq 1$ of the x axis.

Since $F(z) = f(z)$ for all z in the domain of definition of F except on the ray $r_1 > 0$, $\theta_1 = 0$, we need only show that F is analytic on that ray. To do this, we form the product of the branches of $(z - 1)^{1/2}$ and $(z + 1)^{1/2}$ which are given by equation (3). That is, we consider the function

$$G(z) = \sqrt{r_1 r_2} \exp \frac{i(\Theta_1 + \Theta_2)}{2}$$

where $r_1 = |z - 1|$, $r_2 = |z + 1|$, $\Theta_1 = \text{Arg}(z - 1)$, $\Theta_2 = \text{Arg}(z + 1)$ and where

$$r_k > 0, \qquad -\pi < \Theta_k < \pi \qquad\qquad (k = 1, 2).$$

Observe that G is analytic in the entire z plane except for the ray $r_1 \geq 0$, $\Theta_1 = \pi$. Now $F(z) = G(z)$ when the point z lies above or on the ray $r_1 > 0$, $\Theta_1 = 0$; for then $\theta_k = \Theta_k$ ($k = 1, 2$). When z lies below that ray, $\theta_k = \Theta_k + 2\pi$ ($k = 1, 2$). Consequently, $\exp(i\theta_k/2) = -\exp(i\Theta_k/2)$; and this means that

$$\exp\frac{i(\theta_1 + \theta_2)}{2} = \left(\exp\frac{i\theta_1}{2}\right)\left(\exp\frac{i\theta_2}{2}\right) = \exp\frac{i(\Theta_1 + \Theta_2)}{2}.$$

So again $F(z) = G(z)$. Since $F(z) = G(z)$ in a domain containing the ray $r_1 > 0$, $\Theta_1 = 0$ and since $G(z)$ is analytic in that domain, $F(z)$ is analytic there. Hence $F(z)$ is *analytic everywhere except on the line segment $P_2 P_1$ in Fig. 60.*

The function F defined in equation (7) cannot itself be extended to a function which is analytic at points on the line segment $P_2 P_1$. For the value on the right in equation (7) jumps from $i\sqrt{r_1 r_2}$ to numbers near $-i\sqrt{r_1 r_2}$ as the point z moves downward across that line segment. Hence the extension would not even be continuous there.

The transformation $w = F(z)$ is, as we shall see, a one to one mapping of the domain D_z consisting of all points in the z plane except those on the line segment $P_2 P_1$ onto the domain D_w consisting of the entire w plane with the exception of the segment $-1 \leq v \leq 1$ of the v axis (Fig. 61).

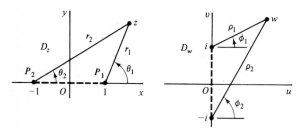

Figure 61. $w = F(z)$.

Before verifying this, we note that if $z = iy$ ($y > 0$), then $r_1 = r_2 > 1$ and $\theta_1 + \theta_2 = \pi$; hence the positive y axis is mapped by $w = F(z)$ onto that part of the v axis for which $v > 1$. The negative y axis is, moreover, mapped onto that part of the v axis for which $v < -1$. Each point in the upper half $y > 0$ of the domain D_z is mapped into the upper half $v > 0$ of the w plane, and each point in the lower half $y < 0$ of the domain D_z is mapped into the lower half $v < 0$ of the w plane. The ray $r_1 > 0$, $\theta_1 = 0$ is mapped onto the posi ve real axis in the w plane, and the ray $r_2 > 0$, $\theta_2 = \pi$ is mapped onto the negative real axis there.

To show that the transformation $w = F(z)$ is *one to one*, observe that if $F(z_1) = F(z_2)$, then $z_1^2 - 1 = z_2^2 - 1$. From this it follows that $z_1 = z_2$ or $z_1 = -z_2$. However, because of the manner in which F maps the upper and lower halves of the domain D_z, as well as the portions of the real axis lying in D_z, the case $z_1 = -z_2$ is impossible. Thus, if $F(z_1) = F(z_2)$, then $z_1 = z_2$; and F is one to one.

We show that F maps the domain D_z *onto* the domain D_w by finding a function H mapping D_w into D_z with the property that if $z = H(w)$, then $w = F(z)$. This will show that for any point w in D_w, there exists a point z in D_z such that $F(z) = w$; and the mapping is therefore onto. The mapping H will be the inverse of F.

To find H, we first note that if w is a value of $(z^2 - 1)^{1/2}$ for a specific z, then $w^2 = z^2 - 1$; and z is therefore a value of $(w^2 + 1)^{1/2}$ for that w. The function H will be a branch of the double-valued function

$$(w^2 + 1)^{1/2} = (w - i)^{1/2}(w + i)^{1/2} \qquad\qquad (w \neq \pm i).$$

Following our procedure for obtaining the function $F(z)$, we write $w - i = \rho_1 \exp(i\phi_1)$ and $w + i = \rho_2 \exp(i\phi_2)$. (See Fig. 61.) With the restrictions

$$\rho_k > 0, \qquad \frac{-\pi}{2} \leq \phi_k < \frac{3\pi}{2} \quad (k = 1, 2), \qquad \text{and} \qquad \rho_1 + \rho_2 > 2,$$

we then write

(8) $$H(w) = \sqrt{\rho_1 \rho_2} \exp \frac{i(\phi_1 + \phi_2)}{2},$$

the domain of definition being D_w. The transformation $z = H(w)$ maps points of D_w lying above or below the u axis onto points above or below the x axis, respectively. It maps the positive u axis into that part of the positive x axis where $x > 1$ and the negative u axis into that part of the negative x axis where $x < -1$. If $z = H(w)$, then $z^2 = w^2 + 1$; and so $w^2 = z^2 - 1$. Since z is in D_z and since $F(z)$ and $-F(z)$ are the two values of $(z^2 - 1)^{1/2}$ for a point in D_z, we see that $w = F(z)$ or $w = -F(z)$. But it is evident from the manner in which F and H map the upper and lower halves of their domains of definition, including the portions of the real axes lying in those domains, that $w = F(z)$.

Branches of double-valued functions

(9) $$w = (z^2 + Az + B)^{1/2} = [(z - z_0)^2 - z_1^2]^{1/2} \qquad\qquad (z_1 \neq 0)$$

where $A = -2z_0$ and $B = z_0^2 - z_1^2$, and mappings by those branches, can be treated with the aid of the results found for F above and the successive transformations

(10) $$Z = \frac{z - z_0}{z_1}, \qquad W = (Z^2 - 1)^{1/2}, \qquad w = z_1 W.$$

EXERCISES

1. Sketch the region onto which the sector $r \leq 1$, $0 \leq \theta \leq \pi/4$ is mapped by the transformation

    (a) $w = z^2$; (b) $w = z^3$; (c) $w = z^4$.

2. Find a region in the z plane whose image under the transformation $w = z^2$ is the square domain in the w plane bounded by the lines $u = 1$, $u = 2$, $v = 1$, and $v = 2$.

3. Verify that the closed region bounded by hyperbolas in Fig. 2, Appendix 2, is mapped in the manner shown there by the transformation $w = z^2$.

4. Show that each branch of a hyperbola $2xy = c_2$ $(c_2 > 0)$ is mapped onto the line $v = c_2$ by the transformation $w = z^2$, as indicated in Fig. 57 (Sec. 67).

5. Find and sketch, showing corresponding orientations, the images of the hyperbolas $x^2 - y^2 = c_1$ $(c_1 < 0)$ and $2xy = c_2$ $(c_2 < 0)$ under the transformation $w = z^2$.

6. Show, indicating corresponding orientations, that the transformation $w = z^2$ is a one to one mapping of (a) the lines $x = c_1$ $(c_1 > 0)$ onto the parabolas $v^2 = -4c_1^2(u - c_1^2)$; (b) the lines $y = c_2$ $(c_2 > 0)$ onto the parabolas $v^2 = 4c_2^2(u + c_2^2)$. (Note that all these parabolas have foci at the point $w = 0$.)

7. Show that the principal branch of the function $z^{1/2}$ maps the domain to the right of the y axis and bounded by the segment $-2 \leq y \leq 2$ of that axis and the parabola $y^2 = -4(x - 1)$ onto the triangular domain bounded by the lines $v = u$, $v = -u$, and $u = 1$. Show corresponding parts of the boundaries of the two regions. [See Exercise 6(a).]

8. Show that the branch $w = \sqrt{r}\,\exp(i\theta/2)$ $(r > 0, 0 < \theta < 2\pi)$ of the multiple-valued function $z^{1/2}$ is a one to one mapping of the domain between the two parabolas

$$y^2 = 4a^2(x + a^2), \qquad y^2 = 4b^2(x + b^2) \qquad (0 < a < b)$$

onto the strip $a < v < b$. [See Exercise 6(b).]

9. Determine the image of the domain $r > 0$, $-\pi < \Theta < \pi$ in the z plane under each of the transformations $w = F_k(z)$ $(k = 0, 1, 2, 3)$ where $F_k(z)$ are the four branches of $z^{1/4}$ given by equation (2), Sec. 69, when $n = 4$. Use these branches to determine the fourth roots of i.

10. The branch F of $(z^2 - 1)^{1/2}$ in Example 2, Sec. 69, was defined in terms of the coordinates r_1, r_2, θ_1, θ_2. Explain geometrically why the conditions $r_1 > 0$, $0 < \theta_1 + \theta_2 < \pi$ describe the quadrant $x > 0$, $y > 0$ of the z plane. Then show that the transformation $w = F(z)$ maps that quadrant onto the quadrant $u > 0$, $v > 0$ of the w plane.

Suggestion: To show that the quadrant $x > 0$, $y > 0$ in the z plane is described, note that $\theta_1 + \theta_2 = \pi$ at each point on the positive y axis and that $\theta_1 + \theta_2$ decreases as a point z moves to the right along a ray $\theta_2 = c$ $(0 < c < \pi/2)$.

11. For the transformation $w = F(z)$ of the first quadrant of the z plane onto the first quadrant of the w plane (Exercise 10), show that

$$u = \frac{1}{\sqrt{2}}\sqrt{r_1 r_2 + x^2 - y^2 - 1} \qquad \text{and} \qquad v = \frac{1}{\sqrt{2}}\sqrt{r_1 r_2 - x^2 + y^2 + 1},$$

where $r_1^2 r_2^2 = (x^2 + y^2 + 1)^2 - 4x^2$, and that the image of the portion of the hyperbola $x^2 - y^2 = 1$ in the first quadrant is the ray $v = u$ $(u > 0)$.

12. Show that in Exercise 11 the domain D that lies under the hyperbola and in the first quadrant of the z plane is described by the conditions $r_1 > 0$, $0 < \theta_1 + \theta_2 < \pi/2$. Then show that the image of D is the octant $0 < v < u$. Sketch the domain D and its image.

13. Let F be the branch of $(z^2 - 1)^{1/2}$ defined in Example 2, Sec. 69, and let $z_0 = r_0 \exp(i\theta_0)$ be a fixed point, where $r_0 > 0$ and $0 \leq \theta_0 < 2\pi$. Show that a branch F_0 of $(z^2 - z_0^2)^{1/2}$, whose branch cut is the line segment between the points z_0 and $-z_0$, can be written $F_0(z) = z_0 F(Z)$ where $Z = z/z_0$.

14. Write $z - 1 = r_1 \exp(i\theta_1)$ and $z + 1 = r_2 \exp(i\Theta_2)$, where $0 < \theta_1 < 2\pi$ and $-\pi < \Theta_2 < \pi$, to define a branch of the function

(a) $(z^2 - 1)^{1/2}$; (b) $\left(\dfrac{z - 1}{z + 1}\right)^{1/2}$.

In each case the branch cut should consist of the two rays $\theta_1 = 0$ and $\Theta_2 = \pi$.

15. Using the notation in Sec. 69, show that the function

$$w = \left(\frac{z-1}{z+1}\right)^{1/2} = \sqrt{\frac{r_1}{r_2}} \exp \frac{i(\theta_1 - \theta_2)}{2}$$

is a branch with the same domain of definition D_z and the same branch cut as the function $w = F(z)$ in that section. Show that this transformation maps D_z onto the right half plane $\rho > 0$, $-\pi/2 < \phi < \pi/2$, where the point $w = 1$ is the image of the point $z = \infty$. Also show that the inverse transformation is

$$z = \frac{1 + w^2}{1 - w^2} \qquad \qquad (\text{Re } w > 0).$$

16. Show that the transformation in Exercise 15 maps the region outside the unit circle $|z| = 1$ in the upper half of the z plane onto the region in the first quadrant of the w plane between the line $v = u$ and the u axis. Sketch the two regions.

17. Write $z = r \exp(i\Theta)$, $z - 1 = r_1 \exp(i\Theta_1)$, and $z + 1 = r_2 \exp(i\Theta_2)$, where the values of all three angles lie between $-\pi$ and π. Then define a branch of the function $[z(z^2 - 1)]^{1/2}$ whose branch cut consists of the two segments $x \leq -1$ and $0 \leq x \leq 1$ of the x axis.

70. The Transformation $w = \exp z$

As already pointed out in Sec. 22, the transformation $w = e^z$, or $\rho e^{i\phi} = e^x e^{iy}$ where $z = x + iy$ and $w = \rho e^{i\phi}$, can be written

(1) $$\rho = e^x, \qquad \phi = y.$$

The image of a typical point $z = (c_1, y)$ on a vertical line $x = c_1$ has polar coordinates $\rho = \exp c_1$ and $\phi = y$ in the w plane. That image moves counterclockwise around the circle shown in Fig. 62 as z moves up the line. The image of the line is evidently the entire circle; and each point on the circle is the image of an infinite number of points, spaced 2π units apart, along the line.

A horizontal line $y = c_2$ is mapped in a one to one manner onto the ray $\phi = c_2$. As a point $z = (x, c_2)$ moves along that line from left to right, the coordinate $\rho = e^x$ of the image point on the ray increases through all positive values, as indicated in Fig. 62.

Vertical and horizontal line *segments* are mapped onto portions of circles and rays, respectively, and images of various regions can be readily obtained from these observations.

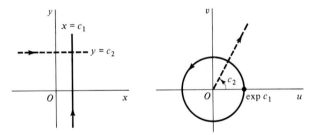

Figure 62. $w = \exp z$.

Example 1. Let us show that the transformation $w = e^z$ maps the rectangular region $a \leq x \leq b$, $c \leq y \leq d$ onto the region $e^a \leq \rho \leq e^b$, $c \leq \phi \leq d$. The two regions and corresponding parts of their boundaries are indicated in Fig. 63. The vertical line segment AD is mapped onto the arc $\rho = e^a$, $c \leq \phi \leq d$, which is labeled $A'D'$. The images of vertical line segments to the right of AD and joining the horizontal parts of the boundary are larger arcs; eventually, the image of the line segment BC is the arc $\rho = e^b$, $c \leq \phi \leq d$, labeled $B'C'$. The mapping is one to one if $d - c < 2\pi$. In particular, if $c = 0$ and $d = \pi$, then $0 \leq \phi \leq \pi$; and the rectangular region is mapped onto half of a circular ring, as shown in Fig. 8, Appendix 2.

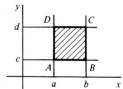 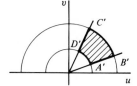

Figure 63. $w = \exp z$.

Example 2. When $w = e^z$, the image of the infinite strip $0 < y < \pi$ is the upper half plane $\rho > 0$, $0 < \phi < \pi$ since the images of the horizontal lines $y = c_2$ $(0 < c_2 < \pi)$ are rays $\phi = c_2$. Corresponding parts of the boundaries of the two regions are shown in Fig. 6 of Appendix 2. The boundary points $z = 0$ and $z = \pi i$ correspond to the points $w = 1$ and $w = -1$, respectively.

This mapping of a strip onto a half plane also follows from the fact that vertical line segments $x = c_1$, $0 < y < \pi$ are transformed into semicircles $\rho = \exp c_1$, $0 < \phi < \pi$.

By considering the images of horizontal lines in any horizontal strip $\alpha < y \leq \alpha + 2\pi$, we see that the image of the strip is the entire w plane with the exception of the origin $w = 0$. Each nonzero point w is the image of just one point in the strip; and, according to equations (1), a point $z = x + iy$ in the strip corresponds to a specified point $w = \rho e^{i\phi}$ $(\rho > 0, \alpha < \phi \leq \alpha + 2\pi)$ if $x = \text{Log } \rho$ and $y = \phi$. Hence the inverse transformation which maps w back onto z is given by the equation

$$z = \text{Log } \rho + i\phi \qquad (\rho > 0, \alpha < \phi \leq \alpha + 2\pi),$$

or (Sec. 25)

$$(2) \qquad z = \log w \qquad (|w| > 0, \alpha < \arg w \leq \alpha + 2\pi).$$

Note that if $\alpha = -\pi$, then $z = \text{Log } w$, where Log w denotes principal values of the multiple-valued function log w. Thus

$$\text{Log}(e^z) = z \qquad (-\pi < \text{Im } z \leq \pi),$$

as seen earlier in Sec. 26.

Interchanging the letters z and w in the equation $w = e^z$ and in equation (2), we arrive at the fact that the branch

$$(3) \qquad w = \log z \qquad (|z| > 0, \alpha < \arg z < \alpha + 2\pi)$$

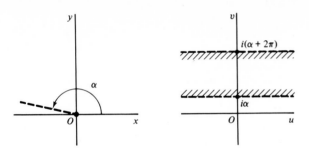

Figure 64. $w = \log z$ $(|z| > 0, \alpha < \arg z < \alpha + 2\pi)$.

of $\log z$ maps the cut z plane (Fig. 64) in a one to one manner onto the strip $\alpha < v < \alpha + 2\pi$ in the w plane, where $w = u + iv$. The inverse mapping is then $z = e^w$.

71. The Transformation $w = \sin z$

Since (Sec. 23)

$$\sin z = \sin x \cosh y + i \cos x \sinh y,$$

the transformation $w = \sin z$ can be written

(1) $$u = \sin x \cosh y, \qquad v = \cos x \sinh y.$$

The following examples illustrate various mapping properties of this transformation.

Example 1. The transformation $w = \sin z$ is a one to one mapping of the semi-infinite strip

$$-\frac{\pi}{2} \le x \le \frac{\pi}{2}, \, y \ge 0$$

in the z plane onto the upper half $v \ge 0$ of the w plane.

To verify this, we first show that the boundary of the strip is mapped in a one to one manner onto the real axis in the w plane, as indicated in Fig. 65. The image of the line segment BA is found by writing $x = \pi/2$ in equations (1) and restricting y to be nonnegative. Since $u = \cosh y$ and $v = 0$ when $x = \pi/2$, a typical point $(\pi/2, y)$ on BA is mapped into the point $(\cosh y, 0)$ in the w plane; and that image must move to the right from B' along the u axis as $(\pi/2, y)$ moves upward from B. A point $(x, 0)$ on the horizontal segment DB has image $(\sin x, 0)$, which moves to the right from D' to B' as x increases from $x = -\pi/2$ to $x = \pi/2$, or as $(x, 0)$ goes from D to B. Finally, as a point $(-\pi/2, y)$ on the line segment DE moves upward from D, its image $(-\cosh y, 0)$ moves to the left from D'.

One way to see how the interior of the strip is mapped onto the upper half $v > 0$ of the w plane is to examine the images of certain vertical half lines. If $0 < c_1 < \pi/2$, points on the line $x = c_1$ are transformed into points on the curve

(2) $$u = \sin c_1 \cosh y, \qquad v = \cos c_1 \sinh y \qquad (-\infty < y < \infty),$$

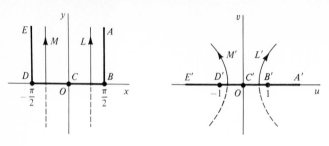

Figure 65. $w = \sin z$.

which is the right-hand branch of the hyperbola

$$(3) \qquad \frac{u^2}{\sin^2 c_1} - \frac{v^2}{\cos^2 c_1} = 1$$

with foci at the points

$$w = \pm\sqrt{\sin^2 c_1 + \cos^2 c_1} = \pm 1.$$

As a point on the line moves upward, its image on the hyperbola also moves upward. In particular, there is a one to one mapping of the top half ($y > 0$) of the line onto the top half ($v > 0$) of the hyperbola's branch. Such a half line L and its image L' are shown in Fig. 65. If $-\pi/2 < c_1 < 0$, the line $x = c_1$ is mapped onto the left-hand branch of the same hyperbola; and, as before, there is a one to one correspondence between points on the top half of the line and those on the top half of the hyperbola's branch. That half line and its image are labeled M and M' in Fig. 65.

The line $x = 0$, or the y axis, needs to be considered separately. According to equations (1), the image of each point $(0, y)$ is $(0, \sinh y)$. Hence the y axis is mapped onto the v axis in a one to one manner, the positive y axis corresponding to the positive v axis.

Now each point in the interior $-\pi/2 < x < \pi/2$, $y > 0$ of the strip lies on one of the above-mentioned half lines, and it is important to notice that the images of those half lines are distinct and constitute the entire half plane $v > 0$. More precisely, if the upper half L of a line $x = c_1$ ($0 < c_1 < \pi/2$) is thought of as moving to the left toward the positive y axis, the right-hand branch of the hyperbola containing L' is opening up wider and its vertex $(\sin c_1, 0)$ is tending toward the origin $w = 0$. Hence L' tends to become the positive v axis. On the other hand, as L approaches the segment BA of the boundary of the strip, the branch of the hyperbola closes down around the segment $B'A'$ of the u axis and the vertex $(\sin c_1, 0)$ tends toward the point $w = 1$. Similar statements can be made regarding the half line M and its image M' in Fig. 65. We may conclude that the image of each point in the interior of the strip lies in the upper half plane $v > 0$ and, furthermore, that each point in the half plane is the image of exactly one point in the interior of the strip.

This completes our demonstration that the transformation $w = \sin z$ is a one to one mapping of the strip $-\pi/2 \le x \le \pi/2$, $y \ge 0$ onto the half plane $v \ge 0$. The final result is shown in Fig. 9, Appendix 2. The right-hand half of the strip is mapped onto the first quadrant of the w plane, as shown in Fig. 10, Appendix 2.

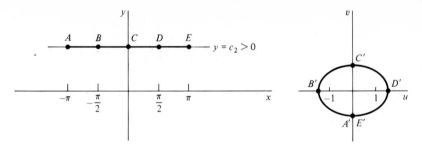

Figure 66. $w = \sin z$.

Another convenient way to find the images of certain regions when $w = \sin z$ is to consider the images of *horizontal* line segments $-\pi \leq x \leq \pi$, $y = c_2$, where $c_2 > 0$. According to equations (1), the image of such a line segment is the curve with parametric representation

(4) $u = \sin x \cosh c_2,$ $v = \cos x \sinh c_2$ $(-\pi \leq x \leq \pi)$.

That curve is readily seen to be the ellipse

(5) $$\frac{u^2}{\cosh^2 c_2} + \frac{v^2}{\sinh^2 c_2} = 1,$$

whose foci lie at the points

$$w = \pm\sqrt{\cosh^2 c_2 - \sinh^2 c_2} = \pm 1.$$

The image of a point (x, c_2) moving to the right from point A to point E in Fig. 66 makes one circuit around the ellipse in the clockwise direction. Note that when smaller values of the positive number c_2 are taken, the ellipse becomes smaller but retains the same foci $(\pm 1, 0)$. The ellipse, in fact, tends to become the interval $-1 \leq u \leq 1$ on the u axis, that line segment being the image of the interval $-\pi \leq x \leq \pi$ on the x axis. In the limiting case $c_2 = 0$, the mapping is not, however, one to one, as it is when $c_2 > 0$.

We now show how these observations can be used.

Example 2. A rectangular region $-\pi \leq x \leq \pi$, $a \leq y \leq b$ lying above the x axis is mapped by $w = \sin z$ onto a region bounded by two ellipses, as shown in Fig. 67. Both of the vertical edges $x = \pm\pi$, $a \leq y \leq b$ are mapped onto the segment

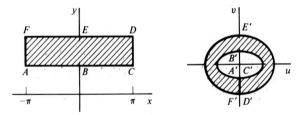

Figure 67. $w = \sin z$.

$$-\sinh b \leq v \leq -\sinh a$$

of the v axis. The interior of the rectangular region is mapped in a one to one manner onto the interior of an elliptic ring which has a cut lying along the negative v axis. As a point z describes the boundary of the rectangular region in the counterclockwise direction from point A in Fig. 67, the image of z moves around the smaller ellipse, then along the cut and around the larger ellipse, and back again along the cut to the starting point. So, while the mapping of the interior of the region is one to one, the mapping of its boundary is not.

Example 3. The transformation $w = \sin z$ is a one to one mapping of the entire rectangular region $-\pi/2 \leq x \leq \pi/2, 0 \leq y \leq b$, including its boundary, onto the semielliptic region shown in Fig. 11 of Appendix 2. Details of the verification are left as an exercise.

EXERCISES

1. Show that the lines $ay = x$ $(a \neq 0)$ are mapped onto the spirals $\rho = \exp(a\phi)$ under the transformation $w = \exp z$, where $w = \rho \exp(a\phi)$.

2. By considering the images of horizontal line segments, verify that the image of the rectangular region $a \leq x \leq b, c \leq y \leq d$ under the transformation $w = \exp z$ is the region $e^a \leq \rho \leq e^b, c \leq \phi \leq d$, as shown in Fig. 63 (Sec. 70).

3. Verify the mapping of the region and boundary shown in Fig. 7 of Appendix 2, where the transformation is $w = \exp z$.

4. Find the image of the semi-infinite strip $x \geq 0, 0 \leq y \leq \pi$ under the transformation $w = \exp z$, and label corresponding portions of the boundaries.

5. (a) By writing $z = re^{i\theta}$ and examining images of rays from the origin in the z plane, give a direct verification that the branch

$$w = \log z = \text{Log } r + i\theta \qquad (r > 0, \alpha < \theta < \alpha + 2\pi)$$

   of the logarithmic function is a one to one mapping of the cut plane in Fig. 64 (Sec. 70) onto the strip shown in that figure.

   (b) Verify that the principal branch

$$w = \text{Log } z \qquad (r > 0, -\pi < \Theta < \pi)$$

   of the logarithmic function is a one to one mapping of the right half plane $x > 0$ onto the strip $-\pi/2 < v < \pi/2$ and the upper half plane $y > 0$ onto the strip $0 < v < \pi$.

6. Show that the transformation $w = \sin z$ maps the top half $(y > 0)$ of the line $x = c_1$ $(-\pi/2 < c_1 < 0)$ in a one to one manner onto the top half $(v > 0)$ of the left-hand branch of hyperbola (3), Sec. 71, as indicated in Fig. 65 of that section.

7. Show that under the transformation $w = \sin z$, a line $x = c_1$ $(\pi/2 < c_1 < \pi)$ is mapped onto the right-hand branch of hyperbola (3), Sec. 71. Note that the mapping is one to one and that the upper and lower halves of the line are mapped onto the *lower* and *upper* halves, respectively, of the branch.

8. Vertical half lines were used in Example 1, Sec. 71, to show that the transformation $w = \sin z$ is a one to one mapping of the open region $-\pi/2 < x < \pi/2, y > 0$ onto the

half plane $v > 0$. Verify that result by using, instead, horizontal line segments $-\pi/2 < x < \pi/2$, $y = c_2$ ($c_2 > 0$), whose images are semiellipses.

9. (a) Show that under the transformation $w = \sin z$, the images of the line segments forming the boundary of the rectangular region $0 \leqq x \leqq \pi/2$, $0 \leqq y \leqq 1$ are the line segments and the arc $D'E'$ indicated in Fig. 68. The arc $D'E'$ is a quarter of the ellipse

$$\frac{u^2}{\cosh^2 1} + \frac{v^2}{\sinh^2 1} = 1.$$

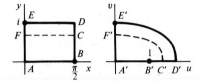

Figure 68. $w = \sin z$.

(b) Complete the mapping indicated in Fig. 68 by using images of horizontal line segments to prove that the transformation $w = \sin z$ establishes a one to one correspondence between the interior points of the regions $ABDE$ and $A'B'D'E'$.

10. Verify the mapping by $w = \sin z$ shown in Fig. 11, Appendix 2.

72. Successive Transformations

Other mappings can be obtained by applying familiar transformations successively.

Example 1. We need only recall the identity (Sec. 23)

$$\cos z = \sin\left(z + \frac{\pi}{2}\right)$$

to see that the transformation

$$w = \cos z$$

can be written successively as

$$Z = z + \frac{\pi}{2}, \qquad w = \sin Z.$$

Hence the cosine transformation is the same as the sine transformation preceded by a translation to the right through $\pi/2$ units.

Example 2. According to Sec. 24, the transformation

$$w = \sinh z$$

can be written $w = -i \sin(iz)$, or

$$Z = iz, \qquad W = \sin Z, \qquad w = -iW.$$

It is, therefore, a combination of the sine transformation and rotations through right angles. The transformation

$$w = \cosh z$$

is, likewise, essentially a cosine transformation since $\cosh z = \cos(iz)$.

Example 3. The transformation

$$w = (\sin z)^{1/2},$$

where the fractional power denotes the principal branch of the square root function, can be written

$$Z = \sin z, \qquad w = Z^{1/2} \qquad (|Z| > 0, \; -\pi < \text{Arg } Z < \pi).$$

As noted at the end of Example 1 in Sec. 71, the first transformation maps the semi-infinite strip $0 \leq x \leq \pi/2$, $y \geq 0$ onto the first quadrant $X \geq 0$, $Y \geq 0$ in the Z plane; the second transforms that quadrant into an octant in the w plane. These successive transformations are illustrated in Fig. 69.

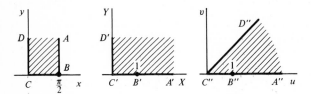

Figure 69. $w = (\sin z)^{1/2}$.

Example 4. To give another illustration of successive transformations, we consider first the linear fractional transformation

$$Z = \frac{z - 1}{z + 1}.$$

By writing $z = x + iy$ and $Z = X + iY$, one can easily see that it maps the half plane $y \geq 0$ onto the half plane $Y \geq 0$. For when the number z is real, so is the number Z; and since the image of the real axis $y = 0$ is either a circle or a line, it must be the real axis $Y = 0$. Furthermore, for any point Z in the finite Z plane,

$$Y = \text{Im } Z = \text{Im}\left[\frac{(z - 1)(\bar{z} + 1)}{(z + 1)(\bar{z} + 1)}\right] = \frac{2y}{|z + 1|^2} \qquad (z \neq -1).$$

The numbers y and Y thus have the same sign, and this means that points above the x axis correspond to points above the X axis and points below the x axis correspond to points below the X axis.

By recalling [Exercise 5(b), Sec. 71] that the transformation $w = \text{Log } Z$ maps the upper half plane $Y > 0$ onto the horizontal strip $0 < v < \pi$ in the w plane, we may conclude, then, that the transformation

$$w = \text{Log } \frac{z - 1}{z + 1}$$

maps the half plane $y > 0$ onto the strip $0 < v < \pi$. Corresponding points on the boundaries are shown in Fig. 19, Appendix 2.

73. Table of Transformations of Regions

Appendix 2 consists of a set of figures showing transformations of a number of simple and useful regions by means of elementary functions. In each case there is a one to one correspondence between the interior points of the given region and the interior points of the image of that region. Corresponding parts of boundaries are indicated by lettering. Some mappings that have not been discussed in the text are shown. Several of them can be derived by means of the Schwarz-Christoffel transformation, to be developed in Chap. 10.

EXERCISES

1. The equation $w = \log(z - 1)$ can be written

$$Z = z - 1, \qquad w = \log Z.$$

   Find a branch of $\log Z$ such that the cut z plane consisting of all points except those on the segment $x \geq 1$ of the real axis is mapped by $w = \log(z - 1)$ onto the strip $0 < v < 2\pi$ in the w plane.

2. Use the identity $\cosh z = \cos(iz)$ to show that the transformation $w = \cosh z$ maps the segment $0 \leq y \leq \pi$ of the y axis onto the segment $-1 \leq u \leq 1$ of the u axis.

3. (a) Describe the transformation $w = \cosh z$ in terms of the sine function. (See Examples 1 and 2 in Sec. 72.)

   (b) Apply the result in part (a), along with the mapping by $\sin z$ shown in Fig. 10, Appendix 2, to show that the transformation $w = \cosh z$ maps the semi-infinite strip $x \geq 0$, $0 \leq y \leq \pi/2$ in the z plane onto the first quadrant $u \geq 0$, $v \geq 0$ of the w plane. Indicate corresponding parts of the boundaries of the two regions.

4. By referring to Fig. 10, Appendix 2, show that the transformation $w = \sin^2 z$ maps the strip $0 \leq x \leq \pi/2$, $y \geq 0$ onto the half plane $v \geq 0$. Indicate corresponding parts of the boundaries.

5. Show that under the transformation $w = (\sin z)^{1/4}$, where the principal branch of the fractional power is taken, the semi-infinite strip $-\pi/2 < x < \pi/2$, $y > 0$ is mapped onto the part of the first quadrant lying between the line $v = u$ and the u axis. Label corresponding parts of the boundaries.

6. Recall (Example 4, Sec. 72) that the linear fractional transformation $Z = (z - 1)/(z + 1)$ maps the x axis onto the X axis and the half planes $y > 0$ and $y < 0$ onto the half planes $Y > 0$ and $Y < 0$, respectively. Show that, in particular, it maps the segment $-1 \leq x \leq 1$ of the x axis onto the segment $X \leq 0$ of the X axis. Then show that when the principal branch of the square root is used, the composite function

$$w = Z^{1/2} = \left(\frac{z-1}{z+1}\right)^{1/2}$$

   maps the z plane, except for the segment $-1 \leq x \leq 1$ of the x axis, onto the half plane $u > 0$. (Compare Exercise 15, Sec. 69.)

7. Verify the mapping shown in Fig. 16, Appendix 2, where $w = z + (1/z)$. (Compare the Example in Sec. 10.)

8. Observe that the transformation $w = \cosh z$ can be expressed as a composition of the mappings

$$Z = e^z, \qquad W = Z + \frac{1}{Z}, \qquad w = \frac{1}{2}W.$$

By appealing to Figs. 7 and 16 in Appendix 2, show that when $w = \cosh z$, the strip $x \leq 0, 0 \leq y \leq \pi$ in the z plane is mapped onto the lower half $v \leq 0$ of the w plane. Indicate corresponding parts of the boundaries.

9. By noting that the transformation $w = \sin z$ can be written

$$Z = i\left(z + \frac{\pi}{2}\right), \qquad W = \cosh Z, \qquad w = -W$$

and using the result obtained in Exercise 8, show that when $w = \sin z$, the strip $-\pi/2 \leq x \leq \pi/2, y \geq 0$ is mapped onto the half plane $v \geq 0$, as shown in Fig. 9, Appendix 2. (This mapping was verified in a different way in Example 1, Sec. 71.)

EIGHT

CONFORMAL MAPPING

In this chapter we introduce and develop the concept of a conformal mapping, with emphasis on connections between such mappings and harmonic functions. Applications to physical problems follow in the next chapter.

74. Basic Properties

Let C be a smooth arc (Sec. 30), represented by the equation $z = z(t)$ $(a \leq t \leq b)$, and let $f(z)$ be a function defined at all points on C. The equation $w = f[z(t)]$ $(a \leq t \leq b)$ is a parametric representation of the image Γ of C under the transformation $w = f(z)$.

Suppose that C passes through a point $z_0 = z(t_0)$ $(a \leq t_0 \leq b)$ where f is analytic and that $f'(z_0) \neq 0$. According to the chain rule given in Exercise 11, Sec. 30, if $w(t) = f[z(t)]$, then

(1) $$w'(t_0) = f'[z(t_0)]z'(t_0);$$

and this means that (see Sec. 5)

(2) $$\arg w'(t_0) = \arg f'[z(t_0)] + \arg z'(t_0).$$

Equation (2) is useful in relating the directions of C and Γ at the points z_0 and $w_0 = f(z_0)$, respectively.

To be specific, let ψ_0 denote a value of $\arg f'(z_0)$, and let θ_0 be the angle of inclination of a directed line tangent to C at z_0. According to Sec. 30, θ_0 is a value of $\arg z'(t_0)$, and it follows from equation (2) that the quantity

$$\phi_0 = \psi_0 + \theta_0$$

is a value of $\arg w'(t_0)$ and is therefore the angle of inclination of a directed line tangent

217

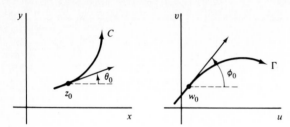

Figure 70. $\phi_0 = \psi_0 + \theta_0$.

to Γ at the point $w_0 = f(z_0)$ (Fig. 70). Hence the angle of inclination of the directed line at w_0 differs from the angle of inclination of the directed line at z_0 by the *angle of rotation*

$$(3) \qquad\qquad \psi_0 = \arg f'(z_0).$$

Now let C_1 and C_2 be two smooth arcs passing through z_0, and let θ_1 and θ_2 be angles of inclination of directed lines tangent to C_1 and C_2, respectively, at z_0. We know from the preceding paragraph that the quantities

$$\phi_1 = \psi_0 + \theta_1 \qquad \text{and} \qquad \phi_2 = \psi_0 + \theta_2$$

are angles of inclination of directed lines tangent to the image curves Γ_1 and Γ_2, respectively, at the point $w_0 = f(z_0)$. Thus $\phi_2 - \phi_1 = \theta_2 - \theta_1$; that is, the angle $\phi_2 - \phi_1$ from Γ_1 to Γ_2 is the same in *magnitude* and *sense* as the angle $\theta_2 - \theta_1$ from C_1 to C_2. Those angles are denoted by α in Fig. 71.

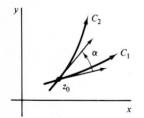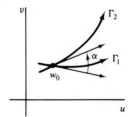

Figure 71

Because of this angle-preserving property, a transformation $w = f(z)$ is said to be *conformal* at a point z_0 if f is analytic there and $f'(z_0) \neq 0$. Such a transformation is actually conformal at each point in a neighborhood of z_0. For f must be analytic in a neighborhood of z_0 (Sec. 19); and since f' is continuous at z_0 (Sec. 39), it follows from Exercise 10, Sec. 15, that there is also a neighborhood of that point throughout which $f'(z) \neq 0$.

A transformation $w = f(z)$, defined on a domain D, is referred to as a conformal transformation, or *conformal mapping,* when it is conformal at each point in D. That is, the mapping is conformal in D if f is analytic in D and its derivative f' has no zeros there. Each of the elementary functions studied in Chap. 7 can be used to define a transformation which is conformal in some domain.

Example 1. The mapping $w = e^z$ is conformal throughout the entire z plane since $d(e^z)/dz = e^z \neq 0$ for each z. Consider any two lines $x = c_1$ and $y = c_2$ in the z plane, the first directed upward and the second directed to the right. According to Sec. 70, their images under the mapping $w = e^z$ are a positively oriented circle centered at the origin and a ray from the origin, respectively. As illustrated in Fig. 62 (Sec. 70), the angle between the lines at their point of intersection is a right angle in the negative direction, and the same is true of the angle between the circle and the ray at the corresponding point in the w plane. The conformality of the mapping $w = e^z$ is also illustrated in Figs. 7 and 8 of Appendix 2.

Example 2. Consider two smooth arcs which are level curves $u(x, y) = c_1$ and $v(x, y) = c_2$ of the real and imaginary components, respectively, of a function

$$f(z) = u(x, y) + iv(x, y),$$

and suppose that they intersect at a point z_0 where f is analytic and $f'(z_0) \neq 0$. The transformation $w = f(z)$ is conformal at z_0 and maps these arcs into the lines $u = c_1$ and $v = c_2$, which are orthogonal at the point $w_0 = f(z_0)$. According to our theory, then, the arcs must be orthogonal at z_0. This has already been verified and illustrated in Exercises 16 and 17 of Sec. 20.

A mapping that preserves the magnitude of the angle between two smooth arcs but not necessarily the sense is called an *isogonal mapping*.

Example 3. The transformation $w = \bar{z}$, which is a reflection in the real axis, is isogonal but not conformal. If it is followed by a conformal transformation, the resulting transformation $w = f(\bar{z})$ is also isogonal but not conformal.

Suppose f is not a constant function and is analytic at a point z_0. If $f'(z_0) = 0$, then z_0 is called a *critical point* of the transformation $w = f(z)$.

Example 4. The point $z = 0$ is a critical point of the transformation

$$w = 1 + z^2.$$

If we write $z = r \exp(i\theta)$, then

$$w = 1 + r^2 \exp(i2\theta),$$

or $w = 1 + \rho \exp(i\phi)$ where $\rho = r^2$, $\phi = 2\theta$. Hence a ray $\theta = \alpha$ from the point $z = 0$ is mapped onto the ray from the point $w = 1$ whose angle of inclination is 2α. Moreover, the angle between any two rays drawn from the critical point $z = 0$ is doubled by the transformation.

More generally, it can be shown that if z_0 is a critical point of a transformation $w = f(z)$, there is an integer m ($m \geq 2$) such that the angle between any two smooth arcs passing through z_0 is multiplied by m under that transformation. The integer m is the smallest positive integer such that $f^{(m)}(z_0) \neq 0$. Verification of these facts is left to the exercises.

75. Further Properties and Examples

Another property of a transformation $w = f(z)$ which is conformal at a point z_0 is obtained by considering the modulus of $f'(z_0)$. From the definition of derivative and property (8), Sec. 12, of limits, we know that

$$(1) \qquad |f'(z_0)| = \left| \lim_{z \to z_0} \frac{f(z) - f(z_0)}{z - z_0} \right| = \lim_{z \to z_0} \frac{|f(z) - f(z_0)|}{|z - z_0|}.$$

Now $|z - z_0|$ is the length of a line segment joining z_0 and z, and $|f(z) - f(z_0)|$ is the length of the line segment joining the points $f(z_0)$ and $f(z)$ in the w plane. Evidently, then, if z is near the point z_0, the ratio

$$\frac{|f(z) - f(z_0)|}{|z - z_0|}$$

of the two lengths is approximately the number $|f'(z_0)|$. Note that $|f'(z_0)|$ represents an expansion if it is greater than unity and a contraction if it is less than unity.

Although the angle of rotation $\arg f'(z)$ (Sec. 74) and the *scale factor* $|f'(z)|$ vary, in general, from point to point, it follows from the continuity of f' that their values are approximately $\arg f'(z_0)$ and $|f'(z_0)|$ at points z near z_0. Hence the image of a small region in a neighborhood of z_0 *conforms* to the original region in the sense that it has approximately the same shape. A large region may, however, be transformed into a region that bears no resemblance to the original one.

Example 1. When $f(z) = z^2$, the transformation

$$w = f(z) = x^2 - y^2 + i2xy$$

is conformal at the point $z = 1 + i$, where the lines $y = x$ and $x = 1$ intersect. Since the image of a point (x, y) in the z plane is the point w whose cartesian coordinates are

$$u = x^2 - y^2 \qquad \text{and} \qquad v = 2xy,$$

the line $y = x$ is transformed into the curve represented by the equations

$$(2) \qquad u = 0, \qquad v = 2y^2 \qquad (-\infty < y < \infty).$$

That is, the image of the line $y = x$ is the upper half $v \geqq 0$ of the v axis. The line $x = 1$ is transformed into the curve

$$(3) \qquad u = 1 - y^2, \qquad v = 2y \qquad (-\infty < y < \infty),$$

which is the parabola $v^2 = -4(u - 1)$ (Fig. 72).

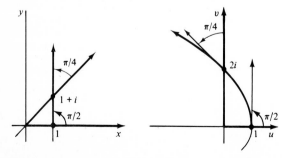

Figure 72. $w = z^2$.

If the direction of increasing y is taken as the positive sense on the two lines in the z plane, the angle from the first to the second at the point $z = 1 + i$ is $\pi/4$. In view of the second of equations (2), when $y > 0$ and y increases along the line $y = x$, the coordinate v also increases. So the positive sense of the first image is upward when $y > 0$. This is also true of the parabola, as we see from the second of equations (3). If u and v are the variables in representation (3) for the parabola, then

$$\frac{dv}{du} = \frac{dv/dy}{du/dy} = \frac{2}{-2y} = -\frac{2}{v}.$$

In particular, $dv/du = -1$ when $v = 2$. Consequently, the angle from the first image curve to the second at the point $w = f(1 + i) = 2i$ is $\pi/4$, as required by the conformality of the mapping.

As anticipated, the angle of rotation $\pi/4$ at the point $z = 1 + i$ is a value of

$$\arg[f'(1 + i)] = \arg[2(1 + i)] = \frac{\pi}{4} + 2n\pi \qquad (n = 0, \pm1, \pm2, \ldots).$$

The scale factor at that point is the number

$$|f'(1 + i)| = |2(1 + i)| = 2\sqrt{2}.$$

To illustrate how the angle of rotation and the scale factor can change from point to point, we note that they are 0 and 2, respectively, at the point $z = 1$ since $f'(1) = 2$. See Fig. 72, where the x axis is transformed into the nonnegative half of the u axis and u increases with x when $x > 0$.

A transformation $w = f(z)$ which is conformal at a point z_0 has a *local inverse* there. That is, if $w_0 = f(z_0)$, there exist rectangular domains R and S, centered at z_0 and w_0, respectively, such that for each point w in S there is a unique point z in R where $f(z) = w$. This correspondence defines the inverse transformation $z = g(w)$, where $f[g(w)] = w$ when w is in S. The function g is, in addition, analytic in S, with derivative

(4)
$$g'(w) = \frac{1}{f'(z)}.$$

Note that $g(w_0) = z_0$ since z_0 lies in R and must then be the unique point z in that domain such that $f(z) = w_0$. Note, too, how expression (4) shows that the transformation $z = g(w)$ is itself conformal at w_0.

Let us verify the existence of such an inverse, which is a direct consequence of results in advanced calculus.* As noted in Sec. 74, the conformality of the transformation $w = f(z)$ at z_0 implies that there is some neighborhood of z_0 where the transformation is conformal and, consequently, where f is analytic. Hence if we write $z_0 = x_0 + iy_0$ and

$$f(z) = u(x, y) + iv(x, y),$$

*The results from advanced calculus to be used here appear in, for instance, A.E. Taylor and W.R. Mann, "Advanced Calculus," 3d ed., pp. 241–247, 1983.

we know that there is a neighborhood of the point (x_0, y_0) throughout which the functions $u(x, y)$ and $v(x, y)$, along with their partial derivatives of all orders, are continuous (see Sec. 39).

Now the pair of equations

$$(5) \qquad u = u(x, y), \qquad v = v(x, y)$$

represents a transformation from the neighborhood just mentioned into the uv plane. Moreover, the determinant

$$J = \begin{vmatrix} u_x & u_y \\ v_x & v_y \end{vmatrix} = u_x v_y - v_x u_y,$$

which is known as the *Jacobian* of the transformation, is nonzero at the point (x_0, y_0). For, in view of the Cauchy-Riemann equations $u_x = v_y$ and $u_y = -v_x$, we can write J as

$$J = (u_x)^2 + (v_x)^2 = |f'(z)|^2 ;$$

and $f'(z_0) \neq 0$ since the transformation $w = f(z)$ is conformal at z_0. The above continuity conditions on the functions $u(x, y)$ and $v(x, y)$ and their derivatives, together with this condition on the Jacobian, are sufficient to ensure the existence of a local inverse of transformation (5) at (x_0, y_0). That is, if

$$(6) \qquad u_0 = u(x_0, y_0) \qquad \text{and} \qquad v_0 = v(x_0, y_0),$$

there are rectangular domains R and S, centered at the points (x_0, y_0) and (u_0, v_0), respectively, where for each point (u, v) in S there is a unique point (x, y) in R such that equations (5) hold. This establishes the existence of the inverse transformation

$$(7) \qquad x = x(u, v), \qquad y = y(u, v)$$

from S back into the domain R. The domains R and S can be taken small enough that R is contained in the neighborhood in which the transformation $w = f(z)$ is conformal, and this means that $f'(z) \neq 0$ anywhere in R. Since $J = |f'(z)|^2$, then, J is nonzero at the points in R which correspond to the points in S. Also, the functions (7) are continuous and have continuous first-order partial derivatives which satisfy the equations

$$(8) \qquad x_u = \frac{1}{J} v_y, \qquad x_v = -\frac{1}{J} u_y, \qquad y_u = -\frac{1}{J} v_x, \qquad y_v = \frac{1}{J} u_x$$

throughout S.

If we write $w = u + iv$ and

$$(9) \qquad g(w) = x(u, v) + iy(u, v),$$

the transformation $z = g(w)$ is the desired local inverse of the transformation $w = f(z)$ at z_0. For if the xy plane is regarded as the z plane, transformations (5) and (7) become

$$u + iv = u(x, y) + iv(x, y) \qquad \text{and} \qquad x + iy = x(u, v) + iy(u, v);$$

and since these last two equations are the same as

$$w = f(z) \qquad \text{and} \qquad z = g(w),$$

where $z = x + iy$ and $w = u + iv$, it is now clear that for each point w in the domain S there is a unique point $z = g(w)$ in R such that $f(z) = w$. Equations (8) can be used to show that g is analytic in S. Details are left to the exercises, where expression (4) for $g'(w)$ is also derived.

Example 2. If $f(z) = e^z$, the transformation $w = f(z)$ is conformal everywhere in the z plane and, in particular, at the point $z_0 = 2\pi i$. The image of this choice of z_0 is the point $w_0 = 1$. When points in the w plane are expressed in the form $w = \rho e^{i\phi}$, the local inverse can be obtained by writing $g(w) = \log w$ where $\log w$ denotes the branch

$$\log w = \text{Log } \rho + i\phi \qquad (\rho > 0, \pi < \phi < 3\pi)$$

of the logarithmic function, restricted to any rectangular domain that is centered at w_0 and does not include the origin. Observe that

$$g(1) = \text{Log } 1 + i2\pi = 2\pi i$$

and that when w is in that rectangular domain,

$$f[g(w)] = \exp(\log w) = w.$$

Also,

$$g'(w) = \frac{d}{dw} \log w = \frac{1}{w} = \frac{1}{\exp z}$$

in accordance with equation (4).

Note that if the point $z_0 = 0$ is chosen, one can use the principal branch

$$\text{Log } w = \text{Log } \rho + i\phi \qquad (\rho > 0, -\pi < \phi < \pi)$$

of the logarithmic function to define g. In this case, $g(1) = 0$.

EXERCISES

1. Determine the angle of rotation at the point $z = 2 + i$ when the transformation is $w = z^2$, and illustrate it for some particular curve. Show that the scale factor of the transformation at that point is $2\sqrt{5}$.

2. What angle of rotation is produced by the transformation $w = 1/z$ at the point (a) $z = 1$; (b) $z = i$?

   Ans. (a) π; (b) 0.

3. Show that under the transformation $w = 1/z$, the images of the lines $y = x - 1$ and $y = 0$ are the circle $u^2 + v^2 - u - v = 0$ and the line $v = 0$, respectively. Sketch all four curves, determine corresponding directions along them, and verify the conformality of the mapping at the point $z = 1$.

4. Show that the angle of rotation at a nonzero point $z_0 = r_0 \exp(i\theta_0)$ under the transformation $w = z^n$ ($n = 1, 2, \ldots$) is $(n - 1)\theta_0$. Determine the scale factor of the transformation at that point.

   Ans. nr_0^{n-1}.

5. Show that the transformation $w = \sin z$ is conformal at all points except $z = (2n + 1)\pi/2$ $(n = 0, \pm 1, \pm 2, \ldots)$. Note that this is in agreement with the mapping of directed line segments shown in Figs. 9, 10, and 11 of Appendix 2.

6. Find a local inverse of the transformation $w = z^2$ at the point
   (a) $z_0 = 2$; (b) $z_0 = -2$; (c) $z_0 = -i$.

   Ans. (a) $w^{1/2} = \sqrt{\rho}\, e^{i\phi/2}$ $(\rho > 0, -\pi < \phi < \pi)$;

   (c) $w^{1/2} = \sqrt{\rho}\, e^{i\phi/2}$ $(\rho > 0, 2\pi < \phi < 4\pi)$.

7. In Sec. 75 it was pointed out that the components $x(u, v)$ and $y(u, v)$ of the inverse function $g(w)$ defined in equation (9) there are continuous and have continuous first-order partial derivatives in the domain S. Use equations (8), Sec. 75, to show that the Cauchy-Riemann equations $x_u = y_v$, $x_v = -y_u$ hold in S. Then conclude that $g(w)$ is analytic in that domain.

8. Show that if $z = g(w)$ is a local inverse of a conformal transformation $w = f(z)$ at a point z_0, then

$$g'(w) = \frac{1}{f'(z)}$$

at points w in the domain S, where g is analytic (Exercise 7).

   Suggestion: Start with the fact that $f[g(w)] = w$, and apply the chain rule for differentiating composite functions.

9. Let C be a smooth arc lying in a domain D throughout which a transformation $w = f(z)$ is conformal, and let Γ denote the image of C under that transformation. Show that Γ is also a smooth arc.

10. Suppose that a function f is analytic at z_0 and that

$$f'(z_0) = f''(z_0) = \cdots = f^{(m-1)}(z_0) = 0, \qquad f^{(m)}(z_0) \neq 0$$

for some positive integer m $(m \geq 1)$. Also, write $w_0 = f(z_0)$.

   (a) Use the Taylor series for f about the point z_0 to show that there is a neighborhood of z_0 in which the difference $f(z) - w_0$ can be written

$$f(z) - w_0 = (z - z_0)^m \frac{f^{(m)}(z_0)}{m!}[1 + g(z)],$$

   where $g(z)$ is continuous at z_0 and $g(z_0) = 0$.

   (b) Let Γ be the image of a smooth arc C under the transformation $w = f(z)$, as shown in Fig. 70, and note that the angles of inclination θ_0 and ϕ_0 in that figure are limits of $\arg(z - z_0)$ and $\arg[f(z) - w_0]$, respectively, as z approaches z_0 along arc C. Then use the result in part (a) to show that θ_0 and ϕ_0 are related by the equation

$$\phi_0 = m\theta_0 + \arg f^{(m)}(z_0).$$

   (c) Let α denote the angle between two smooth arcs C_1 and C_2 passing through z_0, as shown on the left in Fig. 71. Show how it follows from the relation obtained in part (b) that the corresponding angle between the image curves Γ_1 and Γ_2 at the point $w_0 = f(z_0)$ is $m\alpha$. (Note that the transformation is conformal at z_0 when $m = 1$ and that z_0 is a critical point when $m \geq 2$.)

76. Harmonic Conjugates

We saw in Sec. 20 that if a function

$$f(z) = u(x, y) + iv(x, y)$$

is analytic in a domain D, then the real-valued functions u and v are harmonic in that domain. That is, they have continuous first- and second-order partial derivatives in D and satisfy Laplace's equation there:

$$(1) \qquad u_{xx} + u_{yy} = 0, \qquad v_{xx} + v_{yy} = 0.$$

We had seen earlier that the first-order partial derivatives of u and v satisfy the Cauchy-Riemann equations

$$(2) \qquad u_x = v_y, \qquad u_y = -v_x;$$

and, as pointed out in Sec. 20, v is called a harmonic conjugate of u.

Suppose now that $u(x, y)$ is any given harmonic function defined on a *simply connected* (Sec. 36) domain D. In this section we show that $u(x, y)$ always has a harmonic conjugate $v(x, y)$ in D by deriving an expression for $v(x, y)$.

To accomplish this, we first recall some important facts about line integrals in advanced calculus.* Suppose that $P(x, y)$ and $Q(x, y)$ have continuous first-order partial derivatives in a simply connected domain D of the xy plane, and let (x_0, y_0) and (x, y) be any two points in D. If $P_y = Q_x$ everywhere in D, the line integral

$$\int_C P(X, Y)\, dX + Q(X, Y)\, dY$$

from (x_0, y_0) to (x, y) is independent of the contour C that is taken so long as the contour lies entirely in D. Furthermore, when the point (x_0, y_0) is kept fixed and (x, y) is allowed to vary throughout D, the integral represents a single-valued function

$$(3) \qquad F(x, y) = \int_{(x_0, y_0)}^{(x, y)} P(X, Y)\, dX + Q(X, Y)\, dY$$

of x and y whose first-order partial derivatives are given by the equations

$$(4) \qquad F_x(x, y) = P(x, y), \qquad F_y(x, y) = Q(x, y).$$

Note that the value of F is changed by an additive constant when a different point (x_0, y_0) is taken.

Returning to the given harmonic function $u(x, y)$, observe how it follows from Laplace's equation $u_{xx} + u_{yy} = 0$ that

$$(-u_y)_y = (u_x)_x$$

everywhere in D. Also, the second-order partial derivatives of u are continuous in D; and this means that the first-order partial derivatives of $-u_y$ and u_x are continuous there.

---

*See, for example, W. Kaplan, "Advanced Mathematics for Engineers," pp. 546–550, 1981.

Thus if (x_0, y_0) is a fixed point in D, the function

(5)
$$v(x, y) = \int_{(x_0, y_0)}^{(x, y)} -u_Y(X, Y)\, dX + u_X(X, Y)\, dY$$

is well defined for all (x, y) in D; and, according to equations (4),

(6)
$$v_x(x, y) = -u_y(x, y), \qquad v_y(x, y) = u_x(x, y).$$

These are the Cauchy-Riemann equations. Since the first-order partial derivatives of u are continuous, it is evident from equations (6) that those derivatives of v are also continuous. Hence (Sec. 17) $u(x, y) + iv(x, y)$ is an analytic function in D, and v is therefore a harmonic conjugate of u.

The function v defined by equation (5) is, of course, not the only harmonic conjugate of u. For the function $v(x, y) + c$, where c is any real constant, is also a harmonic conjugate of u.

Example. Consider the function $u(x, y) = xy$, which is harmonic throughout the entire xy plane. According to equation (5), the function

$$v(x, y) = \int_{(0, 0)}^{(x, y)} -X\, dX + Y\, dY$$

is a harmonic conjugate of $u(x, y)$. The integral here is readily evaluated by inspection; it can also be evaluated by integrating first along the horizontal path from the point $(0, 0)$ to the point $(x, 0)$ and then along the vertical path from $(x, 0)$ to the point (x, y). The result is

$$v(x, y) = -\frac{1}{2} x^2 + \frac{1}{2} y^2,$$

and the corresponding analytic function is then

$$f(z) = xy - \frac{i}{2}(x^2 - y^2) = -\frac{i}{2} z^2.$$

77. Transformations of Harmonic Functions

The problem of finding a function which is harmonic in a specified domain and which satisfies prescribed conditions on the boundary of that domain is prominent in applied mathematics. If the values of the function are prescribed along the boundary, the problem is known as a boundary value problem of the *first kind,* or a *Dirichlet problem.* If the values of the normal derivative of the function are prescribed on the boundary, the boundary value problem is one of the *second kind,* or a *Neumann problem.* Modifications and combinations of those types of boundary conditions also arise.

The domains most frequently encountered in the applications are simply connected; and since a function which is harmonic in a simply connected domain always has a harmonic conjugate (Sec. 76), solutions of boundary value problems for such domains are the real or imaginary parts of analytic functions.

Example 1. It is easy to verify that the function

(1) $$H(x, y) = e^{-y} \sin x$$

satisfies the conditions

(2) $$H_{xx}(x, y) + H_{yy}(x, y) = 0,$$

(3) $$H(0, y) = 0, \qquad H(\pi, y) = 0,$$

(4) $$H(x, 0) = \sin x, \qquad \lim_{y \to \infty} H(x, y) = 0,$$

which make up a Dirichlet problem for the vertical strip $0 < x < \pi$, $y > 0$. Observe that since

$$-ie^{iz} = e^{-y} \sin x - ie^{-y} \cos x,$$

the stated solution $H(x, y)$ is the real part of the entire function $-ie^{iz}$. It is also the imaginary part of the entire function e^{iz}.

Sometimes the solution of a given boundary value problem can be *discovered* by identifying it as the real or imaginary part of an analytic function. But the success of this procedure depends on the simplicity of the problem and on one's familiarity with the real and imaginary parts of a variety of analytic functions. The following theorem is an important aid.

Theorem. *Suppose that an analytic function*

(5) $$w = f(z) = u(x, y) + iv(x, y)$$

maps a domain D_z in the z plane onto a domain D_w in the w plane. If $h(u, v)$ is a harmonic function defined on D_w, then the function

(6) $$H(x, y) = h[u(x, y), v(x, y)]$$

is harmonic in D_z.

Our proof of the theorem is for the case in which the domain D_w is simply connected. According to Sec. 76, the given harmonic function $h(u, v)$ has a harmonic conjugate $g(u, v)$. Hence the function

$$\Phi(w) = h(u, v) + ig(u, v)$$

is analytic in D_w. Since the function $f(z)$ is analytic in D_z, the composite function $\Phi[f(z)]$ is also analytic in D_z. Consequently, the real part $h[u(x, y), v(x, y)]$ of this composition is harmonic in D_z.

The proof of the theorem for the general case in which D_w is not necessarily simply connected can be accomplished directly by means of the chain rule for partial derivatives. The computations are, however, somewhat involved (see Exercise 8, Sec. 78).

Example 2. The function $h(u, v) = e^{-v} \sin u$ is harmonic in the domain D_w consisting of all points in the upper half plane $v > 0$ (see Example 1). If the transformation is $w = z^2$, then $u(x, y) = x^2 - y^2$ and $v(x, y) = 2xy$; moreover, the domain D_z in the

z plane consisting of the points in the first quadrant $x > 0$, $y > 0$ is mapped onto the domain D_w (Sec. 67). Hence the function

$$H(x, y) = e^{-2xy} \sin(x^2 - y^2)$$

is harmonic in D_z.

Example 3. Consider the function $h(u, v) = \text{Im } w = v$, which is harmonic in the strip $-\pi/2 < v < \pi/2$, and recall [Exercise 5(b), Sec. 71] that the transformation $w = \text{Log } z$ maps the right half plane $x > 0$ onto that strip. Writing

$$\text{Log } z = \text{Log } \sqrt{x^2 + y^2} + i \arctan \frac{y}{x},$$

where $-\pi/2 < \arctan(y/x) < \pi/2$, we find that the function

$$H(x, y) = \arctan \frac{y}{x}$$

is harmonic in the half plane $x > 0$.

78. Transformations of Boundary Conditions

The conditions that a function or its normal derivative have prescribed values along the boundary of a domain in which it is harmonic are the most common, although not the only, important types of boundary conditions. In this section we show that certain of these conditions remain unaltered under the change of variables associated with a conformal transformation. These results are used in Chap. 9 to solve boundary value problems. The basic technique there is to transform a given boundary value problem in the xy plane into a simpler one in the uv plane and then to use the theorems of this and the previous section to write the solution of the original problem in terms of the solution obtained for the simpler one.

Theorem. Suppose that a transformation

(1) $$w = f(z) = u(x, y) + iv(x, y)$$

is conformal on a smooth arc C, and let Γ be the image of C under that transformation. If along Γ a function $h(u, v)$ satisfies either of the conditions

(2) $$h = h_0 \quad \text{or} \quad \frac{dh}{dn} = 0,$$

where h_0 is a real constant and dh/dn denotes derivatives normal to Γ, then along C the function

(3) $$H(x, y) = h[u(x, y), v(x, y)]$$

satisfies the corresponding condition

(4) $$H = h_0 \quad \text{or} \quad \frac{dH}{dN} = 0,$$

where dH/dN denotes derivatives normal to C.

To show that the condition $h = h_0$ on Γ implies that $H = h_0$ on C, we note from equation (3) that the value of H at any point (x, y) on C is the same as the value of h at the image (u, v) of (x, y) under transformation (1). Since the image point (u, v) lies on Γ and since $h = h_0$ along that curve, it follows that $H = h_0$ along C.

Suppose, on the other hand, that $dh/dn = 0$ on Γ. From calculus we know that

$$
(5) \qquad\qquad \frac{dh}{dn} = (\text{grad } h) \cdot \mathbf{n}
$$

where grad h denotes the gradient of h at a point (u, v) on Γ and $\mathbf{n}$ is a unit vector normal to Γ at (u, v). Since $dh/dn = 0$ at (u, v), equation (5) tells us that grad h is orthogonal to $\mathbf{n}$ at (u, v). That is, grad h is tangent to Γ there (Fig. 73). But gradients are orthogonal to level curves; and, because grad h is tangent to Γ, we see that Γ is orthogonal to a level curve $h(u, v) = h_1$ passing through (u, v).

Now, according to equation (3), the level curve $H(x, y) = h_1$ in the z plane can be written

$$
h[u(x, y), v(x, y)] = h_1;
$$

and so it is evidently transformed into the level curve $h(u, v) = h_1$ under transformation (1). Furthermore, since C is transformed into Γ and Γ is orthogonal to the level curve $h(u, v) = h_1$, as demonstrated in the previous paragraph, it follows from the conformality of transformation (1) on C that C is orthogonal to the level curve $H(x, y) = h_1$ at the point (x, y) corresponding to (u, v). Because gradients are orthogonal to level curves, this means that grad H is tangent to C at (x, y) (see Fig. 73). Consequently, if $\mathbf{N}$ denotes a unit vector normal to C at (x, y), grad H is orthogonal to $\mathbf{N}$. That is,

$$
(6) \qquad\qquad (\text{grad } H) \cdot \mathbf{N} = 0.
$$

Finally, since

$$
\frac{dH}{dN} = (\text{grad } H) \cdot \mathbf{N},
$$

we may conclude from equation (6) that $dH/dN = 0$ at points on C.

In this discussion we have tacitly assumed that grad $h \neq \mathbf{0}$. If grad $h = \mathbf{0}$, it follows from the identity

$$
|\text{grad } H(x, y)| = |\text{grad } h(u, v)| \, |f'(z)|,
$$

derived in Exercise 10(a) below, that grad $H = \mathbf{0}$; hence dh/dn and the corresponding

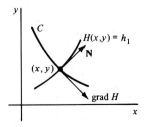

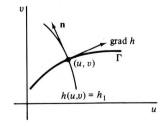

Figure 73

normal derivative dH/dN are both zero. We also assumed that (a) grad h and grad H always exist and (b) the level curve $H(x, y) = h_1$ is a smooth arc at the point (x, y) when grad $h \neq \mathbf{0}$ at (u, v). Condition (b) ensures that angles between arcs are preserved by transformation (1) when it is conformal. In all our applications, both conditions (a) and (b) will be satisfied.

Example. Consider, for instance, the function $h(u, v) = v + 2$. The transformation

$$w = iz^2 = -2xy + i(x^2 - y^2)$$

is conformal when $z \neq 0$. It maps the half line $y = x$ $(x > 0)$ onto the negative u axis, where $h = 2$, and the positive x axis onto the positive v axis, where the normal derivative h_u is 0 (Fig. 74). According to the above theorem, the function

$$H(x, y) = x^2 - y^2 + 2$$

must satisfy the conditions $H = 2$ along the half line $y = x$ $(x > 0)$ and $H_y = 0$ along the positive x axis, as one can verify directly.

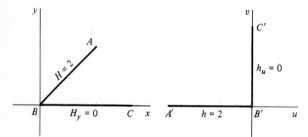

Figure 74. $w = iz^2$.

A boundary condition that is not of one of the two types mentioned in the theorem may be transformed into a condition that is substantially different from the original one (see Exercise 6). New boundary conditions for the transformed problem can be obtained for a particular transformation in any case. It is interesting to note that under a conformal transformation the ratio of a directional derivative of H along a smooth arc C in the z plane to the directional derivative of h along the image curve Γ at the corresponding point in the w plane is $|f'(z)|$; usually this ratio is not constant along a given arc. (See Exercise 10.)

EXERCISES

1. Use expression (5), Sec. 76, to find a harmonic conjugate of the harmonic function $u(x, y) = x^3 - 3xy^2$. Write the resulting analytic function in terms of the complex variable z.

2. Let $u(x, y)$ be harmonic in a simply connected domain D. By appealing to results in Secs. 76 and 39, show that its partial derivatives of all orders are continuous throughout D.

3. The transformation $w = \exp z$ maps the horizontal strip $0 < y < \pi$ onto the upper half plane $v > 0$, as shown in Fig. 6 of Appendix 2; and the function $h(u, v) = \text{Re}(w^2) = u^2 - v^2$ is harmonic in that half plane. With the aid of the theorem in Sec. 77, show that the function $H(x, y) = e^{2x} \cos 2y$ is harmonic in the strip. Verify this result directly.

4. Under the transformation $w = \exp z$, the image of the segment $0 \leq y \leq \pi$ of the y axis is the semicircle $u^2 + v^2 = 1$, $v \geq 0$. Also, the function

$$h(u, v) = \text{Re}\left(2 - w + \frac{1}{w}\right) = 2 - u + \frac{u}{u^2 + v^2}$$

is harmonic everywhere in the w plane except for the origin; and it assumes the value $h = 2$ on the semicircle. Write an explicit expression for the function $H(x, y)$ defined in the theorem of Sec. 78. Then illustrate the theorem by showing directly that $H = 2$ along the segment of the y axis.

5. The transformation $w = z^2$ maps the positive x and y axes and the origin in the z plane onto the u axis in the w plane. Consider the harmonic function

$$h(u, v) = \text{Re}(e^{-w}) = e^{-u} \cos v,$$

and observe that its normal derivative h_v along the u axis is zero. Then illustrate the theorem in Sec. 78 when $f(z) = z^2$ by showing directly that the normal derivative of the function $H(x, y)$ defined in that theorem is zero along both positive axes in the z plane. Note that the transformation $w = z^2$ is not conformal at the origin.

6. Replace the function $h(u, v)$ in Exercise 5 by the harmonic function

$$h(u, v) = \text{Re}(-2iw + e^{-w}) = 2v + e^{-u} \cos v.$$

Then show that $h_v = 2$ along the u axis but that $H_y = 4x$ along the positive x axis and $H_x = 4y$ along the positive y axis. This illustrates how a condition of the type $dh/dn = h_0 \neq 0$ is *not* necessarily transformed into a condition of the type $dH/dN = h_0$.

7. Show that if a function $H(x, y)$ is a solution of a Neumann problem (Sec. 77), then $H(x, y) + A$, where A is any real constant, is also a solution of that problem.

8. Suppose that an analytic function $w = f(z) = u(x, y) + iv(x, y)$ maps a domain D_z in the z plane onto a domain D_w in the w plane; and let a function $h(u, v)$, with continuous first- and second-order partial derivatives, be defined on D_w. Use the chain rule for partial derivatives to show that if $H(x, y) = h[u(x, y), v(x, y)]$, then

$$H_{xx}(x, y) + H_{yy}(x, y) = [h_{uu}(u, v) + h_{vv}(u, v)]|f'(z)|^2.$$

Conclude that the function $H(x, y)$ is harmonic in D_z when $h(u, v)$ is harmonic in D_w. This proves the theorem in Sec. 77 even when the domain D_w is multiply connected.

Suggestion: In the simplifications, it is important to note that since f is analytic, the Cauchy-Riemann equations $u_x = v_y$, $u_y = -v_x$ hold and that the functions u and v both satisfy Laplace's equation. Also, the continuity conditions on the derivatives of h ensure that $h_{vu} = h_{uv}$.

9. Let $p(u, v)$ be a function which has continuous first- and second-order partial derivatives and which satisfies *Poisson's equation,*

$$p_{uu}(u, v) + p_{vv}(u, v) = \Phi(u, v),$$

in a domain D_w of the w plane, where Φ is a prescribed function. Show how it follows from the identity obtained in Exercise 8 that if an analytic function $w = f(z) = u(x, y) + iv(x, y)$ maps a domain D_z onto the domain D_w, the function

$$P(x, y) = p[u(x, y), v(x, y)]$$

satisfies the Poisson equation

$$P_{xx}(x, y) + P_{yy}(x, y) = \Phi[u(x, y), v(x, y)] |f'(z)|^2$$

in D_z.

10. Suppose that $w = f(z) = u(x, y) + iv(x, y)$ is a conformal mapping of a smooth arc C onto a smooth arc Γ in the w plane. Let the function $h(u, v)$ be defined on Γ, and write

$$H(x, y) = h[u(x, y), v(x, y)].$$

(a) From calculus, the x and y components of grad H are the partial derivatives H_x and H_y, respectively; likewise, grad h has components h_u and h_v. By applying the chain rule for partial derivatives and using the Cauchy-Riemann equations, show that if (x, y) is a point on C and (u, v) is its image on Γ, then

$$|\text{grad } H(x, y)| = |\text{grad } h(u, v)| \, |f'(z)|.$$

(b) Show that the angle from the arc C to grad H at a point (x, y) on C is equal to the angle from Γ to grad h at the image (u, v) of the point (x, y).

(c) Let s and σ denote distance along the arcs C and Γ, respectively; and let $\mathbf{t}$ and $\boldsymbol{\tau}$ denote unit tangent vectors at a point (x, y) on C and its image (u, v), in the direction of increasing distance. With the aid of the results in parts (a) and (b) and using the fact that

$$\frac{dH}{ds} = (\text{grad } H) \cdot \mathbf{t} \qquad \text{and} \qquad \frac{dh}{d\sigma} = (\text{grad } h) \cdot \boldsymbol{\tau},$$

show that the directional derivative along the arc Γ is transformed as follows:

$$\frac{dH}{ds} = \frac{dh}{d\sigma} |f'(z)|.$$

APPLICATIONS OF CONFORMAL MAPPING

We now use conformal mapping to solve a number of physical problems involving Laplace's equation in two independent variables. Problems in heat conduction, electrostatic potential, and fluid flow will be treated. Since these problems are intended to illustrate methods, they will be kept on a fairly elementary level.

79. Steady Temperatures

In the theory of heat conduction, the *flux* across a surface within a solid body at a point on that surface is the quantity of heat flowing in a specified direction normal to the surface per unit time per unit area at the point. Flux is therefore measured in such units as calories per second per square centimeter. It is denoted here by Φ, and it varies with the normal derivative of the temperature T at the point on the surface:

$$(1) \qquad\qquad \Phi = -K\frac{dT}{dN} \qquad\qquad (K > 0).$$

The constant K is known as the *thermal conductivity* of the material of the solid, which is assumed to be homogeneous.

The points in the solid are assigned cartesian coordinates in three-dimensional space, and we restrict our attention to those cases in which the temperature T varies with only the x and y coordinates. Since T does not vary with the coordinate along the axis perpendicular to the xy plane, the flow of heat is then two-dimensional and parallel to that plane. We assume, moreover, that the flow is in a steady state; that is, T does not vary with time.

It is assumed that no thermal energy is created or destroyed within the solid. That is, no heat sources or sinks are present there. Also, the temperature function $T(x, y)$ and all its partial derivatives of the first and second order are continuous at each point interior to the solid. This statement and expression (1) for the flux of heat are postulates for the mathematical theory of heat conduction, postulates that also apply at points within a solid containing a continuous distribution of sources or sinks.

Consider now an element of volume which is interior to the solid and which has the shape of a rectangular prism of unit height perpendicular to the xy plane with base Δx by Δy in that plane (Fig. 75). The time rate of flow of heat toward the right across the left-hand face is $-KT_x(x, y)\,\Delta y$; and toward the right across the right-hand face it is $-KT_x(x + \Delta x, y)\,\Delta y$. Subtracting the first rate from the second, we obtain the rate of heat loss from the element through those two faces. This resultant rate can be written

$$-K\left[\frac{T_x(x + \Delta x, y) - T_x(x, y)}{\Delta x}\right]\Delta x\,\Delta y,$$

or

$$(2) \qquad\qquad -KT_{xx}(x, y)\,\Delta x\,\Delta y$$

if Δx is very small. All the expressions here are, of course, approximations whose accuracy increases as Δx and Δy are made smaller.

In like manner, the resultant rate of heat loss through the other two faces perpendicular to the xy plane is found to be

$$(3) \qquad\qquad -KT_{yy}(x, y)\,\Delta x\,\Delta y.$$

Heat enters or leaves the element only through these four faces, and the temperatures within the element are steady. Hence the sum of expressions (2) and (3) is zero; that is,

$$(4) \qquad\qquad T_{xx}(x, y) + T_{yy}(x, y) = 0.$$

The temperature function thus satisfies Laplace's equation at each interior point of the solid.

In view of equation (4) and the continuity of the temperature function and its partial derivatives, T is a harmonic function of x and y in the domain representing the interior of the solid body.

The surfaces $T(x, y) = c_1$, where c_1 is any real constant, are the *isotherms* within the solid. They can also be considered as curves in the xy plane; then $T(x, y)$ can be interpreted as the temperature at a point (x, y) in a thin sheet of material in that plane, with the faces of the sheet thermally insulated. The isotherms are the level curves of the function T.

The gradient of T is perpendicular to the isotherm at each point, and the maximum flux at a point is in the direction of the gradient there. If $T(x, y)$ denotes temperatures in a thin sheet and if S is a harmonic conjugate of the function T, then a curve

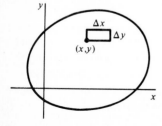

Figure 75

$S(x, y) = c_2$ has the gradient of T as a tangent vector at each point where the analytic function $T(x, y) + iS(x, y)$ is conformal. The curves $S(x, y) = c_2$ are called *lines of flow*.

If the normal derivative dT/dN is zero along any part of the boundary of the sheet, the flux of heat across that part is zero. That is, the part is thermally insulated and is therefore a line of flow.

The function T may also denote the concentration of a substance that is diffusing through a solid. In that case, K is the diffusion constant. The above discussion and the derivation of equation (4) apply as well to steady-state diffusion.

80. Steady Temperatures in a Half Plane

Let us find an expression for the steady temperatures $T(x, y)$ in a thin semi-infinite plate $y \geq 0$ whose faces are insulated and whose edge $y = 0$ is kept at temperature zero except for the segment $-1 < x < 1$, where it is kept at temperature unity (Fig. 76). The function $T(x, y)$ is to be bounded; this condition is natural if we consider the given plate as the limiting case of the plate $0 \leq y \leq y_0$ whose upper edge is kept at a fixed temperature as y_0 is increased. In fact, it would be physically reasonable to stipulate that $T(x, y)$ approach zero as y tends to infinity.

The boundary value problem to be solved can be written

$$(1) \qquad\qquad T_{xx}(x, y) + T_{yy}(x, y) = 0 \qquad (-\infty < x < \infty, \ y > 0),$$

$$(2) \qquad\qquad T(x, 0) = \begin{cases} 1 & \text{when } |x| < 1, \\ 0 & \text{when } |x| > 1; \end{cases}$$

also, $|T(x, y)| < M$ where M is some positive constant.

This is a Dirichlet problem for the upper half of the xy plane. Our method of solution will be to obtain a new Dirichlet problem for a region in the uv plane. That region will be the image of the half plane under a transformation $w = f(z)$ which is analytic in the domain $y > 0$ and which is conformal along the boundary $y = 0$ except at the points $(\pm 1, 0)$, where it is undefined. It will be a simple matter to discover a bounded harmonic function satisfying the new problem. The two theorems in Chap. 8 will then be applied to transform the solution of the problem in the uv plane into a solution of the original problem in the xy plane. Specifically, a harmonic function of

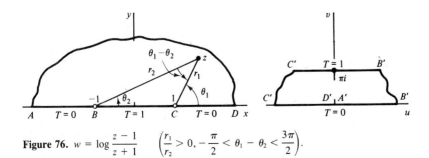

Figure 76. $w = \log \dfrac{z - 1}{z + 1}$ $\left(\dfrac{r_1}{r_2} > 0, \ -\dfrac{\pi}{2} < \theta_1 - \theta_2 < \dfrac{3\pi}{2} \right)$.

u and v will be transformed into a harmonic function of x and y, and the boundary conditions in the uv plane will be preserved on corresponding portions of the boundary in the xy plane. There should be no confusion if we use the same symbol T to denote the different temperature functions in the two planes.

Let us write $z - 1 = r_1 \exp(i\theta_1)$ and $z + 1 = r_2 \exp(i\theta_2)$, where $-\pi/2 < \theta_k < 3\pi/2$ $(k = 1, 2)$. The transformation

$$(3) \qquad w = \log \frac{z-1}{z+1} = \mathrm{Log}\, \frac{r_1}{r_2} + i(\theta_1 - \theta_2) \qquad \left(\frac{r_1}{r_2} > 0, -\frac{\pi}{2} < \theta_1 - \theta_2 < \frac{3\pi}{2}\right)$$

is defined on the upper half plane $y \geq 0$, except for the points $z = \pm 1$, since $0 \leq \theta_1 - \theta_2 \leq \pi$ in that region. (See Fig. 76.) Now the value of the logarithm is the principal value when $0 \leq \theta_1 - \theta_2 \leq \pi$, and we note from Fig. 19 of Appendix 2 that the upper half plane $y > 0$ is mapped onto the strip $0 < v < \pi$ in the w plane. Indeed, it was that figure which suggested transformation (3) here. The segment of the x axis between $z = -1$ and $z = 1$, where $\theta_1 - \theta_2 = \pi$, is mapped onto the upper edge of the strip; and the rest of the x axis, where $\theta_1 - \theta_2 = 0$, is mapped onto the lower edge. The required analyticity and conformality conditions are evidently satisfied by transformation (3).

A bounded harmonic function of u and v that is zero on the edge $v = 0$ of the strip and unity on the edge $v = \pi$ is clearly

$$(4) \qquad T = \frac{1}{\pi} v;$$

it is harmonic since it is the imaginary part of the entire function $(1/\pi)w$. Changing to x and y coordinates by means of the equation

$$(5) \qquad w = \mathrm{Log}\, \left|\frac{z-1}{z+1}\right| + i \arg\left(\frac{z-1}{z+1}\right),$$

we find that

$$v = \arg\left[\frac{(z-1)(\bar{z}+1)}{(z+1)(\bar{z}+1)}\right] = \arg\left[\frac{x^2 + y^2 - 1 + i2y}{(x+1)^2 + y^2}\right],$$

or

$$v = \arctan\left(\frac{2y}{x^2 + y^2 - 1}\right).$$

The range of the arctangent function here is from 0 to π since

$$\arg\left(\frac{z-1}{z+1}\right) = \theta_1 - \theta_2$$

and $0 \leq \theta_1 - \theta_2 \leq \pi$. Expression (4) now takes the form

$$(6) \qquad T = \frac{1}{\pi} \arctan\left(\frac{2y}{x^2 + y^2 - 1}\right) \qquad (0 \leq \arctan t \leq \pi).$$

Since the function (4) is harmonic in the strip $0 < v < \pi$ and since transformation (3) is analytic in the half plane $y > 0$, we may apply the theorem in Sec. 77 to conclude that the function (6) is harmonic in that half plane. The boundary conditions for the two harmonic functions are the same on corresponding parts of the boundaries because they are of the type $h = h_0$, treated in the theorem of Sec. 78. The bounded function (6) is therefore the desired solution of the original problem. One can, of course, verify directly that the function (6) satisfies Laplace's equation and has the values tending to those indicated on the left in Fig. 76 as the point (x, y) approaches the x axis from above.

The isotherms $T(x, y) = c_1$ $(0 < c_1 < 1)$ are arcs of the circles

$$x^2 + (y - \cot \pi c_1)^2 = \csc^2 \pi c_1$$

with centers on the y axis and passing through the points $(\pm 1, 0)$.

Finally, we note that since the product of a harmonic function by a constant is also harmonic, the function

$$T = \frac{T_0}{\pi} \arctan\left(\frac{2y}{x^2 + y^2 - 1}\right) \qquad (0 \le \arctan t \le \pi)$$

represents the steady temperatures in the given half plane when the temperature $T = 1$ along the segment $-1 < x < 1$ of the x axis is replaced by any fixed temperature $T = T_0$.

81. A Related Problem

Consider a semi-infinite slab in three-space, bounded by the planes $x = \pm \pi/2$ and $y = 0$, when the first two surfaces are kept at temperature zero and the last at temperature unity. We wish to find an expression for the temperature $T(x, y)$ at any interior point of the slab. The problem is also that of finding temperatures in a thin plate having the form of a semi-infinite strip $-\pi/2 \le x \le \pi/2$, $y \ge 0$ when the faces of the plate are perfectly insulated (Fig. 77).

The boundary value problem to be solved here is

(1) $$T_{xx}(x, y) + T_{yy}(x, y) = 0 \qquad \left(-\frac{\pi}{2} < x < \frac{\pi}{2}, y > 0\right),$$

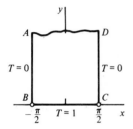

Figure 77

(2)
$$T\left(-\frac{\pi}{2}, y\right) = T\left(\frac{\pi}{2}, y\right) = 0 \qquad (y > 0),$$

(3)
$$T(x, 0) = 1 \qquad \left(-\frac{\pi}{2} < x < \frac{\pi}{2}\right),$$

where $T(x, y)$ is bounded.

In view of Example 1 in Sec. 71, as well as Fig. 9 of Appendix 2, the mapping

(4)
$$w = \sin z$$

transforms this boundary value problem into the one posed in Sec. 80 (Fig. 76). Hence, recalling solution (6) in that section, we write

(5)
$$T = \frac{1}{\pi} \arctan\left(\frac{2v}{u^2 + v^2 - 1}\right) \qquad (0 \le \arctan t \le \pi).$$

The change of variables indicated in equation (4) can be written

$$u = \sin x \cosh y, \qquad v = \cos x \sinh y;$$

and the harmonic function (5) becomes

$$T = \frac{1}{\pi} \arctan\left(\frac{2 \cos x \sinh y}{\sin^2 x \cosh^2 y + \cos^2 x \sinh^2 y - 1}\right).$$

The denominator here reduces to $\sinh^2 y - \cos^2 x$, and the quotient can be written

$$\frac{2 \cos x \sinh y}{\sinh^2 y - \cos^2 x} = \frac{2(\cos x / \sinh y)}{1 - (\cos x / \sinh y)^2} = \tan 2\alpha,$$

where $\tan \alpha = \cos x / \sinh y$. Hence $T = (2/\pi)\alpha$; that is,

(6)
$$T = \frac{2}{\pi} \arctan\left(\frac{\cos x}{\sinh y}\right) \qquad \left(0 \le \arctan t \le \frac{\pi}{2}\right).$$

The arctangent function here has the range 0 to $\pi/2$ since its argument is nonnegative.

Since $\sin z$ is entire and the function (5) is harmonic in the half plane $v > 0$, the function (6) is harmonic in the strip $-\pi/2 < x < \pi/2$, $y > 0$. Also, the function (5) satisfies the boundary condition $T = 1$ when $|u| < 1$ and $v = 0$ and the condition $T = 0$ when $|u| > 1$ and $v = 0$. The function (6) thus satisfies boundary conditions (2) and (3). Moreover, $|T(x, y)| \le 1$ throughout the strip. Expression (6) is therefore the temperature formula sought.

The isotherms $T(x, y) = c_1$ $(0 < c_1 < 1)$ are the portions of the surfaces

$$\cos x = \tan\left(\frac{\pi c_1}{2}\right) \sinh y$$

within the slab, each surface passing through the points $(\pm\pi/2, 0)$ in the xy plane. If K is the thermal conductivity, the flux of heat into the slab through the surface lying in the plane $y = 0$ is

$$-KT_y(x, 0) = \frac{2K}{\pi \cos x} \qquad \left(-\frac{\pi}{2} < x < \frac{\pi}{2}\right);$$

the flux outward through the surface lying in the plane $x = \pi/2$ is

$$-KT_x\left(\frac{\pi}{2}, y\right) = \frac{2K}{\pi \sinh y} \qquad (y > 0).$$

The boundary value problem posed in this section can also be solved by the method of separation of variables. That method is more direct, but it gives the solution in the form of an infinite series.*

82. Temperatures in a Quadrant

Let us find the steady temperatures in a thin plate having the form of a quadrant if a segment at the end of one edge is insulated, if the rest of that edge is kept at a fixed temperature, and if the second edge is kept at another fixed temperature. The surfaces are insulated, and so the problem is two-dimensional.

The temperature scale and the unit of length can be chosen so that the boundary value problem for the temperature function T becomes

(1) $$T_{xx}(x, y) + T_{yy}(x, y) = 0 \qquad (x > 0, y > 0),$$

(2) $$\begin{cases} T_y(x, 0) = 0 & \text{when} & 0 < x < 1, \\ T(x, 0) = 1 & \text{when} & x > 1, \end{cases}$$

(3) $$T(0, y) = 0 \qquad (y > 0),$$

where $T(x, y)$ is bounded in the quadrant. The plate and its boundary conditions are shown on the left in Fig. 78.

Conditions (2) prescribe the value of the normal derivative of the function T over a part of a boundary line and the value of the function itself over the rest of that line. The separation-of-variables method mentioned at the end of Sec. 81 is not adapted to such problems with different types of conditions along the same boundary line.

As indicated in Fig. 10 of Appendix 2, the transformation

(4) $$z = \sin w$$

is a one to one mapping of the strip $0 \leq u \leq \pi/2$, $v \geq 0$ onto the quadrant $x \geq 0$, $y \geq 0$. Observe now that the existence of an inverse is ensured by the fact that the given

*Essentially the same problem is treated in the authors' "Fourier Series and Boundary Value Problems," 3d ed., exercises 3 and 4, p. 141, 1978. Also, a short discussion of the uniqueness of solutions to boundary value problems can be found in chap. 10 of that book.

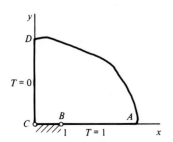
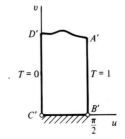

Figure 78

transformation is both one to one and onto. Since transformation (4) is conformal throughout the strip except at the point $w = \pi/2$, the inverse transformation must be conformal throughout the quadrant except at the point $z = 1$. That inverse transformation maps the segment $0 < x < 1$ of the x axis onto the base of the strip and the rest of the boundary onto the sides of the strip as shown in Fig. 78.

Since the inverse of transformation (4) is conformal in the quadrant, except when $z = 1$, the solution to the given problem can be obtained by finding a function that is harmonic in the strip and satisfies the boundary conditions shown on the right in Fig. 78. Observe that these boundary conditions are of the types $h = h_0$ and $dh/dn = 0$ in the theorem of Sec. 78.

The required temperature function T for the new boundary value problem is clearly

$$(5) \qquad T = \frac{2}{\pi} u,$$

the function $(2/\pi)u$ being the real part of the entire function $(2/\pi)w$. We must now express T in terms of x and y.

To obtain u in terms of x and y, we first note that according to equation (4),

$$(6) \qquad x = \sin u \cosh v, \qquad y = \cos u \sinh v.$$

When $0 < u < \pi/2$, both $\sin u$ and $\cos u$ are nonzero; and, consequently,

$$(7) \qquad \frac{x^2}{\sin^2 u} - \frac{y^2}{\cos^2 u} = 1.$$

Now it is convenient to observe that for each fixed u, hyperbola (7) has foci at the points

$$z = \pm\sqrt{\sin^2 u + \cos^2 u} = \pm 1$$

and that the length of the transverse axis, which is the line segment joining the two vertices, is $2 \sin u$. Thus the absolute value of the difference of the distances between the foci and a point (x, y) lying on the part of the hyperbola in the first quadrant is

$$\sqrt{(x + 1)^2 + y^2} - \sqrt{(x - 1)^2 + y^2} = 2 \sin u.$$

It follows directly from equations (6) that this relation also holds when $u = 0$ or $u = \pi/2$. In view of equation (5), the required temperature function is therefore

$$(8) \qquad T = \frac{2}{\pi} \arcsin \frac{1}{2}\left[\sqrt{(x + 1)^2 + y^2} - \sqrt{(x - 1)^2 + y^2}\right],$$

where, since $0 \leq u \leq \pi/2$, the arcsine function has the range 0 to $\pi/2$.

If we wish to verify that this function satisfies boundary conditions (2), we must remember that $\sqrt{(x - 1)^2}$ denotes $x - 1$ when $x > 1$ and $1 - x$ when $0 < x < 1$, the square roots being positive. Note, too, that the temperature at any point along the insulated part of the lower edge of the plate is

$$T(x, 0) = \frac{2}{\pi} \arcsin x \qquad\qquad (0 < x < 1).$$

It can be seen from equation (5) that the isotherms $T(x, y) = c_1$ $(0 < c_1 < 1)$ are the parts of the confocal hyperbolas (7), where $u = \pi c_1/2$, which lie in the first quadrant. Since the function $(2/\pi)v$ is a harmonic conjugate of the function (5), the lines of flow are quarters of the confocal ellipses obtained by holding v constant in equations (6).

EXERCISES

1. In the problem of the semi-infinite plate shown on the left in Fig. 76, obtain a harmonic conjugate of the temperature function $T(x, y)$ from equation (5), Sec. 80, and find the lines of flow of heat. Show that those lines of flow consist of the upper half of the y axis and the upper halves of certain circles on either side of that axis, the circles having their centers on the segment AB or CD of the x axis.

2. Show that if the function T of Sec. 80 is not required to be bounded, the harmonic function (4) of that section can be replaced by the harmonic function

$$T = \text{Im}\left(\frac{1}{\pi}w + A \cosh w\right) = \frac{1}{\pi}v + A \sinh u \sin v,$$

where A is an arbitrary real constant. Conclude that the solution of the Dirichlet problem for the strip in the uv plane (Fig. 76) would not, then, be unique.

3. Suppose that the condition that T be bounded is omitted from the problem for temperatures in the semi-infinite slab of Sec. 81 (Fig. 77). Show that an infinite number of solutions are then possible by noting the effect of adding to the solution found there the imaginary part of the function $A \sin z$, where A is an arbitrary real constant.

4. Use the function Log z to find an expression for the bounded steady temperatures in a plate having the form of a quadrant $x \geq 0, y \geq 0$ if its faces are perfectly insulated and its edges have temperatures $T(x, 0) = 0$ and $T(0, y) = 1$ (Fig. 79). Find the isotherms and lines of flow, and draw some of them.

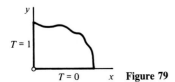

$T = 1$ $T = 0$ x **Figure 79**

Ans. $T = (2/\pi) \arctan(y/x)$.

5. Find the steady temperatures in a solid whose shape is that of a long cylindrical wedge if its boundary planes $\theta = 0$ and $\theta = \theta_0$ $(0 < r < r_0)$ are kept at constant temperatures zero and T_0, respectively, and if its surface $r = r_0$ $(0 < \theta < \theta_0)$ is perfectly insulated (Fig. 80).

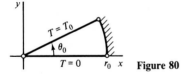

$T = T_0$ θ_0 $T = 0$ r_0 x **Figure 80**

Ans. $T = (T_0/\theta_0) \arctan(y/x)$.

6. Find the bounded steady temperatures $T(x, y)$ in the semi-infinite solid $y \geq 0$ if $T = 0$ on the part $x < -1$, $y = 0$ of the boundary, if $T = 1$ on the part $x > 1$, $y = 0$, and if the strip $-1 < x < 1$, $y = 0$ of the boundary is insulated (Fig. 81).

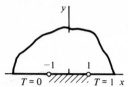

$T = 0$ ///////// $T = 1$ x **Figure 81**

$$Ans. \quad T = \frac{1}{2} + \frac{1}{\pi} \arcsin \frac{1}{2}\left[\sqrt{(x + 1)^2 + y^2} - \sqrt{(x - 1)^2 + y^2}\right]$$

$$\left(-\frac{\pi}{2} \leq \arcsin t \leq \frac{\pi}{2}\right).$$

7. Find the bounded steady temperatures in the solid $x \geq 0$, $y \geq 0$ when the boundary surfaces are kept at fixed temperatures except for insulated strips of equal width at the corner, as shown in Fig. 82.

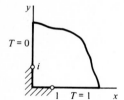

$T = 0$

i

/////// 1 $T = 1$ x **Figure 82**

$$Ans. \quad T = \frac{1}{2} + \frac{1}{\pi} \arcsin \frac{1}{2}\left[\sqrt{(x^2 - y^2 + 1)^2 + 4x^2y^2} - \sqrt{(x^2 - y^2 - 1)^2 + 4x^2y^2}\right]$$

$$\left(-\frac{\pi}{2} \leq \arcsin t \leq \frac{\pi}{2}\right).$$

8. Solve the following Dirichlet problem for a semi-infinite strip (Fig. 83):

$$H_{xx}(x, y) + H_{yy}(x, y) = 0 \qquad \left(0 < x < \frac{\pi}{2}, y > 0\right),$$

$$H(x, 0) = 0 \qquad \left(0 < x < \frac{\pi}{2}\right),$$

$$H(0, y) = 1, \qquad H\left(\frac{\pi}{2}, y\right) = 0 \qquad \qquad (y > 0),$$

where $0 \leq H(x, y) \leq 1$.

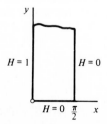

$H = 1$ $H = 0$

$H = 0$ $\frac{\pi}{2}$ x **Figure 83**

Suggestion: This problem can be transformed into the one of Exercise 4.

$$Ans. \quad H = \frac{2}{\pi} \arctan\left(\frac{\tanh y}{\tan x}\right).$$

9. Derive an expression for the temperatures $T(r, \theta)$ in a semicircular plate $r \leq 1, 0 \leq \theta \leq \pi$ with insulated faces if $T = 1$ along the radial edge $\theta = 0$ $(0 < r < 1)$ and $T = 0$ on the rest of the boundary.

   Suggestion: This problem can be transformed into the one of Exercise 8.

$$Ans. \quad T = \frac{2}{\pi} \arctan\left(\frac{1-r}{1+r} \cot \frac{\theta}{2}\right).$$

10. Solve the boundary value problem for the plate $x \geq 0$, $y \geq 0$ in the z plane where the faces are insulated and the boundary conditions are those indicated in Fig. 84.

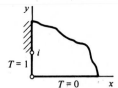

$T = 1$ $T = 0$ x **Figure 84**

   Suggestion: Using the mapping $w = i/z = i\bar{z}/|z|^2$, transform this problem into the one posed in Sec. 82 (Fig. 78).

11. The portions $x < 0$, $y = 0$ and $x < 0$, $y = \pi$ of the edges of an infinite plate $0 \leq y \leq \pi$ are thermally insulated, as are the faces of the plate. The conditions $T(x, 0) = 1$ and $T(x, \pi) = 0$ are maintained when $x > 0$ (Fig. 85). Find the steady temperatures in the plate.

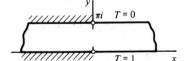

      πi $T = 0$

$T = 1$ x **Figure 85**

   Suggestion: This problem can be transformed to the one of Exercise 6.

12. Consider a thin plate with insulated faces whose shape is the upper half of the region enclosed by an ellipse with foci $(\pm 1, 0)$. The temperature on the elliptical part of its boundary is $T = 1$. The temperature along the segment $-1 < x < 1$ of the x axis is $T = 0$, and the rest of the boundary along the x axis is insulated. With the aid of Fig. 11 in Appendix 2, find the lines of flow of heat.

13. According to Sec. 41 and Exercise 4, Sec. 42, if a function $f(z) = u(x, y) + iv(x, y)$ is continuous in a closed bounded region R and analytic and not constant in the interior of R, then the function $u(x, y)$ reaches its maximum and minimum values on the boundary of R, and never in the interior. By interpreting $u(x, y)$ as a steady temperature, state a physical reason why that property of maximum and minimum values should hold true.

83. Electrostatic Potential

In an electrostatic force field, the *field intensity* at a point is a vector representing the force exerted on a unit positive charge placed at that point. The electrostatic *potential* is a scalar function of the space coordinates such that at each point its directional derivative in any direction is the negative of the component of the field intensity in that direction.

For two stationary charged particles, the magnitude of the force of attraction or repulsion exerted by one particle on the other is directly proportional to the product of the charges and inversely proportional to the square of the distance between those particles. From this inverse-square law it can be shown that the potential at a point due to a single particle in space is inversely proportional to the distance between the point and the particle. In any region free of charges, the potential due to a distribution of charges outside that region can be shown to satisfy Laplace's equation for three-dimensional space.

If conditions are such that the potential V is the same in all planes parallel to the xy plane, then in regions free from charges V is a harmonic function of just the two variables x and y:

$$V_{xx}(x, y) + V_{yy}(x, y) = 0.$$

The field intensity vector at each point is parallel to the xy plane with x and y components $-V_x(x, y)$ and $-V_y(x, y)$, respectively. That vector is therefore the negative of the gradient of $V(x, y)$.

A surface along which $V(x, y)$ is constant is an equipotential surface. The tangential component of the field intensity vector at a point on a conducting surface is zero in the static case since charges are free to move on such a surface. Hence $V(x, y)$ is constant along the surface of a conductor, and that surface is an *equipotential*.

If U is a harmonic conjugate of V, the curves $U(x, y) = c_2$ in the xy plane are called *flux lines*. When such a curve intersects an equipotential curve $V(x, y) = c_1$ at a point where the derivative of the analytic function $V(x, y) + iU(x, y)$ is not zero, the two curves are orthogonal at that point and the field intensity is tangent to the flux line there.

Boundary value problems for the potential V are the same mathematical problems as those for steady temperatures T; and, as in the case of steady temperatures, the methods of complex variables are limited to two-dimensional problems. The problem posed in Sec. 81 (Fig. 77), for instance, can be interpreted as that of finding the two-dimensional electrostatic potential in the empty space $-\pi/2 < x < \pi/2, y > 0$ bounded by the conducting planes $x = \pm\pi/2$ and $y = 0$, insulated at their intersections, when the first two surfaces are kept at potential zero and the third at potential unity. Problems of this type arise in electronics. If the space charge inside a vacuum tube is small, the space is sometimes considered free of charge and the potential there can be assumed to satisfy Laplace's equation.

The potential in the steady flow of electricity in a plane conducting sheet is also a harmonic function at points free from sources and sinks. Gravitational potential is a further example of a harmonic function in physics.

84. Potential in a Cylindrical Space

A long hollow circular cylinder is made out of a thin sheet of conducting material, and the cylinder is split along two of its elements to form two equal parts. These parts are separated by slender strips of insulating material and are used as electrodes, one of which is grounded at potential zero and the other kept at a different fixed potential. We take the coordinate axes and units of length and potential difference as indicated on the left in Fig. 86. We then interpret the electrostatic potential $V(x, y)$ over any cross section of the enclosed space that is distant from the ends of the cylinder as a harmonic function inside the circle $x^2 + y^2 = 1$ in the xy plane; also, $V = 0$ on the upper half of the circle and $V = 1$ on the lower half.

A linear fractional transformation that maps the upper half plane onto the interior of the unit circle centered at the origin, the positive real axis onto the upper half of the circle, and the negative real axis onto the lower half of the circle is verified in Exercise 10, Sec. 66. The result is given in Fig. 13 of Appendix 2; interchanging z and w there, we find that the inverse of the transformation

$$(1) \qquad\qquad z = \frac{i - w}{i + w}$$

gives us a new problem for V in a half plane, indicated on the right in Fig. 86.

Observe now that the imaginary part of the function

$$(2) \qquad\qquad \frac{1}{\pi} \operatorname{Log} w = \frac{1}{\pi} \operatorname{Log} \rho + \frac{i}{\pi} \phi \qquad (\rho > 0, 0 \le \phi \le \pi)$$

is a bounded function of u and v that assumes the required constant values on the two parts $\phi = 0$ and $\phi = \pi$ of the u axis. The desired harmonic function for the half plane is therefore

$$(3) \qquad\qquad V = \frac{1}{\pi} \arctan\left(\frac{v}{u}\right),$$

where the values of the arctangent function range from 0 to π.

The inverse of transformation (1) is

$$(4) \qquad\qquad w = i\frac{1 - z}{1 + z},$$

from which u and v can be expressed in terms of x and y. Equation (3) then becomes

$$(5) \qquad\qquad V = \frac{1}{\pi} \arctan\left(\frac{1 - x^2 - y^2}{2y}\right) \qquad (0 \le \arctan t \le \pi).$$

The function (5) is the potential function for the space enclosed by the cylindrical

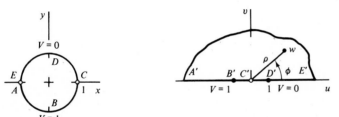

Figure 86

electrodes since it is harmonic inside the circle and it assumes the required values on the semicircles. If we wish to verify this solution, we must note that

$$\lim_{\substack{t \to 0 \\ t > 0}} \arctan t = 0 \quad \text{and} \quad \lim_{\substack{t \to 0 \\ t < 0}} \arctan t = \pi.$$

The equipotential curves $V(x, y) = c_1$ $(0 < c_1 < 1)$ in the circular region are arcs of the circles

$$x^2 + (y + \tan \pi c_1)^2 = \sec^2 \pi c_1,$$

with each circle passing through the points $(\pm 1, 0)$. Also, the segment of the x axis between those points is the equipotential $V(x, y) = 1/2$. A harmonic conjugate U of V is $-(1/\pi) \operatorname{Log} \rho$, or the imaginary part of the function $-(i/\pi) \operatorname{Log} w$. In view of equation (4), U may be written

$$U = -\frac{1}{\pi} \operatorname{Log} \left| \frac{1 - z}{1 + z} \right|.$$

From this equation it can be seen that the flux lines $U(x, y) = c_2$ are arcs of circles with centers on the x axis. The segment of the y axis between the electrodes is also a flux line.

EXERCISES

1. The harmonic function (3) of Sec. 84 is bounded in the half plane $v \geq 0$ and satisfies the boundary conditions indicated on the right in Fig. 86. Show that if the imaginary part of Ae^w, where A is any real constant, is added to that function, the resulting function satisfies all the requirements except the boundedness condition.

2. Prove that transformation (4) of Sec. 84 maps the upper half of the circular region shown on the left in Fig. 86 onto the first quadrant of the w plane and the diameter CE onto the positive v axis. Then find the electrostatic potential V in the space enclosed by the half cylinder $x^2 + y^2 = 1, y \geq 0$ and the plane $y = 0$ when $V = 0$ on the cylindrical surface and $V = 1$ on the planar surface (Fig. 87).

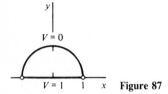

Figure 87

$$\textit{Ans.} \quad V = \frac{2}{\pi} \arctan \left(\frac{1 - x^2 - y^2}{2y} \right).$$

3. Find the electrostatic potential $V(r, \theta)$ in the space $0 < r < 1, 0 < \theta < \pi/4$, bounded by the half planes $\theta = 0$ and $\theta = \pi/4$ and the portion $0 \leq \theta \leq \pi/4$ of the cylindrical surface $r = 1$, when $V = 1$ on the planar surfaces and $V = 0$ on the cylindrical one. (See Exercise 2.) Verify that the function obtained satisfies the boundary conditions.

4. Note that all branches of log z have the same real component, which is harmonic everywhere except at the origin. Then write an expression for the electrostatic potential $V(x, y)$ in the space between two coaxial conducting cylindrical surfaces $x^2 + y^2 = 1$ and $x^2 + y^2 = r_0^2 \ (r_0 \neq 1)$ when $V = 0$ on the first surface and $V = 1$ on the second.

$$Ans. \quad V = \frac{\text{Log}(x^2 + y^2)}{2 \ \text{Log} \ r_0}.$$

5. Find the bounded electrostatic potential $V(x, y)$ in the space $y > 0$ bounded by an infinite conducting plane $y = 0$ one strip $(-a < x < a, y = 0)$ of which is insulated from the rest of the plane and kept at potential $V = 1$, while $V = 0$ on the rest (Fig. 88). Verify that the function obtained satisfies the boundary conditions.

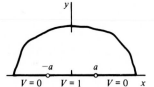

$V = 0 \qquad V = 1 \qquad V = 0 \qquad x$ **Figure 88**

$$Ans. \quad V = \frac{1}{\pi} \arctan\left(\frac{2ay}{x^2 + y^2 - a^2}\right) \quad (0 \leq \arctan t \leq \pi).$$

6. Derive an expression for the electrostatic potential in the semi-infinite space indicated in Fig. 89, bounded by two half planes and a half cylinder, when $V = 1$ on the cylindrical surface and $V = 0$ on the planar surfaces. Draw some of the equipotential curves in the xy plane.

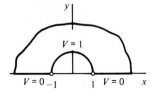

$V = 0 \ _{-1} \qquad _1 \ V = 0 \qquad x$ **Figure 89**

$$Ans. \quad V = \frac{2}{\pi} \arctan\left(\frac{2y}{x^2 + y^2 - 1}\right).$$

7. Find the potential V in the space between the planes $y = 0$ and $y = \pi$ when $V = 0$ on the part of each of those planes where $x > 0$ and $V = 1$ on the parts where $x < 0$ (Fig. 90). Check the result with the boundary conditions.

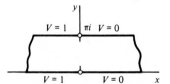

$V = 1 \qquad V = 0 \qquad x$ **Figure 90**

$$Ans. \quad V = \frac{1}{\pi} \arctan\left(\frac{\sin y}{\sinh x}\right) \quad (0 \leq \arctan t \leq \pi).$$

8. Derive an expression for the electrostatic potential V in the space interior to a long cylinder $r = 1$ when $V = 0$ on the first quadrant ($r = 1$, $0 < \theta < \pi/2$) of the cylindrical surface and $V = 1$ on the rest ($r = 1$, $\pi/2 < \theta < 2\pi$) of that surface. (See Exercise 14, Sec. 66, and Fig. 55 there.) Show that $V = 3/4$ on the axis of the cylinder. Check the result with the boundary conditions.

9. Using Fig. 20 of Appendix 2, find a temperature function $T(x, y)$ which is harmonic in the shaded domain of the xy plane shown there and which assumes the values $T = 0$ along the arc ABC and $T = 1$ along the line segment DEF. Verify that the function obtained satisfies the required boundary conditions. (See Exercise 2.)

10. The Dirichlet problem

$$V_{xx}(x, y) + V_{yy}(x, y) = 0 \qquad (0 < x < a, 0 < y < b)$$
$$V(x, 0) = 0, \qquad V(x, b) = 1 \qquad (0 < x < a),$$
$$V(0, y) = V(a, y) = 0 \qquad (0 < y < b)$$

for $V(x, y)$ in a rectangle can be solved by the method of separation of variables.* The solution is

$$V = \frac{4}{\pi} \sum_{n=1}^{\infty} \frac{\sinh(m\pi y/a)}{m \sinh(m\pi b/a)} \sin \frac{m\pi x}{a} \qquad (m = 2n - 1).$$

Accepting this result and adapting it to a problem in the uv plane, find the potential $V(r, \theta)$ in the space $1 < r < r_0, 0 < \theta < \pi$ when $V = 1$ on the part of the boundary where $\theta = \pi$ and $V = 0$ on the rest of the boundary. (See Fig. 91.)

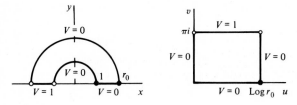

Figure 91. $w = \log z \quad \left(r > 0, -\dfrac{\pi}{2} < \theta < \dfrac{3\pi}{2} \right).$

$$Ans. \quad V = \frac{4}{\pi} \sum_{n=1}^{\infty} \frac{\sinh(\alpha_n \theta)}{\sinh(\alpha_n \pi)} \frac{\sin(\alpha_n \operatorname{Log} r)}{2n - 1} \qquad \left[\alpha_n = \frac{(2n - 1)\pi}{\operatorname{Log} r_0} \right].$$

11. With the aid of the solution of the Dirichlet problem for the rectangle $0 \leq x \leq a$, $0 \leq y \leq b$ that was used in Exercise 10, find the potential function $V(r, \theta)$ for the space $1 < r < r_0, 0 < \theta < \pi$ when $V = 1$ on the part of the boundary $r = r_0, 0 < \theta < \pi$ and $V = 0$ on the rest of the boundary (Fig. 92).

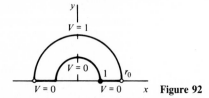

x **Figure 92**

*See the authors' "Fourier Series and Boundary Value Problems," 3d ed., pp. 135–137, 1978.

$$\text{Ans.} \quad V = \frac{4}{\pi} \sum_{n=1}^{\infty} \left(\frac{r^m - r^{-m}}{r_0^m - r_0^{-m}} \right) \frac{\sin m\theta}{m} \qquad (m = 2n - 1).$$

85. Two-Dimensional Fluid Flow

Harmonic functions play an important role in hydrodynamics and aerodynamics. Again, we consider only the two-dimensional steady-state type of problem. That is, the motion of the fluid is assumed to be the same in all planes parallel to the xy plane, the velocity being parallel to that plane and independent of time. It is then sufficient to consider the motion of a sheet of fluid in the xy plane.

We let the vector representing the complex number

$$V = p + iq$$

denote the velocity of a particle of the fluid at any point (x, y); hence the x and y components of the velocity vector are $p(x, y)$ and $q(x, y)$, respectively. At points interior to a region of flow in which no sources or sinks of the fluid occur, the real-valued functions $p(x, y)$ and $q(x, y)$ and their first-order partial derivatives are assumed to be continuous.

The *circulation* of the fluid along any contour C is defined as the line integral with respect to arc length s of the tangential component $V_T(x, y)$ of the velocity vector along C:

$$(1) \qquad \int_C V_T(x, y)\, ds .$$

The ratio of the circulation along C to the length of C is therefore a mean speed of the fluid along that contour. It is shown in advanced calculus that such an integral can be written*

$$(2) \qquad \int_C V_T(x, y)\, ds = \int_C p(x, y)\, dx + q(x, y)\, dy.$$

When C is a positively oriented simple closed contour lying in a simply connected domain of flow containing no sources or sinks, Green's theorem allows us to write

$$\int_C p(x, y)\, dx + q(x, y)\, dy = \iint_R [q_x(x, y) - p_y(x, y)]\, dx\, dy ,$$

where R is the closed region consisting of points interior to and on C. Thus

$$(3) \qquad \int_C V_T(x, y)\, ds = \iint_R [q_x(x, y) - p_y(x, y)]\, dx\, dy$$

for such a contour.

A physical interpretation of the integrand on the right in expression (3) for the circulation along the simple closed contour C is readily given. We let C denote a circle

---

*Properties of line integrals in advanced calculus that are used in this and the following section are to be found in, for instance, W. Kaplan, "Advanced Mathematics for Engineers," chap. 10, 1981.

of radius r which is centered at a point (x_0, y_0) and taken counterclockwise. The mean speed along C is then found by dividing the circulation by the circumference $2\pi r$, and the corresponding mean angular speed of the fluid about the center of the circle is found by dividing that mean speed by r:

$$\frac{1}{\pi r^2} \int\int_R \frac{1}{2} [q_x(x, y) - p_y(x, y)] \, dx \, dy.$$

Now this is also an expression for the mean value of the function

$$(4) \qquad \omega(x, y) = \frac{1}{2} [q_x(x, y) - p_y(x, y)]$$

over the circular region R bounded by C. Its limit as r tends to zero is the value of ω at the point (x_0, y_0). Hence the function $\omega(x, y)$, called the *rotation* of the fluid, represents the limiting angular speed of a circular element of the fluid as the circle shrinks to its center (x, y), the point at which ω is evaluated.

If $\omega(x, y) = 0$ at each point in some simply connected domain, the flow is *irrotational* in that domain. We consider only irrotational flows here, and we also assume that the fluid is *incompressible* and *free from viscosity*. Under our assumption of steady irrotational flow of fluids with uniform density ρ, it can be shown that the fluid pressure $P(x, y)$ satisfies the following special case of *Bernoulli's equation*:

$$\frac{P}{\rho} + \frac{1}{2} |V|^2 = \text{constant}.$$

Note that the pressure is greatest where the speed $|V|$ is least.

Let D be a simply connected domain in which the flow is irrotational. According to equation (4), $p_y = q_x$ throughout D. This relation between partial derivatives implies that the line integral

$$\int_C p(X, Y) \, dX + q(X, Y) \, dY$$

along a contour C lying entirely in D and joining any two points (x_0, y_0) and (x, y) in D is actually independent of path. Thus, if (x_0, y_0) is fixed, the function

$$(5) \qquad \phi(x, y) = \int_{(x_0, y_0)}^{(x, y)} p(X, Y) \, dX + q(X, Y) \, dY$$

is well defined on D; and by taking partial derivatives on each side of this equation, we find that

$$(6) \qquad \phi_x(x, y) = p(x, y), \qquad \phi_y(x, y) = q(x, y).$$

From equations (6) we see that the velocity vector $V = p + iq$ is the gradient of ϕ; and the directional derivative of ϕ in any direction represents the component of the velocity of flow in that direction.

The function $\phi(x, y)$ is called the *velocity potential*. From equation (5) it is evident that $\phi(x, y)$ changes by an additive constant when the reference point (x_0, y_0) is changed. The level curves $\phi(x, y) = c_1$ are called *equipotentials*. Because it is the gradient of

$\phi(x, y)$, the velocity vector V is normal to an equipotential at any point where V is not the zero vector.

Just as in the case of the flow of heat, the condition that the incompressible fluid enter or leave an element of volume only by flowing through the boundaries of that element requires that $\phi(x, y)$ must satisfy Laplace's equation

$$\phi_{xx}(x, y) + \phi_{yy}(x, y) = 0$$

in a domain where the fluid is free from sources or sinks. In view of equations (6) and the continuity of the functions p and q and their first-order partial derivatives, it follows that the first- and second-order partial derivatives of ϕ are continuous in such a domain. Hence the velocity potential ϕ *is a harmonic function* in that domain.

86. The Stream Function

According to Sec. 85, the velocity vector

(1) $$V = p(x, y) + iq(x, y)$$

for a simply connected domain in which the flow is irrotational can be written

(2) $$V = \phi_x(x, y) + i\phi_y(x, y) = \operatorname{grad} \phi(x, y),$$

where ϕ is the velocity potential.

When the velocity vector is not the zero vector, it is normal to an equipotential passing through the point (x, y). If, moreover, $\psi(x, y)$ denotes a harmonic conjugate of $\phi(x, y)$ (see Sec. 76), the velocity vector is tangent to a curve $\psi(x, y) = c_2$. The curves $\psi(x, y) = c_2$ are called the *streamlines* of the flow, and the function ψ is the *stream function*. In particular, a boundary across which fluid cannot flow is a streamline.

The analytic function

$$F(z) = \phi(x, y) + i\psi(x, y)$$

is called the *complex potential* of the flow. Note that

$$F'(z) = \phi_x(x, y) + i\psi_x(x, y),$$

or, in view of the Cauchy-Riemann equations,

$$F'(z) = \phi_x(x, y) - i\phi_y(x, y).$$

Expression (2) for the velocity thus becomes

(3) $$V = \overline{F'(z)}.$$

The speed, or magnitude of the velocity, is obtained by writing

$$|V| = |F'(z)|.$$

According to equation (5), Sec. 76, if ϕ is harmonic in a simply connected domain D, a harmonic conjugate of ϕ there can be written

$$\psi(x, y) = \int_{(x_0, y_0)}^{(x, y)} -\phi_Y(X, Y)\, dX + \phi_X(X, Y)\, dY,$$

where the integration is independent of path. With the aid of equations (6), Sec. 85, we can therefore write

(4)
$$\psi(x, y) = \int_C -q(X, Y)\, dX + p(X, Y)\, dY,$$

where C is any contour in D from (x_0, y_0) to (x, y).

Now it is shown in advanced calculus that the right-hand side of equation (4) represents the integral with respect to arc length s along C of the normal component $V_N(x, y)$ of the vector whose x and y components are $p(x, y)$ and $q(x, y)$, respectively. So expression (4) can be written

(5)
$$\psi(x, y) = \int_C V_N(X, Y)\, ds.$$

Physically, then, $\psi(x, y)$ represents the time rate of flow of the fluid across C. More precisely, $\psi(x, y)$ denotes the rate of flow, by volume, across a surface of unit height standing perpendicular to the xy plane on the curve C.

Example. When the complex potential is the function

(6)
$$F(z) = Az,$$

where A is a positive real constant,

(7)
$$\phi(x, y) = Ax, \qquad \psi(x, y) = Ay.$$

The streamlines $\psi(x, y) = c_2$ are the horizontal lines $y = c_2/A$, and the velocity at any point is

$$V = \overline{F'(z)} = A.$$

Here a point (x_0, y_0) at which $\psi(x, y) = 0$ is any point on the x axis. If the point (x_0, y_0) is taken as the origin, then $\psi(x, y)$ is the rate of flow across any contour drawn from the origin to the point (x, y) (Fig. 93). The flow is uniform and to the right. It can be interpreted as the uniform flow in the upper half plane bounded by the x axis, which is a streamline, or as the uniform flow between two parallel lines $y = y_1$ and $y = y_2$.

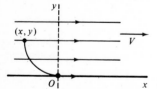

Figure 93

The stream function ψ characterizes a definite flow in a region. The question of whether just one such function exists corresponding to a given region, except possibly for a constant factor or an additive constant, is not examined here. In some of the examples to follow, where the velocity is uniform far from the obstruction, or in

Chap. 10, where sources and sinks are involved, the physical situation indicates that the flow is uniquely determined by the conditions given in the problem.

A harmonic function is not always uniquely determined, even up to a constant factor, by simply prescribing its values on the boundary of a region. In the above example, the function $\psi(x, y) = Ay$ is harmonic in the half plane $y > 0$ and has zero values on the boundary. The function $\psi_1(x, y) = Be^x \sin y$ also satisfies those conditions. However, the streamline $\psi_1(x, y) = 0$ consists not only of the line $y = 0$ but also of the lines $y = n\pi$ $(n = 1, 2, \ldots)$. Here the function $F_1(z) = Be^z$ is the complex potential for the flow in the strip between the lines $y = 0$ and $y = \pi$, both boundaries making up the streamline $\psi_1(x, y) = 0$; if $B > 0$, the fluid flows to the right along the lower boundary and to the left along the upper one.

87. Flows around a Corner and around a Cylinder

In analyzing a flow in the xy, or z, plane, it is often simpler to consider a corresponding flow in the uv, or w, plane. Then, if ϕ is a velocity potential and ψ a stream function for the flow in the uv plane, results in Secs. 77 and 78 can be applied to these harmonic functions. That is, when the domain of flow D_w in the uv plane is the image of a domain D_z under a transformation

$$w = f(z) = u(x, y) + iv(x, y)$$

where f is analytic, the functions

$$\phi[u(x, y), v(x, y)], \qquad \psi[u(x, y), v(x, y)]$$

are harmonic in D_z. These new functions may be interpreted as velocity potential and stream function in the xy plane. A streamline or natural boundary $\psi(u, v) = c_2$ in the uv plane corresponds to a streamline or natural boundary $\psi[u(x, y), v(x, y)] = c_2$ in the xy plane.

In using this technique, it is usually most efficient to first write the complex potential function for the region in the w plane and then obtain from that the velocity potential and stream function for the corresponding region in the xy plane. More precisely, if the potential function in the uv plane is

$$F(w) = \phi(u, v) + i\psi(u, v),$$

then the composite function

$$F[f(z)] = \phi[u(x, y), v(x, y)] + i\psi[u(x, y), v(x, y)]$$

is the desired complex potential in the xy plane.

To avoid an excess of notation, we use the same symbols F, ϕ, and ψ for the complex potential, etc., in both the xy and the uv planes.

Example 1. Consider a flow in the first quadrant $x > 0$, $y > 0$ which comes in downward parallel to the y axis but is forced to turn a corner near the origin, as shown

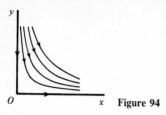

Figure 94

in Fig. 94. To determine the flow, we recall (Sec. 67) that the transformation

$$w = z^2 = x^2 - y^2 + i2xy$$

maps the first quadrant onto the upper half of the uv plane and the boundary of the quadrant onto the entire u axis.

From the example in Sec. 86, we know that the complex potential for a uniform flow to the right in the upper half of the w plane is $F = Aw$, where A is a positive real constant. The potential in the quadrant is therefore

(1) $$F = Az^2 = A(x^2 - y^2) + i2Axy,$$

and it follows that the stream function for the flow there is

(2) $$\psi = 2Axy.$$

This stream function is, of course, harmonic in the first quadrant, and it vanishes on the boundary.

The streamlines are branches of the rectangular hyperbolas

$$2Axy = c_2.$$

According to equation (3), Sec. 86, the velocity of the fluid is

$$V = \overline{2Az} = 2A(x - iy).$$

Observe that the speed

$$|V| = 2A\sqrt{x^2 + y^2}$$

of a particle is directly proportional to its distance from the origin. The value of the stream function (2) at a point (x, y) can be interpreted as the rate of flow across a line segment extending from the origin to that point.

Example 2. Let a long circular cylinder of unit radius be placed in a large body of fluid flowing with a uniform velocity, the axis of the cylinder being perpendicular to the direction of flow. To determine the steady flow around the cylinder, we represent the cylinder by the circle $x^2 + y^2 = 1$ and let the flow distant from it be parallel to the x axis and to the right (Fig. 95). Symmetry shows that the part of the x axis exterior

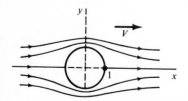

Figure 95

to the circle may be treated as a boundary; so we need to consider only the upper part of the figure as the region of flow.

The boundary of this region of flow, consisting of the upper semicircle and the parts of the x axis exterior to the circle, is mapped onto the entire u axis by the transformation

$$w = z + \frac{1}{z}.$$

The region itself is mapped onto the upper half plane $v \geqq 0$, as indicated in Fig. 17, Appendix 2. The complex potential for the corresponding uniform flow in that half plane is $F = Aw$, where A is a positive real constant. Hence the complex potential for the region exterior to the circle and above the x axis is

(3) $$F = A\left(z + \frac{1}{z}\right).$$

The velocity

(4) $$V = A\left(1 - \frac{1}{z^2}\right)$$

approaches A as $|z|$ increases. Thus the flow is nearly uniform and parallel to the x axis at points distant from the circle, as one would expect. From expression (4) we see that $V(\bar{z}) = \overline{V(z)}$; hence that expression also represents velocities of flow in the lower region, the lower semicircle being a streamline.

According to equation (3), the stream function for the given problem is, in polar coordinates,

(5) $$\psi = A\left(r - \frac{1}{r}\right)\sin\theta.$$

The streamlines

$$A\left(r - \frac{1}{r}\right)\sin\theta = c_2$$

are symmetric to the y axis and have asymptotes parallel to the x axis. Note that when $c_2 = 0$, the streamline consists of the circle $r = 1$ and the parts of the x axis exterior to the circle.

EXERCISES

1. State why the components of velocity can be obtained from the stream function by the equations

$$p(x, y) = \psi_y(x, y), \qquad q(x, y) = -\psi_x(x, y).$$

2. At an interior point of a region of flow and under the conditions that we have assumed, the fluid pressure cannot be less than the pressure at all other points in a neighborhood of that point. Justify this statement with the aid of statements in Secs. 85, 86, and 41.

3. For the flow around a corner described in Example 1, Sec. 87, at what point of the region $x \geq 0$, $y \geq 0$ is the fluid pressure greatest?

4. Show that the speed of the fluid at points on the cylindrical surface in Example 2, Sec. 87, is $2A \left| \sin \theta \right|$ and that the fluid pressure on the cylinder is greatest at the points $z = \pm 1$ and least at the points $z = \pm i$.

5. Write the complex potential for the flow around a cylinder $r = r_0$ when the velocity V at a point z approaches a real constant A as the point recedes from the cylinder.

6. Obtain the stream function $\psi = Ar^4 \sin 4\theta$ for a flow in the angular region $r \geq 0$, $0 \leq \theta \leq \pi/4$ (Fig. 96), and sketch a few of the streamlines in the interior of that region.

x **Figure 96**

7. Obtain the complex potential $F = A \sin z$ for a flow inside the semi-infinite region $-\pi/2 \leq x \leq \pi/2$, $y \geq 0$ (Fig. 97). Write the equations of the streamlines.

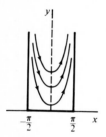

$-\dfrac{\pi}{2}$ $\dfrac{\pi}{2}$ x **Figure 97**

8. Show that if the velocity potential is $\phi = A \, \mathrm{Log} \, r \ (A > 0)$ for flow in the region $r \geq r_0$, the streamlines are the half lines $r \geq r_0$, $\theta = c$ and the rate of flow outward through each complete circle about the origin is $2\pi A$, corresponding to a source of that strength at the origin.

9. Obtain the complex potential $F = A(z^2 + z^{-2})$ for a flow in the region $r \geq 1$, $0 \leq \theta \leq \pi/2$. Write expressions for V and ψ. Note how the speed $\left| V \right|$ varies along the boundary of the region, and verify that $\psi(x, y) = 0$ on the boundary.

10. Suppose that the flow at an infinite distance from the cylinder of unit radius in Example 2, Sec. 87, is uniform in a direction making an angle α with the x axis; that is,

$$\lim_{|z| \to \infty} V = A \exp(i\alpha) \qquad\qquad (A > 0).$$

Find the complex potential.

Ans. $F = A[z \exp(-i\alpha) + z^{-1} \exp(i\alpha)]$.

11. The transformation $z = w + (1/w)$ maps the circle $\left| w \right| = 1$ onto the line segment joining the points $z = -2$ and $z = 2$, and it maps the domain outside that circle onto the rest of the z plane. [See Exercise 5(a), Sec. 12.] Write

$$z - 2 = r_1 \exp(i\theta_1), \qquad z + 2 = r_2 \exp(i\theta_2),$$

and

$$(z^2 - 4)^{1/2} = \sqrt{r_1 r_2} \exp \frac{i(\theta_1 + \theta_2)}{2} \qquad (0 \le \theta_1 < 2\pi, 0 \le \theta_2 < 2\pi);$$

the function $(z^2 - 4)^{1/2}$ is then single-valued and analytic everywhere except on the branch cut consisting of the segment of the x axis joining the points $z = \pm 2$. Show that the inverse of the transformation $z = w + (1/w)$, such that $|w| > 1$ for every point z not on the branch cut, can be written

$$w = \frac{1}{2}[z + (z^2 - 4)^{1/2}] = \frac{1}{4}\left(\sqrt{r_1} \exp \frac{i\theta_1}{2} + \sqrt{r_2} \exp \frac{i\theta_2}{2}\right)^2.$$

The transformation and that inverse establish a one to one correspondence between points in the two domains.

12. With the aid of the results found in Exercises 10 and 11, derive the expression

$$F = A[z \cos \alpha - i(z^2 - 4)^{1/2} \sin \alpha]$$

for the complex potential of the steady flow around a long plate whose width is 4 and whose cross section is the line segment joining the two points $z = \pm 2$ in Fig. 98, assuming that the velocity of the fluid at an infinite distance from the plate is $A \exp(i\alpha)$. The branch $(z^2 - 4)^{1/2}$ is the one described in Exercise 11, and $A > 0$.

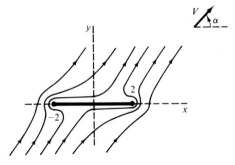

Figure 98

13. Show that if $\sin \alpha \ne 0$ in Exercise 12, the speed of the fluid along the line segment joining the points $z = \pm 2$ is infinite at the ends and is equal to $A|\cos \alpha|$ at the midpoint.

14. For the sake of simplicity, suppose that $0 < \alpha \le \pi/2$ in Exercise 12. Then show that the velocity of the fluid along the upper side of the line segment representing the plate in Fig. 98 is zero at the point $x = 2 \cos \alpha$ and that the velocity along the lower side of the segment is zero at the point $x = -2 \cos \alpha$.

15. A circle with its center at a point x_0 ($0 < x_0 < 1$) on the x axis and passing through the point $z = -1$ is subjected to the transformation $w = z + (1/z)$. Individual nonzero points $z = \exp(i\theta)$ can be mapped geometrically by adding the vector $1/z = (1/r) \exp(-i\theta)$ to the vector z. Indicate by mapping some points that the image of the circle is a profile of the type shown in Fig. 99 and that points exterior to the circle map onto points exterior to the profile. This is a special case of the profile of a *Joukowski airfoil*. (See also Exercises 16 and 17.)

16. (a) Show that the mapping of the circle in Exercise 15 is conformal except at the point $z = -1$.

(b) Let the complex numbers

$$t = \lim_{\Delta z \to 0} \frac{\Delta z}{|\Delta z|}, \qquad \tau = \lim_{\Delta z \to 0} \frac{\Delta w}{|\Delta w|}$$

represent unit vectors tangent to a smooth directed arc at $z = -1$ and that arc's image, respectively, under the transformation $w = z + (1/z)$. Show that $\tau = -t^2$ and hence that the Joukowski profile in Fig. 99 has a cusp at the point $w = -2$, the angle between the tangents at the cusp being zero.

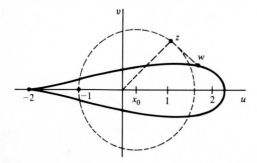

Figure 99

17. The inverse of the transformation $w = z + (1/z)$ used in Exercise 15 is given, with z and w interchanged, in Exercise 11. Find the complex potential for the flow around the airfoil in Exercise 15 when the velocity V of the fluid at an infinite distance from the origin is a real constant A.

18. Note that under the transformation

$$w = e^z + z,$$

both halves, where $x \geqq 0$ and $x \leqq 0$, of the line $y = \pi$ are mapped onto the half line $u \leqq -1$, $v = \pi$. Similarly, the line $y = -\pi$ is mapped onto the half line $u \leqq -1$, $v = -\pi$; and the strip $-\pi \leqq y \leqq \pi$ is mapped onto the w plane. Also note that the change of directions, $\arg(dw/dz)$, under this transformation approaches zero as x tends to $-\infty$. Show that the streamlines of a fluid flowing through the open channel formed by the half lines in the w plane (Fig. 100) are the images of the lines $y = c_2$ in the strip. These streamlines also represent the equipotential curves of the electrostatic field near the edge of a parallel-plate capacitor.

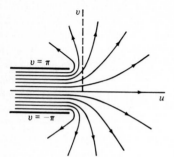

Figure 100

THE SCHWARZ-CHRISTOFFEL
TRANSFORMATION

In this chapter we construct a transformation, known as the Schwarz-Christoffel trans-formation, which maps the x axis and the upper half of the z plane onto a given simple closed polygon and its interior in the w plane. Applications are made to the solution of problems in fluid flow and electrostatic potential theory.

88. Mapping the Real Axis onto a Polygon

We represent the unit vector which is tangent to a smooth arc C at a point z_0 by the complex number t. Let the number τ denote the unit vector tangent to the image Γ of C at the corresponding point w_0 under a transformation $w = f(z)$. We assume that f is analytic at z_0 and that $f'(z_0) \neq 0$. According to Sec. 74,

$$(1) \qquad \qquad \arg \tau = \arg f'(z_0) + \arg t.$$

In particular, if C is a segment of the x axis with positive sense to the right, $t = 1$ and $\arg t = 0$ at each point $z_0 = x$ on C. Equation (1) then becomes

$$(2) \qquad \qquad \arg \tau = \arg f'(x).$$

If $f'(z)$ has a constant argument along that segment, it follows that $\arg \tau$ is constant; that is, the image Γ of C is also a segment of a straight line.

Let us now construct a transformation $w = f(z)$ that maps the whole x axis onto a polygon of n sides, where $x_1, x_2, \ldots, x_{n-1}$, and ∞ are the points on that axis whose images are to be the vertices of the polygon and where

$$x_1 < x_2 < \cdots < x_{n-1}.$$

The vertices are the points $w_j = f(x_j)$ $(j = 1, 2, \ldots, n - 1)$ and $w_n = f(\infty)$. The

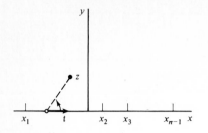
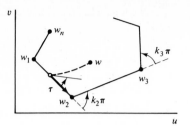

Figure 101

function f should be such that arg $f'(z)$ jumps from one constant value to another at the points $z = x_j$ as the point z traces out the x axis (Fig. 101).

If the function f is chosen such that

(3) $$f'(z) = A(z - x_1)^{-k_1}(z - x_2)^{-k_2} \cdots (z - x_{n-1})^{-k_{n-1}},$$

where A is a complex constant and each k_j is a real constant, the argument of $f'(z)$ changes in the prescribed manner as z describes the real axis; for the argument of the derivative (3) can be written

(4) $$\arg f'(z) = \arg A - k_1 \arg(z - x_1)$$
$$- k_2 \arg(z - x_2) - \cdots - k_{n-1} \arg(z - x_{n-1}).$$

When $z = x$ and $x < x_1$,

$$\arg(z - x_1) = \arg(z - x_2) = \cdots = \arg(z - x_{n-1}) = \pi.$$

When $x_1 < x < x_2$, $\arg(z - x_1) = 0$ and each of the other arguments is π. According to equation (4), then, arg $f'(z)$ increases abruptly by the angle $k_1\pi$ as z moves to the right through the point $z = x_1$. It again jumps in value, by the amount $k_2\pi$, as z passes through the point x_2, etc.

In view of equation (2), the unit vector τ is constant in direction as z moves from x_{j-1} to x_j; w then moves in that fixed direction along a straight line. The direction of τ changes abruptly, by the angle $k_j\pi$, at the image point w_j of x_j, as shown in Fig. 101. Those angles $k_j\pi$ are the exterior angles of the polygon described by the point w.

The exterior angles can be limited to angles between $-\pi$ and π; that is, $-1 < k_j < 1$. We assume that the sides of the polygon never cross one another and that the polygon is given a positive, or counterclockwise, orientation. The sum of the exterior angles of a *closed* polygon is then 2π; and the exterior angle at the vertex w_n, which is the image of the point $z = \infty$, can be written

$$k_n\pi = 2\pi - (k_1 + k_2 + \cdots + k_{n-1})\pi.$$

Thus the numbers k_j must necessarily satisfy the conditions

(5) $$k_1 + k_2 + \cdots + k_{n-1} + k_n = 2, \qquad -1 < k_j < 1 \qquad (j = 1, 2, \ldots, n).$$

Note that $k_n = 0$ if

(6) $$k_1 + k_2 + \cdots + k_{n-1} = 2.$$

In that case, the direction of τ does not change at the point w_n. So w_n is not a vertex, and the polygon has $n - 1$ sides.

The existence of a mapping function f whose derivative is given by equation (3) will now be established.

89. Schwarz-Christoffel Transformation

In our expression

$$(1) \qquad f'(z) = A(z - x_1)^{-k_1}(z - x_2)^{-k_2} \cdots (z - x_{n-1})^{-k_{n-1}}$$

for the derivative of a function that is to map the x axis onto a polygon, let the factors $(z - x_j)^{-k_j}$ represent branches of power functions with branch cuts extending below that axis. To be specific, write

$$(2) \qquad (z - x_j)^{-k_j} = |z - x_j|^{-k_j} \exp(-ik_j\theta_j) \qquad \left(-\frac{\pi}{2} < \theta_j < \frac{3\pi}{2}\right)$$

where $\theta_j = \arg(z - x_j)$ and $j = 1, 2, \ldots, n - 1$. Then $f'(z)$ is analytic everywhere in the half plane $y \geq 0$ except at the $n - 1$ branch points x_j.

If z_0 is a point in that region of analyticity, denoted here by R, then the function

$$(3) \qquad F(z) = \int_{z_0}^{z} f'(s)\, ds$$

is single-valued and analytic throughout that same region, where the path of integration from z_0 to z is any contour lying within R. Moreover, $F'(z) = f'(z)$ (see Sec. 37).

To define the function F at the point $z = x_1$ so that it is continuous there, we note that $(z - x_1)^{-k_1}$ is the only factor in expression (1) that is not analytic at x_1. Hence if $\phi(z)$ denotes the product of the rest of the factors in that expression, $\phi(z)$ is analytic at x_1 and is represented throughout an open disk $|z - x_1| < R_1$ by its Taylor series about x_1. So we can write

$$f'(z) = (z - x_1)^{-k_1} \phi(z)$$

$$= (z - x_1)^{-k_1} \left[\phi(x_1) + \frac{\phi'(x_1)}{1!}(z - x_1) + \frac{\phi''(x_1)}{2!}(z - x_1)^2 + \cdots \right],$$

or

$$(4) \qquad f'(z) = \phi(x_1)(z - x_1)^{-k_1} + (z - x_1)^{1-k_1}\psi(z)$$

where ψ is analytic, and therefore continuous, throughout the entire open disk. Since $1 - k_1 > 0$, the last term on the right in equation (4) thus represents a continuous function of z throughout the upper half of the disk, where $\text{Im } z \geq 0$, if we assign that term the value zero at $z = x_1$. It follows that the integral

$$\int_{Z_1}^{z} (s - x_1)^{1-k_1}\psi(s)\, ds$$

of the last term along a contour from Z_1 to z, where Z_1 and the contour lie in the half disk, is a continuous function of z at $z = x_1$. The integral

$$\int_{Z_1}^{z} (s - x_1)^{-k_1} ds = \frac{1}{1 - k_1} [(z - x_1)^{1-k_1} - (Z_1 - x_1)^{1-k_1}]$$

along the same path also represents a continuous function of z at x_1 if we define the value of the integral there as its limit as z approaches x_1 in the half disk. The integral of the function (4) along the stated path from Z_1 to z is therefore continuous at $z = x_1$, and so then is integral (3) since it can be written as an integral along a contour in R from z_0 to Z_1 plus the integral from Z_1 to z.

The above argument applies at each of the $n - 1$ points x_j to make F continuous throughout the region $y \geqq 0$.

From equation (1) we can show that for a sufficiently large positive number R a positive constant M exists such that if Im $z \geqq 0$,

(5) $$|f'(z)| < \frac{M}{|z|^{2-k_n}} \qquad \text{whenever} \qquad |z| > R.$$

Since $2 - k_n > 1$, this order property of the integrand in equation (3) ensures the existence of the limit of the integral there as z tends to infinity; that is, a number W_n exists such that

(6) $$\lim_{z \to \infty} F(z) = W_n \qquad\qquad (\text{Im } z \geqq 0).$$

Details of the argument are left to Exercises 10 and 11, Sec. 91.

Our mapping function, whose derivative is given by equation (1), can be written $f(z) = F(z) + B$, where B is a complex constant. The resulting transformation,

(7) $$w = A \int_{z_0}^{z} (s - x_1)^{-k_1} (s - x_2)^{-k_2} \cdots (s - x_{n-1})^{-k_{n-1}} ds + B,$$

is the *Schwarz-Christoffel transformation,* named in honor of the two German mathematicians H. A. Schwarz (1843–1921) and E. B. Christoffel (1829–1900) who discovered it independently.

Transformation (7) is continuous throughout the half plane $y \geqq 0$ and is conformal there except for the points x_j. We have assumed that the numbers k_j satisfy conditions (5), Sec. 88. In addition, we suppose that the constants x_j and k_j are such that the sides of the polygon do not cross, so that the polygon is a simple closed contour. Then, according to Sec. 88, as the point z describes the x axis in the positive direction, its image w describes the polygon P in the positive sense; and there is a one to one correspondence between points on that axis and points on P. According to condition (6), the image w_n of the point $z = \infty$ exists and $w_n = W_n + B$.

If z is an interior point of the upper half plane $y \geqq 0$ and x_0 is any point on the x axis other than one of the x_j, then the angle from the vector t at x_0 up to the line segment joining x_0 and z is positive and less than π (Fig. 101). At the image w_0 of x_0, the corresponding angle from the vector τ to the image of the line segment joining x_0 and z has that same value. Thus the images of interior points in the half plane lie to the left of the sides of the polygon taken counterclockwise. A proof that the transformation establishes a one to one correspondence between the interior points of the half plane and the points within the polygon is left to the reader (Exercise 12, Sec. 91).

Given a specific polygon P, let us examine the number of constants in the Schwarz-

Christoffel transformation that must be determined in order to map the x axis onto P. For this purpose, we may write $z_0 = 0$, $A = 1$, and $B = 0$ and simply require that the x axis be mapped onto some polygon P' similar to P. The size and position of P' can then be adjusted to match those of P by introducing the appropriate constants A and B.

The numbers k_j are all determined from the exterior angles at the vertices of P. The $n - 1$ constants x_j remain to be chosen. The image of the x axis is some polygon P' which has the same angles as P. But if P' is to be similar to P, then $n - 2$ connected sides must have a common ratio to the corresponding sides of P; this condition is expressed by means of $n - 3$ equations in the $n - 1$ real unknowns x_j. Thus *two of the numbers x_j, or two relations between them, can be chosen arbitrarily,* provided those $n - 3$ equations in the remaining $n - 3$ unknowns have real-valued solutions.

When a finite point $z = x_n$ on the x axis, instead of the point at infinity, represents the point whose image is the vertex w_n, it follows from Sec. 88 that the Schwarz-Christoffel transformation takes the form

$$(8) \qquad w = A \int_{z_0}^{z} (s - x_1)^{-k_1}(s - x_2)^{-k_2} \cdots (s - x_n)^{-k_n}\, ds + B,$$

where $k_1 + k_2 + \cdots + k_n = 2$. The exponents k_j are determined from the exterior angles of the polygon. But in this case there are n real constants x_j which must satisfy the $n - 3$ equations noted above. Thus *three of the numbers x_j, or three conditions on those n numbers, can be chosen arbitrarily* in transformation (8) of the x axis onto a given polygon.

90. Triangles and Rectangles

The Schwarz-Christoffel transformation is written in terms of the points x_j and not in terms of their images, the vertices of the polygon. No more than three of those points can be chosen arbitrarily; so when the given polygon has more than three sides, some of the points x_j must be determined in order to make the given polygon, or any polygon similar to it, be the image of the x axis. The selection of conditions for the determination of those constants, conditions that are convenient to use, often requires ingenuity.

Another limitation in using the transformation is due to the integration that is involved. Often the integral cannot be evaluated in terms of a finite number of elementary functions. In such cases the solution of problems by means of the transformation can become quite involved.

If the polygon is a triangle with vertices at the points w_1, w_2, and w_3 (Fig. 102), the transformation can be written

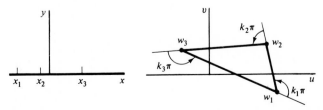

Figure 102

$$(1) \qquad w = A \int_{z_0}^{z} (s - x_1)^{-k_1}(s - x_2)^{-k_2}(s - x_3)^{-k_3}\, ds + B,$$

where $k_1 + k_2 + k_3 = 2$. In terms of the interior angles θ_j,

$$k_j = 1 - \frac{1}{\pi}\theta_j \qquad\qquad (j = 1, 2, 3).$$

Here we have taken all three points x_j as finite points on the x axis. Arbitrary values can be assigned to each of them. The complex constants A and B, which are associated with the size and position of the triangle, can be determined so that the upper half plane is mapped onto the given triangular region.

If we take the vertex w_3 as the image of the point at infinity, the transformation becomes

$$(2) \qquad w = A \int_{z_0}^{z} (s - x_1)^{-k_1}(s - x_2)^{-k_2}\, ds + B,$$

where arbitrary real values can be assigned to x_1 and x_2.

The integrals in equations (1) and (2) do not represent elementary functions unless the triangle is degenerate with one or two of its vertices at infinity. The integral in equation (2) becomes an *elliptic integral* when the triangle is equilateral or when it is a right triangle with one of its angles equal to either $\pi/3$ or $\pi/4$.

Example 1. For an equilateral triangle, $k_1 = k_2 = k_3 = 2/3$. It is convenient to write $x_1 = -1$, $x_2 = 1$, and $x_3 = \infty$ and to use equation (2), where $z_0 = 1$, $A = 1$, and $B = 0$. The transformation then becomes

$$(3) \qquad w = \int_{1}^{z} (s + 1)^{-2/3}(s - 1)^{-2/3}\, ds.$$

The image of the point $z = 1$ is clearly $w = 0$; that is, $w_2 = 0$. When $z = -1$ in the integral, we can write $s = x$, where $-1 < x < 1$. Then $x + 1 > 0$ and $\arg(x + 1) = 0$, while $|x - 1| = 1 - x$ and $\arg(x - 1) = \pi$. Hence

$$(4) \qquad w = \int_{1}^{-1} (x + 1)^{-2/3}(1 - x)^{-2/3} \exp\!\left(-\frac{2\pi i}{3}\right) dx$$

$$= \exp\!\left(\frac{\pi i}{3}\right) \int_{0}^{1} \frac{2\, dx}{(1 - x^2)^{2/3}}.$$

With the substitution $x = \sqrt{t}$, the last integral here reduces to a special case of the one used to define the beta function B (Exercise 10, Sec. 62). Let b denote its value, which is positive:

$$(5) \qquad b = \int_{0}^{1} \frac{2\, dx}{(1 - x^2)^{2/3}} = \int_{0}^{1} t^{-1/2}(1 - t)^{-2/3}\, dt = B\!\left(\frac{1}{2}, \frac{1}{3}\right).$$

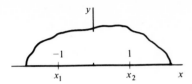 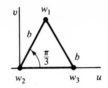

Figure 103

The vertex w_1 is therefore the point (Fig. 103)

(6) $$w_1 = b \exp \frac{\pi i}{3}.$$

The vertex w_3 is on the positive u axis because

$$w_3 = \int_1^\infty (x + 1)^{-2/3}(x - 1)^{-2/3}\, dx = \int_1^\infty \frac{dx}{(x^2 - 1)^{2/3}}.$$

But the value of w_3 is also represented by integral (3) when z tends to infinity along the negative x axis; that is,

$$w_3 = \int_1^{-1} (|x + 1||x - 1|)^{-2/3} \exp\left(-\frac{2\pi i}{3}\right) dx$$

$$+ \int_{-1}^{-\infty} (|x + 1||x - 1|)^{-2/3} \exp\left(-\frac{4\pi i}{3}\right) dx.$$

In view of the first of expressions (4) for w_1, then,

$$w_3 = w_1 + \exp\left(-\frac{4\pi i}{3}\right) \int_{-1}^{-\infty} (|x + 1||x - 1|)^{-2/3}\, dx$$

$$= b \exp \frac{\pi i}{3} + \exp\left(-\frac{\pi i}{3}\right) \int_1^\infty \frac{dx}{(x^2 - 1)^{2/3}},$$

or

$$w_3 = b \exp \frac{\pi i}{3} + w_3 \exp\left(-\frac{\pi i}{3}\right).$$

Solving for w_3, we find that

(7) $$w_3 = b.$$

We have thus verified that the image of the x axis is the equilateral triangle of side b shown in Fig. 103. We can see also that $w = (b/2) \exp(\pi i/3)$ when $z = 0$.

When the polygon is a rectangle, each $k_j = 1/2$. If we choose ± 1 and $\pm a$ as the points x_j whose images are the vertices and write

(8) $$g(z) = (z + a)^{-1/2}(z + 1)^{-1/2}(z - 1)^{-1/2}(z - a)^{-1/2},$$

where $0 \leqq \arg(z - x_j) \leqq \pi$, the Schwarz-Christoffel transformation becomes

(9)
$$w = -\int_0^z g(s)\,ds,$$

except for a transformation $W = Aw + B$ to adjust the size and position of the rectangle. Integral (9) is a constant times the elliptic integral

$$\int_0^z (1 - s^2)^{-1/2}(1 - k^2 s^2)^{-1/2}\,ds \qquad \left(k = \frac{1}{a}\right);$$

but form (8) of the integrand indicates more clearly the appropriate branches of the power functions involved.

Example 2. Let us locate the vertices of the rectangle when $a > 1$. As shown in Fig. 104, $x_1 = -a$, $x_2 = -1$, $x_3 = 1$, and $x_4 = a$. All four vertices can be described in terms of two positive numbers b and c which depend on the value of a in the following manner:

(10)
$$b = \int_0^1 |g(x)|\,dx = \int_0^1 \frac{dx}{\sqrt{(1 - x^2)(a^2 - x^2)}},$$

(11)
$$c = \int_1^a |g(x)|\,dx = \int_1^a \frac{dx}{\sqrt{(x^2 - 1)(a^2 - x^2)}}.$$

If $-1 < x < 0$, then $\arg(x + a) = \arg(x + 1) = 0$ and $\arg(x - 1) = \arg(x - a) = \pi$; hence

$$g(x) = \left[\exp\left(-\frac{\pi i}{2}\right)\right]^2 |g(x)| = -|g(x)|.$$

If $-a < x < -1$, then $g(x) = [\exp(-\pi i/2)]^3 |g(x)| = i|g(x)|$. Thus

$$w_1 = -\int_0^{-a} g(x)\,dx = -\int_0^{-1} g(x)\,dx - \int_{-1}^{-a} g(x)\,dx$$

$$= \int_0^{-1} |g(x)|\,dx - i\int_{-1}^{-a} |g(x)|\,dx = -b + ic.$$

It is left to the exercises to show that

(12)
$$w_2 = -b, \qquad w_3 = b, \qquad w_4 = b + ic.$$

The position and dimensions of the rectangle are shown in Fig. 104.

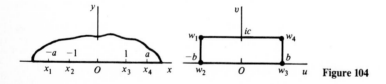

Figure 104

91. Degenerate Polygons

We now apply the Schwarz-Christoffel transformation to some degenerate polygons for which the integrals represent elementary functions. For purposes of illustration, the examples here result in transformations that we have already seen in Chap. 7.

Example 1. Let us map the half plane $y \geq 0$ onto the semi-infinite strip

$$-\frac{\pi}{2} \leq u \leq \frac{\pi}{2}, \, v \geq 0.$$

We consider the strip as the limiting form of a triangle with vertices w_1, w_2, and w_3 (Fig. 105) as the imaginary part of w_3 tends to infinity.

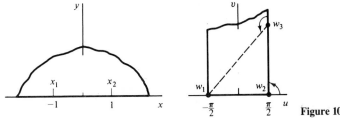

Figure 105

The limiting values of the exterior angles are

$$k_1 \pi = k_2 \pi = \frac{\pi}{2}, \qquad k_3 \pi = \pi.$$

We choose the points $x_1 = -1$, $x_2 = 1$, and $x_3 = \infty$ as the points whose images are the vertices. Then the derivative of the mapping function can be written

$$\frac{dw}{dz} = A(z + 1)^{-1/2}(z - 1)^{-1/2} = A'(1 - z^2)^{-1/2}.$$

Hence $w = A' \sin^{-1} z + B$. If we write $A' = 1/a$ and $B = b/a$, it follows that

$$z = \sin(aw - b).$$

This transformation from the w to the z plane satisfies the conditions $z = -1$ when $w = -\pi/2$ and $z = 1$ when $w = \pi/2$ if $a = 1$ and $b = 0$. The resulting transformation is

$$z = \sin w,$$

which we verified in Sec. 71 as one that maps the strip onto the half plane.

Example 2. Consider the strip $0 < v < \pi$ as the limiting form of a rhombus with vertices at the points $w_1 = \pi i$, w_2, $w_3 = 0$, and w_4 as the points w_2 and w_4 are moved

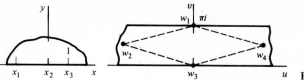

Figure 106

infinitely far to the left and right, respectively (Fig. 106). In the limit, the exterior angles become

$$k_1 \pi = 0, \qquad k_2 \pi = \pi, \qquad k_3 \pi = 0, \qquad k_4 \pi = \pi.$$

We leave x_1 to be determined and choose the values $x_2 = 0$, $x_3 = 1$, and $x_4 = \infty$. The derivative of the Schwarz-Christoffel mapping function then becomes

$$\frac{dw}{dz} = A(z - x_1)^0 z^{-1} (z - 1)^0 = \frac{A}{z};$$

thus

$$w = A \operatorname{Log} z + B.$$

Now $B = 0$ because $w = 0$ when $z = 1$. The constant A must be real because the point w lies on the real axis when $z = x$ and $x > 0$. The point $w = \pi i$ is the image of the point $z = x_1$, where x_1 is a negative number; therefore

$$\pi i = A \operatorname{Log} x_1 = A \operatorname{Log}|x_1| + A \pi i.$$

By identifying real and imaginary parts here, we see that $|x_1| = 1$ and $A = 1$. Hence the transformation becomes

$$w = \operatorname{Log} z;$$

also $x_1 = -1$. We already know from Exercise 5(b), Sec. 71, that this transformation maps the half plane onto the strip.

The procedure used in these two examples is not rigorous because limiting values of angles and coordinates were not introduced in an orderly way. Limiting values were used whenever it seemed expedient to do so. But if we verify the mapping obtained, it is not essential that we justify the steps in our derivation of the mapping function. The formal method used here is shorter and less tedious than rigorous methods.

EXERCISES

1. In transformation (1), Sec. 90, write $B = z_0 = 0$ and

$$A = \exp \frac{3\pi i}{4}, \qquad x_1 = -1, \qquad x_2 = 0, \qquad x_3 = 1,$$

$$k_1 = \frac{3}{4}, \qquad k_2 = \frac{1}{2}, \qquad k_3 = \frac{3}{4}.$$

to map the x axis onto an *isosceles right triangle*. Show that the vertices of that triangle are the points

$$w_1 = bi, \qquad w_2 = 0, \qquad \text{and} \qquad w_3 = b$$

where b is the positive constant

$$b = \int_0^1 (1 - x^2)^{-3/4} x^{-1/2}\, dx.$$

Also show that $2b = B(\tfrac{1}{4}, \tfrac{1}{4})$, where B is the beta function.

2. Obtain expressions (12) in Sec. 90 for the rest of the vertices of the rectangle shown in Fig. 104.

3. Show that when $0 < a < 1$ in equations (8) and (9), Sec. 90, the vertices of the rectangle are those shown in Fig. 104, where b and c now have the values

$$b = \int_0^a |g(x)|\, dx, \qquad c = \int_a^1 |g(x)|\, dx.$$

4. Show that the special case

$$w = i \int_0^z (s + 1)^{-1/2}(s - 1)^{-1/2} s^{-1/2}\, ds$$

of the Schwarz-Christoffel transformation (7), Sec. 89, maps the x axis onto the *square* with vertices

$$w_1 = bi, \qquad w_2 = 0, \qquad w_3 = b, \qquad w_4 = b + ib,$$

where the positive number b is given in terms of the beta function:

$$b = \frac{1}{2} B\left(\frac{1}{4}, \frac{1}{2}\right).$$

5. Use the Schwarz-Christoffel transformation to arrive at the transformation $w = z^m$ $(0 < m < 1)$, which maps the half plane $y \ge 0$ onto the angular region $|w| \ge 0$, $0 \le \arg w \le m\pi$ and the point $z = 1$ into the point $w = 1$. Consider the angular region as the limiting case of the triangular one shown in Fig. 107 as the angle α tends to 0.

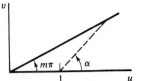

$m\pi$ $\quad$ α

1 $\qquad$ u $\qquad$ **Figure 107**

6. Refer to Fig. 26, Appendix 2. As the point z moves to the right along the negative real axis, its image point w is to move to the right along the entire u axis. As z describes the segment $0 \le x \le 1$ of the real axis, its image point w is to move to the left along the half line $u \ge 1$, $v = \pi i$; and as z moves to the right along that part of the positive real axis where $x \ge 1$, its image point w is to move to the right along the same half line $u \ge 1$, $v = \pi i$. Note the changes in direction of the motion of w at the images of the points $z = 0$ and $z = 1$. These changes suggest that the derivative of a mapping function should be

$$f'(z) = A(z - 0)^{-1}(z - 1),$$

where A is some constant; thus obtain formally the mapping function

$$w = \pi i + z - \text{Log } z,$$

which can be verified as one that maps the half plane Re $z > 0$ as indicated in the figure.

7. As the point z moves to the right along that part of the negative real axis where $x \le -1$, its image point is to move to the right along the negative real axis in the w plane. As z moves on the real axis to the right along the segment $-1 \le x \le 0$ and then along the segment $0 \le x \le 1$, its image point w is to move in the direction of increasing v along the segment $0 \le v \le 1$ of the v axis and then in the direction of decreasing v along the same segment. Finally, as z moves to the right along that part of the positive real axis where $x \ge 1$, its image point is to move to the right along the positive real axis in the w plane. Note the changes in direction of the motion of w at the images of the points $z = -1$, $z = 0$, and $z = 1$. A mapping function whose derivative is

$$f'(z) = A(z + 1)^{-1/2}(z - 0)^1(z - 1)^{-1/2},$$

where A is some constant, is thus indicated. Obtain formally the mapping function

$$w = \sqrt{z^2 - 1}$$

where $0 < \arg \sqrt{z^2 - 1} < \pi$. By considering the successive mappings $Z = z^2$, $W = Z - 1$, and $w = \sqrt{W}$, verify that the resulting transformation maps the half plane Re $z > 0$ onto the half plane Im $w > 0$ with a cut along the segment $0 < v \le 1$ of the v axis.

8. The inverse of the linear fractional transformation

$$Z = \frac{i - z}{i + z}$$

maps the unit disk $|Z| \le 1$ conformally, except at the point $Z = -1$, onto the half plane Im $z \ge 0$. (See Fig. 13, Appendix 2.) Let Z_j be points on the circle $|Z| = 1$ whose images are the points $z = x_j$ ($j = 1, 2, \ldots, n$) which are used in the Schwarz-Christoffel transformation (8), Sec. 89. Show formally, without determining the branches of the power functions, that

$$\frac{dw}{dZ} = A'(Z - Z_1)^{-k_1}(Z - Z_2)^{-k_2} \cdots (Z - Z_n)^{-k_n}$$

where A' is a constant. Thus show that *the transformation*

$$w = A' \int_0^Z (S - Z_1)^{-k_1}(S - Z_2)^{-k_2} \cdots (S - Z_n)^{-k_n} \, dS + B$$

maps the interior of the circle $|Z| = 1$ *onto the interior of a polygon,* the vertices of the polygon being the images of the points Z_j on the circle.

9. In the integral of Exercise 8, let the numbers Z_j ($j = 1, 2, \ldots, n$) be the nth roots of unity. Write $\omega = \exp(2\pi i / n)$ and $Z_1 = 1, Z_2 = \omega, \ldots, Z_n = \omega^{n-1}$. Let each of the numbers k_j ($j = 1, 2, \ldots, n$) have the value $2/n$. The integral in Exercise 8 then becomes

$$w = A' \int_0^Z \frac{dS}{(S^n - 1)^{2/n}} + B.$$

Show that when $A' = 1$ and $B = 0$, this transformation maps the interior of the unit circle $|Z| = 1$ onto the interior of a regular polygon of n sides and that the center of the polygon is the point $w = 0$.

Suggestion: The image of each of the points Z_j $(j = 1, 2, \ldots, n)$ is a vertex of some polygon with an exterior angle of $2\pi/n$ at that vertex. Write

$$w_1 = \int_0^1 \frac{dS}{(S^n - 1)^{2/n}},$$

where the path of the integration is along the positive real axis from $Z = 0$ to $Z = 1$ and the principal value of the nth root of $(S^n - 1)^2$ is to be taken. Then show that the images of the points $Z_2 = \omega, \ldots, Z_n = \omega^{n-1}$ are the points $\omega w_1, \ldots, \omega^{n-1} w_1$, respectively. Thus verify that the polygon is regular and is centered at $w = 0$.

10. Obtain inequality (5), Sec. 89.

   Suggestion: Let R be larger than any of the numbers $|x_j|$ $(j = 1, 2, \ldots, n - 1)$. Note that for R sufficiently large, the inequalities $|z|/2 < |z - x_j| < 2|z|$ hold for each x_j when $|z| > R$. Then use equation (1), Sec. 89, along with conditions (5), Sec. 88.

11. Use condition (5), Sec. 89, and sufficient conditions for the existence of improper integrals of real-valued functions to show that $F(x)$ has some limit W_n as x tends to infinity, where $F(z)$ is defined by equation (3) in that section. Also show that the integral of $f'(z)$ over each arc of a semicircle $|z| = R$, Im $z \geq 0$ approaches 0 as R tends to ∞. Then deduce that

$$\lim_{z \to \infty} F(z) = W_n \qquad\qquad (\text{Im } z \geq 0),$$

as stated in equation (6) of Sec. 89.

12. According to Exercise 14, Sec. 58, the expression

$$N = \frac{1}{2\pi i} \int_C \frac{g'(z)}{g(z)} \, dz$$

can be used to determine the number (N) of zeros of a function g interior to a positively oriented simple closed contour C when $g(z) \neq 0$ on C and when C lies in a simply connected domain D throughout which g is analytic and $g'(z)$ is never zero. In that expression, write $g(z) = f(z) - w_0$, where $f(z)$ is the Schwarz-Christoffel mapping function (7), Sec. 89, and the point w_0 is either interior to or exterior to the polygon P that is the image of the x axis; thus $f(x) \neq w_0$. Let the contour C consist of the upper half of a circle $|z| = R$ and a segment $-R < x < R$ of the x axis that contains all $n - 1$ points x_j, except that a small segment about each point x_j is replaced by the upper half of a circle $|z - x_j| = \rho_j$ with that segment as its diameter. Then the number of points z interior to C such that $f(z) = w_0$ is

$$N_C = \frac{1}{2\pi i} \int_C \frac{f'(z)}{f(z) - w_0} \, dz.$$

Note that $f(z) - w_0$ approaches the nonzero point $W_n - w_0$ when $|z| = R$ and R tends to ∞, and recall the order property (5), Sec. 89, for $|f'(z)|$. Let the ρ_j tend to zero, and prove that the number of points in the upper half of the z plane at which $f(z) = w_0$ is

$$N = \frac{1}{2\pi i} \lim_{R \to \infty} \int_{-R}^R \frac{f'(x)}{f(x) - w_0} \, dx.$$

Deduce that since

$$\int_P \frac{dw}{w - w_0} = \lim_{R \to \infty} \int_{-R}^R \frac{f'(x)}{f(x) - w_0} \, dx,$$

$N = 1$ if w_0 is interior to P and that $N = 0$ if w_0 is exterior to P. Thus show that the mapping of the half plane Im $z > 0$ onto the interior of P is one to one.

92. Fluid Flow in a Channel through a Slit

We now present a further example of the idealized steady flow treated in Chap. 9, an example that will help to show how sources and sinks can be accounted for in problems of fluid flow. In this and the following two sections, the problems are posed in the uv plane, rather than the xy plane. This allows us to refer directly to earlier results in this chapter without interchanging the planes.

Consider the two-dimensional steady flow of fluid between two parallel planes $v = 0$ and $v = \pi$ when the fluid is entering through a narrow slit along the line in the first plane which is perpendicular to the uv plane at the origin (Fig. 108). Let the rate of flow of fluid into the channel through the slit be Q units of volume per unit time for each unit of depth of the channel, where the depth is measured perpendicular to the uv plane. The rate of flow out at either end is then $Q/2$.

The transformation $w = \text{Log } z$, derived in Example 2, Sec. 91, is a one to one mapping of the upper half of the z plane onto the strip in the w plane. The inverse transformation

$$(1) \qquad z = e^w = e^u e^{iv}$$

maps the strip onto the half plane (see Example 2, Sec. 70). Under transformation (1), the image of the u axis is the positive half of the x axis, and the image of the line $v = \pi$ is the negative half of the x axis. Hence the boundary of the strip is transformed into the boundary of the half plane.

The image of the point $w = 0$ is the point $z = 1$. The image of a point $w = u_0$, where $u_0 > 0$, is a point $z = x_0$ where $x_0 > 1$. The rate of flow of fluid across a curve joining the point $w = u_0$ to a point (u,v) within the strip is a stream function $\psi(u,v)$ for the flow (Sec. 86). If u_1 is a negative real number, then the rate of flow into the channel through the slit can be written

$$\psi(u_1, 0) = Q.$$

Now, under a conformal transformation, the function ψ is transformed into a function of x and y that represents the stream function for the flow in the corresponding region of the z plane; that is, the rate of flow is the same across corresponding curves in the two planes. As in Chap. 9, the same symbol ψ is used to represent the different stream

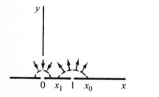

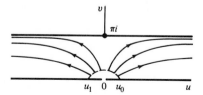

Figure 108

functions in the two planes. Since the image of the point $w = u_1$ is a point $z = x_1$ where $0 < x_1 < 1$, the rate of flow across any curve connecting the points $z = x_0$ and $z = x_1$ and lying in the upper half of the z plane is also equal to Q. Hence there is a source at the point $z = 1$ equal to the source at $w = 0$.

The above argument applies in general to show that *under a conformal transformation, a source or sink at a given point corresponds to an equal source or sink at the image of that point.*

As Re(w) tends to $-\infty$, the image of w approaches the point $z = 0$. A sink of strength $Q/2$ at the latter point corresponds to the sink infinitely far to the left in the strip. To apply the above argument in this case, we consider the rate of flow across a curve connecting the boundaries $v = 0$ and $v = \pi$ of the left-hand part of the strip and the rate of flow across the image of that curve in the z plane.

The sink at the right-hand end of the strip is transformed into a sink at infinity in the z plane.

The stream function ψ for the flow in the upper half of the z plane in this case must be a function whose values are constant along each of the three parts of the x axis. Moreover, its value must increase by Q as the point z moves around the point $z = 1$ from the position $z = x_0$ to the position $z = x_1$, and its value must decrease by $Q/2$ as z moves about the origin in the corresponding manner. We see that the function

$$\psi = \frac{Q}{\pi}\left[\operatorname{Arg}(z - 1) - \frac{1}{2}\operatorname{Arg} z\right]$$

satisfies those requirements. Furthermore, this function is harmonic in the half plane Im $z > 0$ because it is the imaginary component of the function

$$F = \frac{Q}{\pi}\left[\operatorname{Log}(z - 1) - \frac{1}{2}\operatorname{Log} z\right] = \frac{Q}{\pi}\operatorname{Log}(z^{1/2} - z^{-1/2}).$$

The function F is a complex potential function for the flow in the upper half of the z plane. Since $z = e^w$, a complex potential function $F(w)$ for the flow in the channel is

$$F(w) = \frac{Q}{\pi}\operatorname{Log}(e^{w/2} - e^{-w/2}).$$

By dropping an additive constant, we can write

(2) $$F(w) = \frac{Q}{\pi}\operatorname{Log}\left(\sinh\frac{w}{2}\right).$$

We have used the same symbol F to denote three distinct functions, once in the z plane and twice in the w plane.

The velocity vector $\overline{F'(w)}$ is given by the equation

(3) $$V = \frac{Q}{2\pi}\coth\frac{\overline{w}}{2}.$$

From this it can be seen that

$$\lim_{|u| \to \infty} V = \frac{Q}{2\pi}.$$

Also, the point $w = \pi i$ is a *stagnation point*; that is, the velocity is zero there. Hence the fluid pressure along the wall $v = \pi$ of the channel is greatest at points opposite the slit.

The stream function $\psi(u, v)$ for the channel is the imaginary component of the function $F(w)$ given by equation (2). The streamlines $\psi(u, v) = c_2$ are therefore the curves

$$\frac{Q}{\pi} \operatorname{Arg}\left(\sinh \frac{w}{2}\right) = c_2.$$

This equation reduces to

(4) $$\tan \frac{v}{2} = c \tanh \frac{u}{2},$$

where c is any real constant. Some of these streamlines are indicated in Fig. 108.

93. Flow in a Channel with an Offset

To further illustrate the use of the Schwarz-Christoffel transformation, let us find the complex potential for the flow of a fluid in a channel with an abrupt change in its breadth (Fig. 109). We take our unit of length such that the breadth of the wide part of the channel is π units; then $h\pi$, where $0 < h < 1$, represents the breadth of the narrow part. Let the real constant V_0 denote the velocity of the fluid far from the offset in the wide part; that is,

$$\lim_{u \to -\infty} V = V_0$$

where the complex variable V represents the velocity vector. The rate of flow per unit depth through the channel, or the strength of the source on the left and of the sink on the right, is then

(1) $$Q = \pi V_0.$$

The cross section of the channel can be considered as the limiting case of the quadrilaterial with the vertices w_1, w_2, w_3, and w_4 shown in Fig. 109 as the first and last of these vertices are moved infinitely far to the left and to the right, respectively. In the limit, the exterior angles become

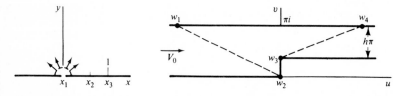

Figure 109

$$k_1\pi = \pi, \qquad k_2\pi = \frac{\pi}{2}, \qquad k_3\pi = -\frac{\pi}{2}, \qquad k_4\pi = \pi.$$

As before, we proceed formally here, using limiting values whenever it is convenient to do so. If we write $x_1 = 0$, $x_3 = 1$, $x_4 = \infty$ and leave x_2 to be determined, where $0 < x_2 < 1$, the derivative of the mapping function becomes

$$(2) \qquad \frac{dw}{dz} = Az^{-1}(z - x_2)^{-1/2}(z - 1)^{1/2}.$$

To simplify the determination of the constants A and x_2 here, we proceed at once to the use of the complex potential of the flow. The source of the flow in the channel infinitely far to the left corresponds to an equal source at $z = 0$ (Sec. 92). The entire boundary of the cross section of the channel is the image of the x axis. In view of equation (1), then, the function

$$(3) \qquad F = V_0 \operatorname{Log} z = V_0 \operatorname{Log} r + iV_0\theta$$

is the potential for the flow in the upper half of the z plane with the required source at the origin. Here the stream function is $\psi = V_0\theta$. It increases in value from 0 to $V_0\pi$ over each semicircle $z = Re^{i\theta}$ ($0 \le \theta \le \pi$), where $R > 0$, as θ varies from 0 to π. [Compare equation (5), Sec. 86, and Exercise 8, Sec. 87.]

The complex conjugate of the velocity V in the w plane can be written

$$\overline{V(w)} = \frac{dF}{dw} = \frac{dF}{dz}\frac{dz}{dw}.$$

Thus, by referring to equations (2) and (3), we can write

$$(4) \qquad \overline{V(w)} = \frac{V_0}{A}\left(\frac{z - x_2}{z - 1}\right)^{1/2}.$$

At the limiting position of the point w_1, which corresponds to $z = 0$, the velocity is the real constant V_0. It therefore follows from equation (4) that

$$V_0 = \frac{V_0}{A}\sqrt{x_2}.$$

At the limiting position of w_4, which corresponds to $z = \infty$, let the real number V_4 denote the velocity. Now it seems plausible that as a vertical line segment spanning the narrow part of the channel is moved infinitely far to the right, V approaches V_4 at each point on that segment. We could establish this conjecture as a fact by first finding w as a function of z from equation (2); but, to shorten our discussion, we assume that this is true. Then, since the flow is steady,

$$\pi hV_4 = \pi V_0 = Q,$$

or $V_4 = V_0/h$. Letting z tend to infinity in equation (4), we find that

$$\frac{V_0}{h} = \frac{V_0}{A}.$$

Thus

(5) $$A = h, \qquad x_2 = h^2,$$

and

(6) $$\overline{V(w)} = \frac{V_0}{h}\left(\frac{z - h^2}{z - 1}\right)^{1/2}.$$

From equation (6) we can see that the magnitude $|V|$ of the velocity becomes infinite at the corner w_3 of the offset since it is the image of the point $z = 1$. Also, the corner w_2 is a stagnation point, a point where $V = 0$. Along the boundary of the channel, the fluid pressure is therefore greatest at w_2 and least at w_3.

To write the relation between the potential and the variable w, we must integrate equation (2), which can now be written

(7) $$\frac{dw}{dz} = \frac{h}{z}\left(\frac{z - 1}{z - h^2}\right)^{1/2}.$$

By substituting a new variable s here, where

$$\frac{z - h^2}{z - 1} = s^2,$$

one can show that equation (7) reduces to

$$\frac{dw}{ds} = 2h\left(\frac{1}{1 - s^2} - \frac{1}{h^2 - s^2}\right).$$

Hence

(8) $$w = h \operatorname{Log}\frac{1 + s}{1 - s} - \operatorname{Log}\frac{h + s}{h - s}.$$

The constant of integration here is zero because when $z = h^2$, s is zero and so, therefore, is w.

In terms of s, the potential F of equation (3) becomes

$$F = V_0 \operatorname{Log}\frac{h^2 - s^2}{1 - s^2};$$

consequently,

(9) $$s^2 = \frac{\exp(F/V_0) - h^2}{\exp(F/V_0) - 1}.$$

By substituting s from this equation into equation (8), we get an implicit relation which defines the potential F as a function of w.

94. Electrostatic Potential about an Edge of a Conducting Plate

Two parallel conducting plates of infinite extent are kept at the electrostatic potential $V = 0$; and a parallel semi-infinite plate, placed midway between them, is kept at

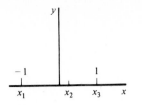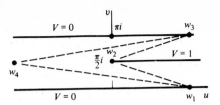

Figure 110

the potential $V = 1$. The coordinate system and the unit of length are chosen so that the plates lie in the planes $v = 0$, $v = \pi$, and $v = \pi/2$ (Fig. 110). Let us determine the potential function $V(u, v)$ in the region between those plates.

The cross section of that region in the uv plane has the limiting form of the quadrilateral bounded by the dashed lines in the figure as the points w_1 and w_3 move out to the right and w_4 to the left. In applying the Schwarz-Christoffel transformation here, we let the point x_4, corresponding to the vertex w_4, be the point at infinity. We choose the points $x_1 = -1$, $x_3 = 1$ and leave x_2 to be determined. The limiting values of the exterior angles of the quadrilateral are

$$k_1 \pi = \pi, \qquad k_2 \pi = -\pi, \qquad k_3 \pi = k_4 \pi = \pi.$$

Thus

$$\frac{dw}{dz} = A(z + 1)^{-1}(z - x_2)(z - 1)^{-1} = A\left(\frac{z - x_2}{z^2 - 1}\right) = \frac{A}{2}\left(\frac{1 + x_2}{z + 1} + \frac{1 - x_2}{z - 1}\right),$$

and so the transformation of the upper half of the z plane into the divided strip in the w plane has the form

$$(1) \qquad w = \frac{A}{2}[(1 + x_2)\, \mathrm{Log}(z + 1) + (1 - x_2)\, \mathrm{Log}(z - 1)] + B.$$

Let A_1, A_2 and B_1, B_2 denote the real and imaginary parts of the constants A and B. When $z = x$, the point w lies on the boundary of the divided strip; and, according to equation (1),

$$(2) \qquad u + iv = \frac{A_1 + iA_2}{2}\{(1 + x_2)[\mathrm{Log}|x + 1| + i\,\arg(x + 1)]$$

$$+ (1 - x_2)[\mathrm{Log}|x - 1| + i\,\arg(x - 1)]\} + B_1 + iB_2.$$

To determine the constants here, we first note that the limiting position of the line segment joining points w_1 and w_4 is the u axis. That segment is the image of the part of the x axis to the left of the point $x_1 = -1$; this is because the line segment joining w_3 and w_4 is the image of the part of the x axis to the right of $x_3 = 1$, and the other two sides of the quadrilateral are the images of the remaining two segments of the x axis. Hence when $v = 0$ and u tends to infinity through positive values, the corresponding point x approaches the point $z = -1$ from the left. Thus

$$\arg(x + 1) = \pi, \qquad \arg(x - 1) = \pi,$$

and $\mathrm{Log}|x + 1|$ tends to $-\infty$. Also, since $-1 < x_2 < 1$, the real part of the quantity inside the braces in equation (2) tends to $-\infty$. Since $v = 0$, it follows that $A_2 = 0$; otherwise, the imaginary part on the right would become infinite. By equating imaginary parts on the two sides, we now see that

$$0 = \frac{A_1}{2}[(1 + x_2)\pi + (1 - x_2)\pi] + B_2.$$

Hence

(3) $$-\pi A_1 = B_2, \qquad A_2 = 0.$$

The limiting position of the line segment joining the points w_1 and w_2 is the half line $v = \pi/2$, $u \geqq 0$. Points on that half line are images of the points $z = x$ where $-1 < x \leqq x_2$; consequently,

$$\arg(x + 1) = 0, \qquad \arg(x - 1) = \pi.$$

Identifying the imaginary parts on the two sides of equation (2) for those points, we see that

(4) $$\frac{\pi}{2} = \frac{A_1}{2}(1 - x_2)\pi + B_2.$$

Finally, the limiting positions of the points on the line segment joining w_3 to w_4 are the points $u + \pi i$, which are the images of the points x where $x > 1$. By identifying the imaginary parts in equation (2) for those points, we find that

$$\pi = B_2.$$

Then, in view of equations (3) and (4),

$$A_1 = -1, \qquad x_2 = 0.$$

Thus $x = 0$ is the point whose image is the vertex $w = \pi i/2$; and, upon substituting these values into equation (2) and identifying real parts, we see that $B_1 = 0$.

Transformation (1) now becomes

(5) $$w = -\frac{1}{2}[\text{Log}(z + 1) + \text{Log}(z - 1)] + \pi i,$$

or

(6) $$z^2 = 1 + e^{-2w}.$$

Under this transformation, the required harmonic function $V(u, v)$ becomes a harmonic function of x and y in the region $y > 0$; and the boundary conditions indicated in Fig. 111 are satisfied. Note that $x_2 = 0$ now. The harmonic function in that half plane assuming those values on the boundary is the imaginary component of the analytic function

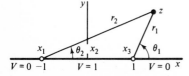

Figure 111

$$\frac{1}{\pi} \text{Log} \frac{z-1}{z+1} = \frac{1}{\pi} \text{Log} \frac{r_1}{r_2} + \frac{i}{\pi}(\theta_1 - \theta_2),$$

where θ_1 and θ_2 range from 0 to π. Writing the tangents of these angles as functions of x and y and simplifying, we find that

(7) $$\tan \pi V = \tan(\theta_1 - \theta_2) = \frac{2y}{x^2 + y^2 - 1}.$$

Equation (6) furnishes expressions for $x^2 + y^2$ and $x^2 - y^2$ in terms of u and v. Then from equation (7) we find that the relation between the potential V and the coordinates u and v can be written

(8) $$\tan \pi V = \frac{1}{s} \sqrt{e^{-4u} - s^2}$$

where

$$s = -1 + \sqrt{1 + 2e^{-2u} \cos 2v + e^{-4u}}.$$

EXERCISES

1. Use the Schwarz-Christoffel transformation to obtain formally the mapping function given with Fig. 22, Appendix 2.

2. Explain why the solution of the problem of flow in a channel with a semi-infinite rectangular obstruction (Fig. 112) is included in the solution of the problem treated in Sec. 93.

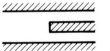

Figure 112

3. Refer to Fig. 29, Appendix 2. As the point z moves to the right along the negative part of the real axis where $x \leq -1$, its image point w is to move to the right along the half line $u \leq 0$, $v = h$. As the point z moves to the right along the segment $-1 \leq x \leq 1$ of the x axis, its image point w is to move in the direction of decreasing v along the segment $0 \leq v \leq h$ of the v axis. Finally, as z moves to the right along the positive part of the real axis where $x \geq 1$, its image point w is to move to the right along the positive real axis. Note the changes in the direction of motion of w at the images of the points $z = -1$ and $z = 1$. These changes indicate that the derivative of a mapping function might be

$$\frac{dw}{dz} = A\left(\frac{z+1}{z-1}\right)^{1/2},$$

where A is some constant. Thus obtain formally the transformation given with the figure. Verify that the transformation, written in the form

$$w = \frac{h}{\pi}\{(z+1)^{1/2}(z-1)^{1/2} + \text{Log}[z + (z+1)^{1/2}(z-1)^{1/2}]\},$$

where $0 \leq \arg(z \pm 1) \leq \pi$, maps the boundary in the manner indicated in the figure.

4. Let $T(u, v)$ denote the bounded steady-state temperatures in the shaded region of the w plane in Fig. 29, Appendix 2, with the boundary conditions $T(u, h) = 1$ when $u < 0$ and $T = 0$ on the rest $(B'C'D')$ of the boundary. In terms of the real parameter α $(0 < \alpha < \pi/2)$, show that the image of each point $z = i \tan \alpha$ on the positive y axis is the point

$$w = \frac{h}{\pi} \left[\text{Log}(\tan \alpha + \sec \alpha) + i \left(\frac{\pi}{2} + \sec \alpha \right) \right]$$

(see Exercise 3) and that the temperature at that point w is

$$T(u, v) = \frac{\alpha}{\pi} \qquad \left(0 < \alpha < \frac{\pi}{2} \right).$$

5. Let $F(w)$ denote the complex potential function for the flow of a fluid over a step in the bed of a deep stream represented by the shaded region of the w plane in Fig. 29, Appendix 2, where the fluid velocity V approaches a real constant V_0 as $|w|$ tends to infinity in that region. The transformation that maps the upper half of the z plane onto that region is noted in Exercise 3. Using the identity $dF/dw = (dF/dz)(dz/dw)$, show that

$$\overline{V(w)} = V_0(z - 1)^{1/2}(z + 1)^{-1/2};$$

and, in terms of the points $z = x$ whose images are the points along the bed of the stream, show that

$$|V| = |V_0| \sqrt{\left| \frac{x - 1}{x + 1} \right|}.$$

Thus note that the speed increases from $|V_0|$ along $A'B'$ until $|V| = \infty$ at B', then diminishes to zero at C', and increases toward $|V_0|$ from C' to D'; note that the speed is $|V_0|$ at the point

$$w = i \left(\frac{1}{2} + \frac{1}{\pi} \right) h$$

between B' and C'.

ELEVEN

INTEGRAL FORMULAS OF THE POISSON TYPE

In this chapter we develop a theory which enables us to obtain solutions to a variety of boundary value problems where those solutions are expressed in terms of definite or improper integrals. Many of the integrals occurring are then readily evaluated.

95. Poisson Integral Formula

Suppose that a function f is analytic within and on a simple closed contour C_0 which is described in the positive sense. The Cauchy integral formula

$$(1) \qquad f(z) = \frac{1}{2\pi i} \int_{C_0} \frac{f(s)}{s - z} \, ds$$

expresses the value of f at any point z interior to C_0 in terms of the values of f at points s on C_0. When C_0 is a circle, we can obtain from formula (1) a corresponding formula for a harmonic function; that is, we can solve the Dirichlet problem for the circle.

Consider the case in which C_0 is the circle $s = r_0 \exp(i\phi)$ $(0 \le \phi \le 2\pi)$, and write $z = r \exp(i\theta)$ where $0 < r < r_0$ (Fig. 113). The inverse of the nonzero point z with respect to the circle is the point z_1 lying on the same ray from the origin as z and satisfying the condition $|z_1| |z| = r_0^2$; that is,

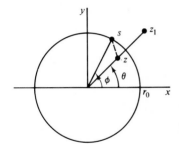

Figure 113

(2)
$$z_1 = \frac{r_0^2}{r}\exp(i\theta) = \frac{r_0^2}{\bar{z}} = \frac{s\bar{s}}{\bar{z}}.$$

Since z_1 is exterior to the circle C_0, it follows from the Cauchy-Goursat theorem that the value of the integral in equation (1) is zero when z is replaced by z_1 in the integrand. Thus, using the stated parametric representation for C_0, we can write

$$f(z) = \frac{1}{2\pi}\int_0^{2\pi}\left(\frac{s}{s-z} - \frac{s}{s-z_1}\right)f(s)\,d\phi$$

where, for convenience, we retain the s instead of writing $r_0\exp(i\phi)$.

Note that in view of the last of expressions (2) for z_1, the factor inside the parentheses here can be written

(3)
$$\frac{s}{s-z} - \frac{1}{1-(\bar{s}/\bar{z})} = \frac{s}{s-z} + \frac{\bar{z}}{\bar{s}-\bar{z}} = \frac{r_0^2 - r^2}{|s-z|^2}.$$

An alternative form of the Cauchy integral formula (1) is therefore

(4)
$$f(re^{i\theta}) = \frac{r_0^2 - r^2}{2\pi}\int_0^{2\pi}\frac{f(r_0e^{i\phi})}{|s-z|^2}\,d\phi$$

when $0 < r < r_0$. This form is also valid when $r = 0$; in that case it reduces directly to formula (1) with $z = 0$.

The quantity $|s-z|$ is the distance between the points s and z, and the law of cosines can be used to write (see Fig. 113)

(5)
$$|s-z|^2 = r_0^2 - 2r_0r\cos(\phi-\theta) + r^2 > 0.$$

Hence, if u is the real part of the analytic function f, it follows from formula (4) that

(6)
$$u(r,\theta) = \frac{1}{2\pi}\int_0^{2\pi}\frac{(r_0^2 - r^2)u(r_0,\phi)}{r_0^2 - 2r_0r\cos(\phi-\theta) + r^2}\,d\phi \qquad (r < r_0).$$

This is the *Poisson integral formula* for the harmonic function u in the open disk bounded by the circle $r = r_0$.

Formula (6) defines a linear integral transformation of $u(r_0,\phi)$ into $u(r,\theta)$. The kernel of the transformation is, except for the factor $1/(2\pi)$, the real-valued function

(7)
$$P(r_0, r, \phi - \theta) = \frac{r_0^2 - r^2}{r_0^2 - 2r_0r\cos(\phi-\theta) + r^2},$$

which is known as the *Poisson kernal*. The function $P(r_0, r, \phi - \theta)$ is also represented by expressions (3), and we see from the last of those expressions that it is always positive. Moreover, since $\bar{z}/(\bar{s}-\bar{z})$ and its complex conjugate $z/(s-z)$ have the same real parts, we find from the second of expressions (3) that

(8)
$$P(r_0, r, \phi - \theta) = \mathrm{Re}\left(\frac{s}{s-z} + \frac{z}{s-z}\right) = \mathrm{Re}\left(\frac{s+z}{s-z}\right).$$

Thus $P(r_0, r, \phi - \theta)$ is a harmonic function of r and θ interior to C_0 for each fixed s

on C_0. From equation (7) we see that $P(r_0, r, \phi - \theta)$ is an even periodic function of $\phi - \theta$, with period 2π; and its value is 1 when $r = 0$.

The Poisson integral formula (6) can now be written

$$(9) \qquad u(r, \theta) = \frac{1}{2\pi} \int_0^{2\pi} P(r_0, r, \phi - \theta) u(r_0, \phi) \, d\phi \qquad (r < r_0).$$

When $f(z) = u(r, \theta) = 1$, equation (9) shows that P has the property

$$(10) \qquad \frac{1}{2\pi} \int_0^{2\pi} P(r_0, r, \phi - \theta) \, d\phi = 1 \qquad (r < r_0).$$

We have assumed that f is analytic not only interior to C_0 but also on C_0 itself and that u is therefore harmonic in a domain which includes all points on that circle. In particular, u is continuous on C_0. The conditions will now be relaxed.

96. Dirichlet Problem for a Disk

Let F be a given piecewise continuous function of θ $(0 \leq \theta \leq 2\pi)$. We shall prove that the function

$$(1) \qquad U(r, \theta) = \frac{1}{2\pi} \int_0^{2\pi} P(r_0, r, \phi - \theta) F(\phi) \, d\phi \qquad (r < r_0),$$

which we call the Poisson integral transform of F, satisfies the following conditions: *U is harmonic inside the circle $r = r_0$ and*

$$(2) \qquad \lim_{\substack{r \to r_0 \\ r < r_0}} U(r, \theta) = F(\theta)$$

for each fixed θ where F is continuous. Thus U is a solution of the Dirichlet problem for the disk $r < r_0$ in the sense that $U(r, \theta)$ approaches the boundary value $F(\theta)$ as the point (r, θ) approaches (r_0, θ) along a radius, except at the finite number of points (r_0, θ) where the discontinuities of F may occur.

Example. Before proving the above statement, let us apply it to find the potential $V(r, \theta)$ inside a cylinder $r = 1$ where the boundary conditions are those illustrated on the left in Fig. 86 (Sec. 84), the potential being zero on one half of the surface and unity on the other half. This problem was solved by conformal mapping in that earlier section. In equation (1) write V for U, $r_0 = 1$, and $F(\phi) = 0$ when $0 < \phi < \pi$ and $F(\phi) = 1$ when $\pi < \phi < 2\pi$ to get

$$V(r, \theta) = \frac{1}{2\pi} \int_\pi^{2\pi} P(1, r, \phi - \theta) \, d\phi = \frac{1}{2\pi} \int_\pi^{2\pi} \frac{(1 - r^2) \, d\phi}{1 + r^2 - 2r \cos(\phi - \theta)}.$$

An antiderivative of $P(1, r, \psi)$ is

$$(3) \qquad \int P(1, r, \psi) \, d\psi = 2 \arctan\left(\frac{1 + r}{1 - r} \tan \frac{\psi}{2}\right)$$

since the integrand here is the derivative with respect to ψ of the function on the right. That function is assigned the value $-\pi$ when $\psi/2 = -\pi/2$ and the value π when $\psi/2 = \pi/2$ so that it can be a continuous function whose values increase from $-\pi$ to 2π as $\psi/2$ increases from $-\pi/2$ to π. This is the required range of values for $\psi/2$ since $\psi = \phi - \theta$ and the angles ϕ and θ range from π to 2π and 0 to 2π, respectively. Thus

$$\pi V(r,\theta) = \arctan\left(\frac{1+r}{1-r}\tan\frac{2\pi-\theta}{2}\right) - \arctan\left(\frac{1+r}{1-r}\tan\frac{\pi-\theta}{2}\right).$$

It is physically evident that the values of $\pi V(r,\theta)$ must range from 0 to π. By simplifying the expression for $\tan(\pi V)$ obtained from this last equation, we find that

$$(4) \qquad\qquad V(r,\theta) = \frac{1}{\pi}\arctan\left(\frac{1-r^2}{2r\sin\theta}\right) \qquad (0 \leq \arctan t \leq \pi).$$

This is the solution obtained in Sec. 84 in terms of cartesian coordinates.

We turn now to the proof that the function U defined in equation (1) satisfies the Dirichlet problem for the disk $r < r_0$, as asserted just prior to this example. First of all, U is harmonic inside the circle $r = r_0$ because P is a harmonic function of r and θ there. More precisely, since F is piecewise continuous, integral (1) can be written as the sum of a finite number of definite integrals each of which has an integrand that is continuous in r, θ, and ϕ. The partial derivatives of those integrands with respect to r and θ are likewise continuous. Since the order of integration and differentiation with respect to r and θ may then be interchanged and since P satisfies Laplace's equation in the polar coordinates r and θ, it follows that U satisfies that equation too.

To establish condition (2), we need to show that if F is continuous at θ, there corresponds to each positive number ε a positive number δ such that

$$(5) \qquad\qquad |U(r,\theta) - F(\theta)| < \varepsilon \qquad \text{whenever} \qquad 0 < r_0 - r < \delta.$$

In view of property (10), Sec. 95, the first inequality here can be written

$$(6) \qquad\qquad \left|\frac{1}{2\pi}\int_0^{2\pi} P(r_0,r,\phi-\theta)[F(\phi)-F(\theta)]\,d\phi\right| < \varepsilon.$$

We shall find it convenient to let F be extended periodically with period 2π so that the integrand is periodic in ϕ with that same period.

Since F is continuous at θ, there is a small number α corresponding to the given positive number ε such that

$$|F(\phi)-F(\theta)| < \frac{\varepsilon}{2} \qquad \text{whenever} \qquad |\phi-\theta| \leq \alpha.$$

Let us write

$$I_1(r) = \frac{1}{2\pi}\int_{\theta-\alpha}^{\theta+\alpha} P(r_0,r,\phi-\theta)[F(\phi)-F(\theta)]\,d\phi,$$

$$I_2(r) = \frac{1}{2\pi}\int_{\theta+\alpha}^{\theta-\alpha+2\pi} P(r_0,r,\phi-\theta)[F(\phi)-F(\theta)]\,d\phi.$$

Then inequality (6) can be written

(7) $$|I_1(r) + I_2(r)| < \varepsilon.$$

Since P is a positive-valued function and because of property (10), Sec. 95,

$$|I_1(r)| \leq \frac{1}{2\pi} \int_{\theta-\alpha}^{\theta+\alpha} P(r_0, r, \phi - \theta) |F(\phi) - F(\theta)| \, d\phi$$

$$< \frac{\varepsilon}{4\pi} \int_0^{2\pi} P(r_0, r, \phi - \theta) \, d\phi = \frac{\varepsilon}{2}$$

whenever $r > r_0$.

Next, recall that $P(r_0, r, \phi - \theta) = (r_0^2 - r^2)/|s - z|^2$, and observe in Fig. 113 that when $r \leq r_0$, the quantity $|s - z|^2$ has a positive minimum value $m(\alpha)$ as ϕ, the argument of s, varies between $\theta + \alpha$ and $\theta - \alpha + 2\pi$. If M denotes an upper bound of $|F(\phi) - F(\theta)|$ for all values of θ and ϕ, it follows that

$$|I_2(r)| \leq \frac{(r_0^2 - r^2)M}{2\pi m(\alpha)} 2\pi < \frac{2Mr_0}{m(\alpha)} (r_0 - r) < \frac{\varepsilon}{2}$$

whenever $r_0 - r < m(\alpha)\varepsilon/(4Mr_0)$. Consequently, if α is selected corresponding to the given ε, the number $m(\alpha)$ is determined; and $|I_1(r)| + |I_2(r)| < \varepsilon$ whenever $r_0 - r < \delta$ if

$$\delta = \frac{m(\alpha)\varepsilon}{4Mr_0}.$$

This is, then, a value of δ such that inequality (7), or (6), is satisfied. Statement (5) is therefore true when δ has this value.

According to expression (1), the value of U at $r = 0$ is

$$\frac{1}{2\pi} \int_0^{2\pi} F(\phi) \, d\phi.$$

Thus *the value of a harmonic function at the center of the circle $r = r_0$ is the average of the boundary values on the circle.*

It is left to the exercises to prove that P and U can be represented by series involving the elementary harmonic functions $r^n \cos n\theta$ and $r^n \sin n\theta$ as follows:

(8) $$P(r_0, r, \phi - \theta) = 1 + 2 \sum_{n=1}^{\infty} \left(\frac{r}{r_0}\right)^n \cos n(\phi - \theta) \qquad (r < r_0),$$

(9) $$U(r, \theta) = \frac{1}{2} a_0 + \sum_{n=1}^{\infty} \left(\frac{r}{r_0}\right)^n (a_n \cos n\theta + b_n \sin n\theta) \qquad (r < r_0),$$

where

(10) $$a_n = \frac{1}{\pi} \int_0^{2\pi} F(\phi) \cos n\phi \, d\phi, \qquad b_n = \frac{1}{\pi} \int_0^{2\pi} F(\phi) \sin n\phi \, d\phi.$$

97. Related Boundary Value Problems

Details of proofs of results given below are left to the exercises. The function F representing boundary values on the circle $r = r_0$ is assumed to be piecewise continuous.

Suppose that $F(2\pi - \theta) = -F(\theta)$. The Poisson integral formula (1) of Sec. 96 then becomes

(1) $$U(r, \theta) = \frac{1}{2\pi} \int_0^\pi [P(r_0, r, \phi - \theta) - P(r_0, r, \phi + \theta)]F(\phi)\,d\phi.$$

This function U has zero values on the horizontal radii $\theta = 0$ and $\theta = \pi$ of the circle, as one would expect when U is interpreted as a steady temperature. Formula (1) therefore solves *the Dirichlet problem for the semicircular region $r < r_0$, $0 < \theta < \pi$* (Fig. 114) *where $U = 0$ on the diameter AB and*

(2) $$\lim_{\substack{r \to r_0 \\ r < r_0}} U(r, \theta) = F(\theta) \qquad\qquad (0 < \theta < \pi)$$

for each fixed θ where F is continuous.

If $F(2\pi - \theta) = F(\theta)$, then

(3) $$U(r, \theta) = \frac{1}{2\pi} \int_0^\pi [P(r_0, r, \phi - \theta) + P(r_0, r, \phi + \theta)]F(\phi)\,d\phi;$$

and $U_\theta(r, \theta) = 0$ when $\theta = 0$ or $\theta = \pi$. Formula (3) thus furnishes a function U that is *harmonic in the semicircular region $r < r_0$, $0 < \theta < \pi$* (Fig. 114) *and satisfies condition (2) as well as the condition that its normal derivative be zero on the diameter AB.*

The analytic function $z = r_0^2/Z$ maps the circle $|Z| = r_0$ in the Z plane onto the circle $|z| = r_0$ in the z plane, and it maps the exterior of the first circle onto the interior of the second. Writing $z = r \exp(i\theta)$ and $Z = R \exp(i\psi)$, we note that $r = r_0^2/R$ and $\theta = 2\pi - \psi$. The harmonic function $U(r, \theta)$ represented by formula (1), Sec. 96, is then transformed into the function

$$U\left(\frac{r_0^2}{R}, 2\pi - \psi\right) = -\frac{1}{2\pi} \int_0^{2\pi} \frac{r_0^2 - R^2}{r_0^2 - 2r_0 R \cos(\phi + \psi) + R^2} F(\phi)\,d\phi,$$

which is harmonic in the domain $R > r_0$. Now, in general, if $u(r, \theta)$ is harmonic, then so is $u(r, -\theta)$ (see Exercise 10). Hence the function $H(R, \psi) = U(r_0^2/R, \psi - 2\pi)$, or

(4) $$H(R, \psi) = -\frac{1}{2\pi} \int_0^{2\pi} P(r_0, R, \phi - \psi)F(\phi)\,d\phi \qquad\qquad (R > r_0),$$

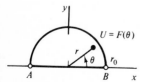

Figure 114

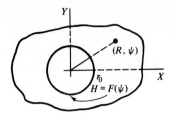

Figure 115

is also harmonic. For each fixed ψ where $F(\psi)$ is continuous, we find from condition (2), Sec. 96, that

(5)
$$\lim_{\substack{R \to r_0 \\ R > r_0}} H(R, \psi) = F(\psi).$$

Thus formula (4) solves *the Dirichlet problem for the region exterior to the circle $R = r_0$ in the Z plane* (Fig. 115). We note that the Poisson kernel is negative there; also,

(6)
$$\frac{1}{2\pi} \int_0^{2\pi} P(r_0, R, \phi - \psi) \, d\phi = -1 \qquad\qquad (R > r_0),$$

(7)
$$\lim_{R \to \infty} H(R, \psi) = \frac{1}{2\pi} \int_0^{2\pi} F(\phi) \, d\phi.$$

EXERCISES

1. Use the Poisson integral formula (1), Sec. 96, to derive the expression

$$V(x, y) = \frac{1}{\pi} \arctan\left[\frac{1 - x^2 - y^2}{(x - 1)^2 + (y - 1)^2 - 1} \right] \qquad (0 \le \arctan t \le \pi)$$

for the electrostatic potential interior to a cylinder $x^2 + y^2 = 1$ when $V = 1$ on the first quadrant $(x > 0, y > 0)$ of the cylindrical surface and $V = 0$ on the rest of that surface. Also, point out why $1 - V$ is the solution to Exercise 8, Sec. 84.

2. Let T denote the steady temperatures in a disk $r \le 1$ with insulated faces when $T = 1$ on the arc $0 < \theta < 2\theta_0$ $(0 < \theta_0 < \pi/2)$ of the edge $r = 1$ and $T = 0$ on the rest of the edge. Use the Poisson integral formula to show that

$$T(x, y) = \frac{1}{\pi} \arctan\left[\frac{(1 - x^2 - y^2)y_0}{(x - 1)^2 + (y - y_0)^2 - y_0^2} \right] \qquad (0 \le \arctan t \le \pi)$$

where $y_0 = \tan \theta_0$. Verify that this function T satisfies the boundary conditions.

3. Let I denote this *finite unit impulse function*:

$$I(h, \theta - \theta_0) = \begin{cases} \dfrac{1}{h} & \text{when } \theta_0 \le \theta \le \theta_0 + h, \\[2mm] 0 & \text{when } 0 \le \theta < \theta_0 \text{ or } \theta_0 + h < \theta \le 2\pi, \end{cases}$$

where h is a positive number and $0 \le \theta_0 < \theta_0 + h < 2\pi$. Note that

$$\int_{\theta_0}^{\theta_0+h} I(h, \theta - \theta_0) \, d\theta = 1.$$

With the aid of a mean-value theorem for definite integrals, show that

$$\int_{\theta_0}^{\theta_0+h} P(r_0, r, \phi - \theta) I(h, \phi - \theta_0) \, d\phi = P(r_0, r, c - \theta) \int_{\theta_0}^{\theta_0+h} I(h, \phi - \theta_0) \, d\phi,$$

where $\theta_0 \leq c \leq \theta_0 + h$, and hence that

$$\lim_{\substack{h \to 0 \\ h > 0}} \int_0^{2\pi} P(r_0, r, \phi - \theta) I(h, \phi - \theta_0) \, d\phi = P(r_0, r, \theta - \theta_0) \qquad (r < r_0).$$

Thus the Poisson kernel $P(r_0, r, \theta - \theta_0)$ is the limit, as h approaches 0 through positive values, of the harmonic function inside the circle $r = r_0$ whose boundary values are represented by the impulse function $2\pi I(h, \theta - \theta_0)$.

4. Show that the expression in Exercise 11, Sec. 53, for the sum of a cosine series can be written

$$1 + 2 \sum_{n=1}^{\infty} a^n \cos n\theta = \frac{1 - a^2}{1 - 2a \cos \theta + a^2} \qquad (-1 < a < 1).$$

Then show that the Poisson kernel has the series representation (8), Sec. 96.

5. Show that the series in representation (8), Sec. 96, for the Poisson kernel converges uniformly with respect to ϕ. Then obtain from formula (1) of that section the series representation (9) for $U(r, \theta)$ there.*

6. Use expressions (9) and (10) of Sec. 96 to find the steady temperatures $T(r, \theta)$ in a solid cylinder $r \leq r_0$ of infinite length if $T(r_0, \theta) = A \cos \theta$. Show that no heat flows across the plane $y = 0$.

$$Ans. \quad T = A(r/r_0) \cos \theta = Ax/r_0.$$

7. Obtain the special case

(a) $H(R, \psi) = \dfrac{1}{2\pi} \displaystyle\int_0^\pi [P(r_0, R, \phi + \psi) - P(r_0, R, \phi - \psi)] F(\phi) \, d\phi;$

(b) $H(R, \psi) = -\dfrac{1}{2\pi} \displaystyle\int_0^\pi [P(r_0, R, \phi + \psi) + P(r_0, R, \phi - \psi)] F(\phi) \, d\phi$

of formula (4), Sec. 97, for the harmonic function H in the unbounded region $R > r_0$, $0 < \psi < \pi$, shown in Fig. 116, if that function satisfies the boundary condition

$$\lim_{\substack{R \to r_0 \\ R > r_0}} H(R, \psi) = F(\psi) \qquad (0 < \psi < \pi)$$

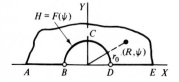

$H = F(\psi)$ C r_0 (R, ψ)

A B D E X **Figure 116**

*This result is obtained when $r_0 = 1$ by the method of separation of variables in the authors' "Fourier Series and Boundary Value Problems," 3d ed., sec. 60, 1978.

on the semicircle and (1) it is zero on the rays BA and DE; (2) its normal derivative is zero on the rays BA and DE.

8. Give the details needed in establishing formula (1) of Sec. 97 as a solution of the Dirichlet problem stated there for the region shown in Fig. 114.

9. Give the details needed in establishing formula (3) of Sec. 97 as a solution of the boundary value problem stated there.

10. Obtain formula (4), Sec. 97, as a solution of the Dirichlet problem for the region exterior to a circle (Fig. 115). To show that $u(r, -\theta)$ is harmonic when $u(r, \theta)$ is harmonic, refer to the polar form of Laplace's equation in Exercise 14 of Sec. 20.

11. State why equation (6), Sec. 97, is valid.

12. Establish the limit (7), Sec. 97.

98. Integral Formulas for a Half Plane

Let f be an analytic function of z throughout the half plane $\operatorname{Im} z \geq 0$ such that, for some positive constants a and M, f satisfies the order property

$$(1) \qquad\qquad |z^a f(z)| < M \qquad\qquad (\operatorname{Im} z \geq 0).$$

For a fixed point z above the real axis, let C_R denote the upper half of a positively oriented circle of radius R centered at the origin, where $R > |z|$ (Fig. 117). Then according to the Cauchy integral formula,

$$(2) \qquad\qquad f(z) = \frac{1}{2\pi i} \int_{C_R} \frac{f(s)\,ds}{s - z} + \frac{1}{2\pi i} \int_{-R}^{R} \frac{f(t)\,dt}{t - z}.$$

We find that the first of these integrals approaches 0 as R tends to ∞ because $|f(s)| < M/R^a$; consequently,

$$(3) \qquad\qquad f(z) = \frac{1}{2\pi i} \int_{-\infty}^{\infty} \frac{f(t)}{t - z}\,dt \qquad\qquad (\operatorname{Im} z > 0).$$

Because of condition (1), the improper integral here converges; and the number to which it converges is the same as its Cauchy principal value (see Sec. 59). Representation (3) is a *Cauchy integral formula for the half plane* $\operatorname{Im} z > 0$.

When the point z lies below the real axis, the right-hand side of equation (2) is zero; hence integral (3) is zero for such a point. Thus, when z is above the real axis, we have the following formula where c is an arbitrary complex constant:

$$(4) \qquad\qquad f(z) = \frac{1}{2\pi i} \int_{-\infty}^{\infty} \left(\frac{1}{t - z} + \frac{c}{t - \bar{z}} \right) f(t)\,dt \qquad\qquad (\operatorname{Im} z > 0).$$

In the two cases $c = -1$ and $c = 1$ this reduces, respectively, to

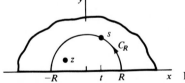

Figure 117

(5) $$f(z) = \frac{1}{\pi} \int_{-\infty}^{\infty} \frac{yf(t)}{|t - z|^2} dt \qquad (y > 0),$$

(6) $$f(z) = \frac{1}{\pi i} \int_{-\infty}^{\infty} \frac{(t - x)f(t)}{|t - z|^2} dt \qquad (y > 0).$$

If $f(z) = u(x, y) + iv(x, y)$, it follows from formulas (5) and (6) that the harmonic functions u and v are represented in the half plane $y > 0$ in terms of the boundary values of u by the formulas

(7) $$u(x, y) = \frac{1}{\pi} \int_{-\infty}^{\infty} \frac{yu(t, 0)}{|t - z|^2} dt = \frac{1}{\pi} \int_{-\infty}^{\infty} \frac{yu(t, 0)}{(t - x)^2 + y^2} dt \qquad (y > 0),$$

(8) $$v(x, y) = \frac{1}{\pi} \int_{-\infty}^{\infty} \frac{(x - t)u(t, 0)}{(t - x)^2 + y^2} dt \qquad (y > 0).$$

Formula (7) is known as the *Poisson integral formula for the half plane,* or the *Schwarz integral formula.* In the next section we relax the conditions for the validity of formulas (7) and (8).

99. Dirichlet Problem for a Half Plane

Let F denote a real-valued function of x that is bounded for all x and continuous except for at most a finite number of finite jumps. When $y \geq \varepsilon$ and $|x| \leq 1/\varepsilon$, where ε is any positive constant, the integral

$$I(x, y) = \int_{-\infty}^{\infty} \frac{F(t)\, dt}{(t - x)^2 + y^2}$$

converges uniformly with respect to x and y, as do the integrals of the partial derivatives of the integrand with respect to x and y. Each of these integrals is the sum of a finite number of improper or definite integrals over intervals where F is continuous; hence the integrand of each component integral is a continuous function of t, x, and y when $y \geq \varepsilon$. Consequently, each partial derivative of $I(x, y)$ is represented by the integral of the corresponding derivative of the integrand whenever $y > 0$.

We write $U(x, y) = yI(x, y)/\pi$. Thus U is the Schwarz integral transform of F, suggested by the second of expressions (7), Sec. 98:

(1) $$U(x, y) = \frac{1}{\pi} \int_{-\infty}^{\infty} \frac{yF(t)}{(t - x)^2 + y^2} dt \qquad (y > 0).$$

Except for the factor $1/\pi$, the kernel here is $y/|t - z|^2$. It is the imaginary component of the function $1/(t - z)$, which is analytic in z when $y > 0$. It follows that the kernel is harmonic, and so it satisfies Laplace's equation in x and y. Because the order of differentiation and integration can be interchanged, the function (1) then satisfies that equation. Consequently, *U is harmonic when $y > 0$.*

To prove that

(2)
$$\lim_{\substack{y \to 0 \\ y > 0}} U(x, y) = F(x)$$

for each fixed x at which F is continuous, we substitute $t = x + y \tan \tau$ in formula (1) and write

(3)
$$U(x, y) = \frac{1}{\pi} \int_{-\pi/2}^{\pi/2} F(x + y \tan \tau) \, d\tau \qquad (y > 0).$$

If
$$G(x, y, \tau) = F(x + y \tan \tau) - F(x)$$

and α is some small positive constant, then

(4)
$$\pi[U(x, y) - F(x)] = \int_{-\pi/2}^{\pi/2} G(x, y, \tau) \, d\tau = I_1(y) + I_2(y) + I_3(y)$$

where

$$I_1(y) = \int_{-\pi/2}^{(-\pi/2)+\alpha} G(x, y, \tau) \, d\tau, \qquad I_2(y) = \int_{(-\pi/2)+\alpha}^{(\pi/2)-\alpha} G(x, y, \tau) \, d\tau,$$

$$I_3(y) = \int_{(\pi/2)-\alpha}^{\pi/2} G(x, y, \tau) \, d\tau.$$

If M denotes an upper bound for $|F(x)|$, then $|G(x, y, \tau)| \leq 2M$. For a given positive number ε we select α so that $6M\alpha < \varepsilon$; then

$$|I_1(y)| \leq 2M\alpha < \frac{\varepsilon}{3} \qquad \text{and} \qquad |I_3(y)| \leq 2M\alpha < \frac{\varepsilon}{3}.$$

We next show that corresponding to ε there is a positive number δ such that

$$|I_2(y)| < \frac{\varepsilon}{3} \qquad \text{whenever} \qquad 0 < y < \delta.$$

Since F is continuous at x, there is a positive number γ such that

$$|G(x, y, \tau)| < \frac{\varepsilon}{3\pi} \qquad \text{whenever} \qquad 0 < y|\tan \tau| < \gamma.$$

Note that the maximum value of $|\tan \tau|$ as τ ranges from $(-\pi/2) + \alpha$ to $(\pi/2) - \alpha$ is $\tan[(\pi/2) - \alpha] = \cot \alpha$. Hence if we write $\delta = \gamma \tan \alpha$, it follows that

$$|I_2(y)| < \frac{\varepsilon}{3\pi} (\pi - 2\alpha) < \frac{\varepsilon}{3} \qquad \text{whenever} \qquad 0 < y < \delta.$$

We have thus shown that

$$|I_1(y)| + |I_2(y)| + |I_3(y)| < \varepsilon \qquad \text{whenever} \qquad 0 < y < \delta.$$

Condition (2) now follows from this result and equation (4).

Formula (1) therefore solves *the Dirichlet problem for the half plane y > 0* with the boundary condition (2). It is evident from form (3) of formula (1) that $|U(x, y)| \leq M$ in the half plane, where M is an upper bound of $|F(x)|$; that is, U is bounded. We note that $U(x, y) = F_0$ when $F(x) = F_0$, where F_0 is a constant.

According to formula (8) of Sec. 98, under certain conditions on F the function

$$(5) \qquad V(x, y) = \frac{1}{\pi} \int_{-\infty}^{\infty} \frac{(x - t)F(t)}{(t - x)^2 + y^2}\, dt \qquad (y > 0)$$

is a harmonic conjugate of the function U given by formula (1). Actually, *formula (5) furnishes a harmonic conjugate of U if F is everywhere continuous, except for at most a finite number of finite jumps, and if F satisfies an order property* $|x^a F(x)| < M$, *where a > 0.* For under those conditions we find that U and V satisfy the Cauchy-Riemann equations when $y > 0$.

Special cases of formula (1) when F is an odd or an even function are left to the exercises.

100. Neumann Problem for a Disk

As in Sec. 95 and Fig. 113, we write $s = r_0 \exp(i\phi)$ and $z = r \exp(i\theta)$, where $r < r_0$. When s is fixed, the function

$$(1) \quad Q(r_0, r, \phi - \theta) = -2r_0 \operatorname{Log}|s - z| = -r_0 \operatorname{Log}[r_0^2 - 2r_0 r \cos(\phi - \theta) + r^2]$$

is harmonic interior to the circle $|z| = r_0$ because it is the real component of $-2r_0 \log(z - s)$ where the branch cut of $\log(z - s)$ is an outward ray from the point s. If, moreover, $r \neq 0$,

$$(2) \qquad Q_r(r_0, r, \phi - \theta) = -\frac{r_0}{r}\left[\frac{2r^2 - 2r_0 r \cos(\phi - \theta)}{r_0^2 - 2r_0 r \cos(\phi - \theta) + r^2}\right]$$

$$= \frac{r_0}{r}[P(r_0, r, \phi - \theta) - 1]$$

where P is the Poisson kernel (7) of Sec. 95.

These observations suggest that the function Q may be used to write an integral representation for a harmonic function U whose normal derivative U_r on the circle $r = r_0$ assumes prescribed values $G(\theta)$.

If G is piecewise continuous and U_0 is an arbitrary constant, the function

$$(3) \qquad U(r, \theta) = \frac{1}{2\pi} \int_0^{2\pi} Q(r_0, r, \phi - \theta)G(\phi)\, d\phi + U_0 \qquad (r < r_0)$$

is harmonic because the integrand is a harmonic function of r and θ. If the mean value of G over the circle $|z| = r_0$ is zero, or

$$(4) \qquad \int_0^{2\pi} G(\phi)\, d\phi = 0,$$

then, in view of equation (2),

$$U_r(r, \theta) = \frac{1}{2\pi} \int_0^{2\pi} \frac{r_0}{r} [P(r_0, r, \phi - \theta) - 1] G(\phi) \, d\phi$$

$$= \frac{r_0}{r} \frac{1}{2\pi} \int_0^{2\pi} P(r_0, r, \phi - \theta) G(\phi) \, d\phi.$$

Now according to equations (1) and (2) of Sec. 96,

$$\lim_{\substack{r \to r_0 \\ r < r_0}} \frac{1}{2\pi} \int_0^{2\pi} P(r_0, r, \phi - \theta) G(\phi) \, d\phi = G(\theta).$$

Hence

(5) $$\lim_{\substack{r \to r_0 \\ r < r_0}} U_r(r, \theta) = G(\theta)$$

for each value of θ where G is continuous.

Since the value of Q is constant when $r = 0$, it follows from equations (3) and (4) that U_0 is the value of U at the center of the circle.

When G is piecewise continuous and satisfies condition (4), the formula

(6) $$U(r, \theta) = -\frac{r_0}{2\pi} \int_0^{2\pi} \text{Log}[r_0^2 - 2r_0 r \cos(\phi - \theta) + r^2] G(\phi) \, d\phi + U_0$$

$$(r < r_0)$$

therefore solves *the Neumann problem for the region interior to the circle* $r = r_0$, where $G(\theta)$ is the normal derivative of the harmonic function $U(r, \theta)$ at the boundary in the sense of condition (5).

The values $U(r, \theta)$ may represent steady temperatures in a disk $r < r_0$ with insulated faces. In that case, condition (5) states that the flux of heat into the disk through its edge is proportional to $G(\theta)$. Condition (4) is the natural physical requirement that the total rate of flow of heat into the disk be zero since temperatures do not vary with time.

A corresponding formula for a harmonic function H in the domain *exterior* to the circle $r = r_0$ can be written in terms of Q as

(7) $$H(R, \psi) = -\frac{1}{2\pi} \int_0^{2\pi} Q(r_0, R, \phi - \psi) G(\phi) \, d\phi + H_0 \qquad (R > r_0),$$

where H_0 is a constant. As before, we assume that G is piecewise continuous and that

(8) $$\int_0^{2\pi} G(\phi) \, d\phi = 0.$$

Then

$$H_0 = \lim_{R \to \infty} H(R, \psi)$$

and

(9)
$$\lim_{\substack{R \to r_0 \\ R > r_0}} H_R(R, \psi) = G(\psi)$$

for each ψ where G is continuous.

The verification of formula (7), as well as special cases of formula (3) that apply to semicircular regions, is left to the exercises.

101. Neumann Problem for a Half Plane

Let $G(x)$ be continuous for all real x, except for at most a finite number of finite jumps, and let it satisfy an order property

(1)
$$|x^a G(x)| < M \qquad\qquad (-\infty < x < \infty),$$

where $a > 1$. For each fixed real number t, the function $\text{Log}|z - t|$ is harmonic in the half plane $\text{Im } z > 0$. Consequently, the function

(2)
$$U(x, y) = \frac{1}{\pi} \int_{-\infty}^{\infty} \text{Log}|z - t| G(t) \, dt + U_0$$

$$= \frac{1}{2\pi} \int_{-\infty}^{\infty} \text{Log}[(t - x)^2 + y^2] G(t) \, dt + U_0 \qquad (y > 0),$$

where U_0 is a real constant, is harmonic in that half plane.

Formula (2) was written with the Schwarz formula (1) of Sec. 99 in mind; for it follows from formula (2) that

(3)
$$U_y(x, y) = \frac{1}{\pi} \int_{-\infty}^{\infty} \frac{y G(t)}{(t - x)^2 + y^2} \, dt \qquad\qquad (y > 0).$$

In view of equations (1) and (2) of Sec. 99, then,

(4)
$$\lim_{\substack{y \to 0 \\ y > 0}} U_y(x, y) = G(x)$$

at each point x where G is continuous.

Integral formula (2) therefore solves *the Neumann problem for the half plane $y > 0$* with boundary condition (4). But we have not presented conditions on G that are sufficient to ensure that the harmonic function U is bounded as $|z|$ increases.

When G is an odd function, formula (2) can be written

(5)
$$U(x, y) = \frac{1}{2\pi} \int_{0}^{\infty} \text{Log}\left[\frac{(t - x)^2 + y^2}{(t + x)^2 + y^2}\right] G(t) \, dt \quad (x > 0, y > 0).$$

This represents a function which is harmonic in the *first quadrant $x > 0$, $y > 0$* and which satisfies the boundary conditions

(6)
$$U(0, y) = 0 \qquad\qquad (y > 0),$$

(7)
$$\lim_{\substack{y \to 0 \\ y > 0}} U_y(x, y) = G(x) \qquad\qquad (x > 0).$$

The kernels of all the integral formulas for harmonic functions presented in this chapter can be described in terms of a single real-valued function of the complex variables $z = x + iy$ and $w = u + iv$:

(8) $$K(z, w) = \text{Log}|z - w| \qquad (z \neq w).$$

This is *Green's function* for the *logarithmic potential* in the z plane. The function is symmetric; that is, $K(w, z) = K(z, w)$. Expressions for kernels used earlier, in terms of K and its derivatives, are given in the exercises.

EXERCISES

1. Obtain as a special case of formula (1), Sec. 99, the formula

$$U(x, y) = \frac{y}{\pi} \int_0^\infty \left[\frac{1}{(t - x)^2 + y^2} - \frac{1}{(t + x)^2 + y^2} \right] F(t)\, dt \qquad (x > 0, y > 0)$$

for a bounded function U which is harmonic in the *first quadrant* and which satisfies the boundary conditions

$$U(0, y) = 0 \qquad\qquad (y > 0),$$
$$\lim_{\substack{y \to 0 \\ y > 0}} U(x, y) = F(x) \qquad\qquad (x > 0, x \neq x_j),$$

where F is bounded for all positive x and continuous except for at most a finite number of finite jumps at the points x_j $(j = 1, 2, \dots, n)$.

2. Obtain as a special case of formula (1), Sec. 99, the formula

$$U(x, y) = \frac{y}{\pi} \int_0^\infty \left[\frac{1}{(t - x)^2 + y^2} + \frac{1}{(t + x)^2 + y^2} \right] F(t)\, dt \qquad (x > 0, y > 0)$$

for a bounded function U which is harmonic in the *first quadrant* and which satisfies the boundary conditions

$$U_x(0, y) = 0 \qquad\qquad (y > 0),$$
$$\lim_{\substack{y \to 0 \\ y > 0}} U(x, y) = F(x) \qquad\qquad (x > 0, x \neq x_j),$$

where F is bounded for all positive x and continuous except possibly for finite jumps at a finite number of points $x = x_j$ $(j = 1, 2, \dots, n)$.

3. Interchange the x and y axes in Sec. 99 to write the solution

$$U(x, y) = \frac{1}{\pi} \int_{-\infty}^\infty \frac{xF(t)}{(t - y)^2 + x^2}\, dt \qquad (x > 0)$$

of the Dirichlet problem for the half plane $x > 0$. Write

$$\lim_{\substack{x \to 0 \\ x > 0}} U(x, y) = \begin{cases} 1 & \text{when } -1 < y < 1, \\ 0 & \text{when } |y| > 1. \end{cases}$$

Then obtain these formulas for U and its harmonic conjugate $-V$:

$$U(x, y) = \frac{1}{\pi}\left(\arctan\frac{y + 1}{x} - \arctan\frac{y - 1}{x}\right),$$

$$V(x, y) = \frac{1}{2\pi}\text{Log}\frac{x^2 + (y + 1)^2}{x^2 + (y - 1)^2},$$

where $-\pi/2 \leq \arctan t \leq \pi/2$. Also, show that

$$V(x, y) + iU(x, y) = \frac{1}{\pi}[\text{Log}(z + i) - \text{Log}(z - i)],$$

where $z = x + iy$.

4. Let $T(x, y)$ denote the bounded steady temperatures in a plate $x > 0, y > 0$ with insulated faces when

$$\lim_{\substack{y \to 0 \\ y > 0}} T(x, y) = F_1(x) \qquad (x > 0)$$

and

$$\lim_{\substack{x \to 0 \\ x > 0}} T(x, y) = F_2(y) \qquad (y > 0)$$

(Fig. 118). Here F_1 and F_2 are bounded and continuous except for at most a finite number

$T = F_2(y)$

$T = F_1(x)$ x **Figure 118**

of finite jumps. Write $x + iy = z$ and show with the aid of the formula in Exercise 1 that

$$T(x, y) = T_1(x, y) + T_2(x, y) \qquad (x > 0, y > 0)$$

where

$$T_1(x, y) = \frac{y}{\pi}\int_0^\infty \left(\frac{1}{|t - z|^2} - \frac{1}{|t + z|^2}\right)F_1(t)\,dt,$$

$$T_2(x, y) = \frac{x}{\pi}\int_0^\infty \left(\frac{1}{|it - z|^2} - \frac{1}{|it + z|^2}\right)F_2(t)\,dt.$$

5. Establish formula (7), Sec. 100, as a solution of the Neumann problem for the region exterior to a circle $r = r_0$, using earlier results found in that section.

6. Obtain as a special case of formula (3), Sec. 100, the formula

$$U(r, \theta) = \frac{1}{2\pi}\int_0^\pi [Q(r_0, r, \phi - \theta) - Q(r_0, r, \phi + \theta)]G(\phi)\,d\phi$$

for a function U which is harmonic in the *semicircular region* $r < r_0$, $0 < \theta < \pi$ and which satisfies the boundary conditions

$$U(r, 0) = U(r, \pi) = 0 \qquad (r < r_0)$$

and

$$\lim_{\substack{r\to r_0 \\ r<r_0}} U_r(r, \theta) = G(\theta) \qquad (0 < \theta < \pi)$$

for each θ where G is continuous.

7. Obtain as a special case of formula (3), Sec. 100, the formula

$$U(r, \theta) = \frac{1}{2\pi} \int_0^\pi [Q(r_0, r, \phi - \theta) + Q(r_0, r, \phi + \theta)]G(\phi)\,d\phi + U_0$$

for a function U which is harmonic in the *semicircular region* $r < r_0$, $0 < \theta < \pi$ and which satisfies the boundary conditions

$$U_\theta(r, 0) = U_\theta(r, \pi) = 0 \qquad (r < r_0)$$

and

$$\lim_{\substack{r\to r_0 \\ r<r_0}} U_r(r, \theta) = G(\theta) \qquad (0 < \theta < \pi)$$

for each θ where G is continuous, provided that

$$\int_0^\pi G(\phi)\,d\phi = 0.$$

8. Let $T(x, y)$ denote the steady temperatures in a plate $x \geq 0$, $y \geq 0$. The faces of the plate are insulated, and $T = 0$ on the edge $x = 0$. The flux of heat into the plate along the segment $0 < x < 1$ of the edge $y = 0$ is a constant A, and the rest of that edge is insulated. Use formula (5), Sec. 101, to show that the flux out of the plate along the edge $x = 0$ is

$$\frac{A}{\pi} \text{Log}\left(1 + \frac{1}{y^2}\right).$$

9. Show that the Poisson kernel is given in terms of Green's function

$$K(z, w) = \text{Log}|z - w| = \frac{1}{2} \text{Log}[\rho^2 - 2\rho r \cos(\phi - \theta) + r^2],$$

where $z = r\exp(i\theta)$ and $w = \rho\exp(i\phi)$, by the equation

$$P(\rho, r, \phi - \theta) = 2\rho\frac{\partial K}{\partial\rho} - 1.$$

10. Show that the kernel used in the Schwarz integral transform (Sec. 99) can be written in terms of Green's function

$$K(z, w) = \text{Log}|z - w| = \frac{1}{2} \text{Log}[(x - u)^2 + (y - v)^2],$$

where $z = x + iy$ and $w = u + iv$, as

$$\frac{y}{|u - z|^2} = \frac{\partial K}{\partial y}\bigg]_{v=0} = -\frac{\partial K}{\partial v}\bigg]_{v=0}.$$

Here K is to be interpreted as a function of the four real variables x, y, u, and v.

TWELVE

FURTHER THEORY OF FUNCTIONS

Many topics in the theory of functions which were not essential to the continuity of the presentation in the earlier chapters were omitted there. Several such topics do, however, deserve a place in an introductory course because of their general interest, and we include them in this chapter.

A. ANALYTIC CONTINUATION

We first consider how the behavior of an analytic function in a domain is determined by its behavior in a smaller set contained in that domain. We then consider the problem of extending the domain of definition of an analytic function.

102. Conditions under which $f(z) \equiv 0$

From Exercise 10, Sec. 15, we know that if a function f is analytic at a point z_0 and $f(z_0) \neq 0$, then there is a neighborhood of that point throughout which $f(z) \neq 0$. Also, in Sec. 53 we proved that the zeros of an analytic function are isolated unless the function is identically zero. Thus, when a function f is analytic at a point z_0, there is a neighborhood $|z - z_0| < \varepsilon$ such that either $f(z) \equiv 0$ throughout that neighborhood or else f has no zeros in it except possibly at the point z_0 itself.

Suppose now that z_0 is an accumulation point (Sec. 8) of an infinite set and that $f(z) = 0$ at each point z of that set. Then every neighborhood of z_0 contains a zero of f other than the point z_0 itself; and if f is analytic at z_0, it follows that there is some neighborhood of z_0 throughout which $f(z) \equiv 0$. All the coefficients $f^{(n)}(z_0)/n!$ $(n = 0, 1, 2, \ldots)$ in the Taylor series for $f(z)$ about z_0 are therefore zero. Hence if this function f is analytic interior to a circle $|z - z_0| = R_0$, we may conclude that $f(z) \equiv 0$ in the open disk $|z - z_0| < R_0$.

In particular, if $f(z) = 0$ at each point z in some *domain* containing z_0, or at each point on some *arc* containing z_0, and if f is analytic in an open disk $|z - z_0| < R_0$, then $f(z)$ is identically zero throughout that disk. This result allows us to prove the following important theorem.

Theorem. *If a function f is analytic throughout a domain D and f(z) = 0 at each point z of a domain or an arc interior to D, then f(z) = 0 at each point of D.*

We first prove the theorem for the case in which $f(z) = 0$ at each point z of a domain D_0 interior to D. Let s_0 be any point in D_0, and let σ be any point which lies in D but not in D_0. Since a domain is connected, there exists a polygonal path C, consisting of a finite number of line segments joined end to end, which lies in D and joins s_0 to σ (Fig. 119).

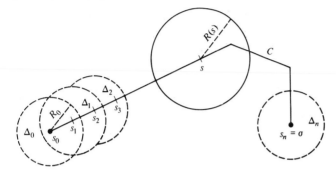

Figure 119

Now the analytic function f has a Taylor series expansion about each point s of C, the radius of the circle of convergence being some positive number $R(s)$. We agree, however, to write $R(s) = 1$ whenever that radius is greater than unity; hence $0 < R(s) \leqq 1$. A circle $|z - s| = R(s)$ may, of course, extend beyond D.

We shall need the fact that $R(s)$ is a continuous function of s. To show this, let s be any point on C and let $s + \Delta s$ be a point on C close enough to s such that $|\Delta s| < R(s)$. It follows that f is analytic in the open disk

$$|z - (s + \Delta s)| < R(s) - |\Delta s|,$$

which is centered at $s + \Delta s$ (Fig. 120). But f may actually be analytic in a larger open disk centered there. Hence $R(s + \Delta s) \geqq R(s) - |\Delta s|$, or

$$(1) \qquad\qquad -[R(s + \Delta s) - R(s)] \leqq |\Delta s|.$$

If $R(s + \Delta s) \leqq R(s)$, inequality (1) can be written

$$(2) \qquad\qquad |R(s + \Delta s) - R(s)| \leqq |\Delta s|.$$

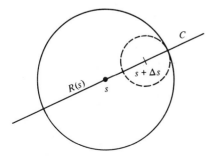

Figure 120

Suppose, on the other hand, that $R(s + \Delta s) > R(s)$. Observe that if z is a point lying in the disk

(3)
$$|z - s| < R(s + \Delta s) - |\Delta s|,$$

then

$$|z - (s + \Delta s)| \leq |z - s| + |\Delta s| < R(s + \Delta s).$$

The function f is therefore analytic at z because that point lies within the circle of convergence about $s + \Delta s$. Consequently, the disk (3) is contained in the disk $|z - s| < R(s)$. That is, $R(s + \Delta s) - |\Delta s| \leq R(s)$; and inequality (2) is again satisfied.

With inequality (2), we see that $|R(s + \Delta s) - R(s)|$ is less than any positive number ε when $|\Delta s|$ is less than both ε and $R(s)$. That is,

$$\lim_{\Delta s \to 0} R(s + \Delta s) = R(s);$$

and the continuity of $R(s)$ is established.

When the polygonal path, or arc, C is given a parametric representation $z = z(t)$ $(a \leq t \leq b)$, the function $R(s)$ can be considered as a real-valued function $R[z(t)]$ of a real variable which is continuous and positive on a closed bounded interval. The function $R(s)$ has, therefore, a positive minimum value R_0. Thus the function f is analytic in the disk $|z - s_0| < R_0$, denoted by Δ_0 in Fig. 119. Since $f(z) = 0$ at each point in the domain D_0 which contains s_0, it follows from the remarks preceding the statement of the theorem that $f(z) = 0$ at each point z in the disk Δ_0.

Let $s_0, s_1, s_2, \ldots, s_n = \sigma$ be a finite sequence of points on C such that

$$\frac{1}{2} R_0 \leq |s_k - s_{k-1}| < R_0 \qquad (k = 1, 2, \ldots, n).$$

As illustrated in Fig. 119, there is about each point s_k an open disk Δ_k of radius R_0 throughout which f is analytic. Because the center s_1 of Δ_1 lies in the domain Δ_0 where $f(z)$ is always zero, $f(z) = 0$ throughout Δ_1. Likewise, the center of Δ_2 lies in the domain Δ_1, and so $f(z) = 0$ throughout Δ_2. Continuing in this manner, we eventually reach Δ_n and find that $f(\sigma) = 0$. The proof of the theorem is thus complete for the case in which $f(z) = 0$ at each point of a domain D_0 interior to D.

Suppose now that $f(z) = 0$ along an arc in D. Then about any point on the arc there is an open disk, or domain, interior to D; and, by the remarks preceding the statement of the theorem, $f(z) = 0$ throughout that disk. Hence, in view of the case just completed, we can conclude that $f(z) = 0$ at each point of D.

103. Permanence of Forms of Functional Identities

Suppose that two functions f and g are analytic in the same domain D and that $f(z) = g(z)$ at each point z of some domain or arc contained in D. The function h defined by the equation $h(z) = f(z) - g(z)$ is also analytic in D, and $h(z) = 0$ throughout the subdomain or along the arc. According to the theorem in Sec. 102, then, $h(z) = 0$ throughout D; that is, $f(z) = g(z)$ when z is in D. We thus arrive at the following result.

Theorem. *A function that is analytic in a domain D is uniquely determined over D by its values over a domain, or along an arc, interior to D.*

Example 1. As an illustration, the function e^z is the only entire function than can assume the values e^x along a segment of the real axis. Moreover, since e^{-z} is entire and since $e^x e^{-x} = 1$ whenever x is real, the function

$$e^z e^{-z} - 1$$

is entire and takes on zero values all along the real axis. Consequently,

$$e^z e^{-z} - 1 = 0$$

everywhere; and the identity $e^{-z} = 1/e^z$ is then valid for all complex numbers z.

Such permanence of forms of other identities between functions, in passing from a real to a complex variable, is established in the same way. We limit our attention in the corollary below to the class of identities that involve only polynomials in the functions.

Corollary 1. *Let $P(w_1, w_2, \ldots, w_n)$ be a polynomial in the n variables w_k, and let f_k $(k = 1, 2, \ldots, n)$ be analytic functions of z in a domain D that contains some interval $a < x < b$ of the x axis. If on that interval the functions f_k satisfy the identity*

(1) $$P[f_1(x), f_2(x), \ldots, f_n(x)] = 0,$$

then

(2) $$P[f_1(z), f_2(z), \ldots, f_n(z)] = 0$$

throughout the domain D.

As for the proof, the left-hand side of equation (2) represents an analytic function of z in the given domain; and it is zero along an arc in that domain, according to identity (1). Hence identity (2) is valid throughout the domain.

Example 2. To illustrate this corollary, we consider the polynomial

$$P(w_1, w_2) = w_1^2 + w_2^2 - 1$$

and the entire functions $f_1(z) = \sin z$ and $f_2(z) = \cos z$. On the real axis,

$$P[f_1(x), f_2(x)] = \sin^2 x + \cos^2 x - 1 = 0.$$

Thus $P[f_1(z), f_2(z)] = \sin^2 z + \cos^2 z - 1 = 0$, or $\sin^2 z + \cos^2 z = 1$, throughout the entire z plane.

Another important consequence of the theorem is the following corollary which was anticipated in Sec. 41.

Corollary 2. *If a function f is analytic and not constant throughout a domain D, then it is not constant over any neighborhood in D.*

To prove this, we suppose that f has a constant value w_0 over some neighborhood $|z - z_0| < \varepsilon$ contained in D. Since the function $g(z) = w_0$ is well defined and analytic throughout D, the theorem tells us that $f(z) = g(z) = w_0$ for every z in D. But this contradicts the fact that f is not constant throughout D.

104. Uniqueness of Analytic Continuation

The *intersection* of two domains D_1 and D_2 is the domain $D_1 \cap D_2$ consisting of all points that are common to both D_1 and D_2. If the two domains have points in common, their *union* $D_1 \cup D_2$, consisting of the totality of points that lie in either D_1 or D_2, is also a domain.

If we have two domains D_1 and D_2 with points in common (Fig. 121) and a function f_1 that is analytic in D_1, there *may* exist a function f_2 which is analytic in D_2 such that $f_2(z) = f_1(z)$ for each z in the intersection $D_1 \cap D_2$. If so, we call f_2 an *analytic continuation* of f_1 into the domain D_2.

Whenever that analytic continuation f_2 exists, it is *unique,* according to the theorem of Sec. 103; for not more than one function can be analytic in D_2 and also assume the value $f_1(z)$ at each point z of the domain $D_1 \cap D_2$ interior to D_2. However, if there is an analytic continuation f_3 of f_2 from D_2 into a domain D_3 which intersects D_1, as indicated in Fig. 121, it is not necessarily true that $f_3(z) = f_1(z)$ for each z in $D_1 \cap D_3$. In Example 3 below we illustrate the fact that such a chain of continuations of a given function from a domain D_1 may lead to a different function defined on D_1.

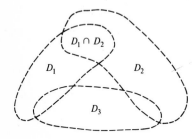

Figure 121

If f_2 is the analytic continuation of f_1 from a domain D_1 into a domain D_2, then the function F defined by the equations

$$F(z) = \begin{cases} f_1(z) & \text{when } z \text{ is in } D_1, \\ f_2(z) & \text{when } z \text{ is in } D_2 \end{cases}$$

is analytic in the union $D_1 \cup D_2$. The function F is the analytic continuation into $D_1 \cup D_2$ of either f_1 or f_2; and f_1 and f_2 are called *elements* of F.

Example 1. Consider first the function f_1 defined by the equation

(1)
$$f_1(z) = \sum_{n=0}^{\infty} z^n.$$

The power series here converges to $1/(1-z)$ when $|z| < 1$ (Sec. 45). It diverges when $|z| \geq 1$ since then the nth term z^n does not approach zero as n tends to infinity. Hence

$$f_1(z) = \frac{1}{1-z} \qquad \text{when} \qquad |z| < 1,$$

and f_1 is not defined when $|z| \geq 1$.

Now the function

$$(2) \qquad\qquad f_2(z) = \frac{1}{1-z} \qquad\qquad (z \neq 1)$$

is defined and analytic everywhere except at the point $z = 1$. Since $f_2(z) = f_1(z)$ inside the circle $|z| = 1$, the function f_2 is the analytic continuation of f_1 into the domain consisting of all points in the z plane except for $z = 1$. It is the only possible analytic continuation of f_1 into that domain, according to remarks at the beginning of the section. In this example, f_1 is also an element of f_2.

It is of interest to note that if we begin with the information that the power series

$$\sum_{n=0}^{\infty} z^n$$

converges when $|z| < 1$ and that its sum is $1/(1-x)$ when $z = x$, then we can conclude that its sum is $1/(1-z)$ whenever $|z| < 1$. For a convergent power series represents an analytic function, and $1/(1-z)$ is the analytic function interior to the circle $|z| = 1$ that assumes the values $1/(1-x)$ along the segment of the x axis inside the circle.

Example 2. Consider the function

$$(3) \qquad\qquad f_1(z) = \int_0^\infty e^{-zt}\, dt.$$

As shown in Exercise 1(c), Sec. 30, this integral exists only when Re $z > 0$, its value being $1/z$. Hence we can write

$$(4) \qquad\qquad f_1(z) = \frac{1}{z} \qquad\qquad (\text{Re } z > 0).$$

The domain of definition Re $z > 0$ is denoted by D_1 in Fig. 122, and f_1 is analytic there.

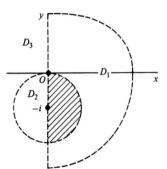

Figure 122

Let f_2 be defined in terms of a geometric series by the equation

$$(5) \qquad f_2(z) = i \sum_{n=0}^{\infty} \left(\frac{z+i}{i} \right)^n \qquad (|z+i| < 1).$$

Within its circle of convergence, the unit circle about the point $z = -i$, the series is convergent. To be specific,

$$(6) \qquad f_2(z) = i \frac{1}{1 - (z+i)/i} = \frac{1}{z}$$

when z is in the domain $|z+i| < 1$, denoted by D_2. Evidently, then, $f_2(z) = f_1(z)$ for each z in the intersection $D_1 \cap D_2$; and f_2 is the analytic continuation of f_1 into D_2.

The function $F(z) = 1/z \ (z \neq 0)$ is the analytic continuation of both f_1 and f_2 into the domain D_3 consisting of all points in the z plane except the origin. The functions f_1 and f_2 are elements of F.

Example 3. Finally, consider this branch of $z^{1/2}$:

$$f_1(z) = \sqrt{r}\, e^{i\theta/2} \qquad (r > 0, 0 < \theta < \pi).$$

An analytic continuation of f_1 across the negative real axis into the lower half plane is

$$f_2(z) = \sqrt{r}\, e^{i\theta/2} \qquad \left(r > 0, \frac{\pi}{2} < \theta < 2\pi \right).$$

An analytic continuation of f_2 across the positive real axis into the first quadrant is then

$$f_3(z) = \sqrt{r}\, e^{i\theta/2} \qquad \left(r > 0, \pi < \theta < \frac{5\pi}{2} \right).$$

Note that $f_3(z) \neq f_1(z)$ in the first quadrant; in fact, $f_3(z) = -f_1(z)$ there.

105. Principle of Reflection

In Chap. 3 we found that some elementary functions $f(z)$ possess the property $f(\bar{z}) = \overline{f(z)}$ for all points z in some domain, and others do not.

Examples. The functions

$$z, \qquad z^2 + 1, \qquad e^z, \qquad \sin z$$

have that property; for when z is replaced by its conjugate, the value of each of these functions changes to the conjugate of the original value. On the other hand, the functions

$$iz, \qquad z^2 + i, \qquad e^{iz}, \qquad (1+i)\sin z$$

do not have the property that the reflection of z in the real axis corresponds to the reflection of $f(z)$ in the real axis.

The following theorem, known as the *reflection principle,* explains these observations.

Theorem. *Let a function f be analytic in some domain D that includes a segment of the x axis and is symmetric to the x axis. If f(x) is real whenever x is a point on that segment, then*

(1) $$\overline{f(\bar{z})} = f(z)$$

whenever z is a point in D. Conversely, if condition (1) is satisfied, then f(x) is real.

Equation (1) represents the same condition on f as the equation

(2) $$\overline{f(\bar{z})} = f(z),$$

where $f(z) = u(x, y) + iv(x, y)$ and

(3) $$\overline{f(\bar{z})} = u(x, -y) - iv(x, -y).$$

When condition (2) is satisfied at a point $(x, 0)$ on the real axis,

$$u(x, 0) - iv(x, 0) = u(x, 0) + iv(x, 0);$$

hence $v(x, 0) = 0$, and $f(x)$ is real. The converse statement in the theorem is therefore true.

To prove the direct statement in the theorem, we first show that the function $\overline{f(\bar{z})}$ is analytic throughout the domain D. We write

$$F(z) = \overline{f(\bar{z})} = U(x, y) + iV(x, y).$$

Then, according to equation (3),

(4) $$U(x, y) = u(x, t), \qquad V(x, y) = -v(x, t)$$

where $t = -y$. Since $f(x + it)$ is an analytic function of $x + it$, the functions $u(x, t)$ and $v(x, t)$, together with their partial derivatives, are continuous throughout D, and the Cauchy-Riemann equations

(5) $$u_x = v_t, \qquad u_t = -v_x$$

are satisfied there. Now, in view of equations (4),

$$U_x = u_x, \qquad V_y = -v_t \frac{dt}{dy} = v_t;$$

and it follows from these and the first of equations (5) that $U_x = V_y$. Similarly, $U_y = -V_x$. These partial derivatives of U and V are continuous, and the function

$$F(z) = \overline{f(\bar{z})}$$

is then analytic in the domain D.

Since $f(x)$ is real, $v(x, 0) = 0$. Hence

$$F(x) = U(x, 0) + iV(x, 0) = u(x, 0);$$

that is, $F(z) = f(z)$ when the point z is on the segment of the x axis within the domain D. It follows from the theorem of Sec. 103 that $F(z) = f(z)$ at each point z of D since both functions are analytic there. Thus condition (2) is established, and the proof of the theorem is complete.

EXERCISES

1. Given that the hyperbolic sine and hyperbolic cosine functions, the exponential function, and the sine and cosine functions are all entire, use Corollary 1 in Sec. 103 to obtain each of these identities for all complex z from the corresponding identities when z is real:

   (a) $\sinh z + \cosh z = e^z$; (b) $\sin 2z = 2 \sin z \cos z$;

   (c) $\cosh^2 z - \sinh^2 z = 1$; (d) $\sin\left(\dfrac{\pi}{2} - z\right) = \cos z$.

2. Show that the function

$$f_2(z) = \frac{1}{z^2 + 1} \qquad\qquad (z \neq \pm i)$$

   is the analytic continuation of the function

$$f_1(z) = \sum_{n=0}^{\infty} (-1)^n z^{2n} \qquad\qquad (|z| < 1)$$

   into the domain consisting of all points in the z plane except $z = \pm i$.

3. Show that the function $f_2(z) = 1/z^2$ $(z \neq 0)$ is the analytic continuation of the function

$$f_1(z) = \sum_{n=0}^{\infty} (n + 1)(z + 1)^n \qquad\qquad (|z + 1| < 1)$$

   into the domain consisting of all points in the z plane except $z = 0$.

4. State why the function

$$f_4(z) = \sqrt{r}\, e^{i\theta/2} \qquad\qquad (r > 0, -\pi < \theta < \pi)$$

   is the analytic continuation of the function (Example 3, Sec. 104)

$$f_1(z) = \sqrt{r}\, e^{i\theta/2} \qquad\qquad (r > 0, 0 < \theta < \pi)$$

   across the positive real axis into the lower half plane.

5. Find the analytic continuation of $\operatorname{Log} z$ from the upper half plane $\operatorname{Im} z > 0$ into the lower half plane across the negative real axis. Note that this analytic continuation is different from $\operatorname{Log} z$ in the lower half plane.

   Ans. $\operatorname{Log} r + i\theta \quad (r > 0, 0 < \theta < 2\pi)$.

6. Find the analytic continuation of the function

$$f(z) = \int_0^{\infty} t e^{-zt}\, dt \qquad\qquad (\operatorname{Re} z > 0)$$

   into the domain consisting of all points in the z plane except the origin.

   Ans. $1/z^2$.

7. Show that the function $1/(z^2 + 1)$ is the analytic continuation of the function

$$f(z) = \int_0^\infty e^{-zt} \sin t \, dt \qquad\qquad (\text{Re } z > 0)$$

into the domain consisting of all points in the z plane except $z = \pm i$.

8. Show that if the condition that $f(x)$ be real in the theorem of Sec. 105 is replaced by the condition that $f(x)$ be pure imaginary, the conclusion is changed to

$$f(\bar{z}) = -\overline{f(z)}.$$

9. Let S denote a set of points in a domain D such that S has an accumulation point in D. Generalize the theorem in Sec. 103 by proving that a function which is analytic in D is uniquely determined by its values on the set S.

B. SINGULAR POINTS AND ZEROS

We now examine further the behavior of functions near their singular points.

106. Poles and Zeros

It was shown in Exercise 12, Sec. 58, that if z_0 is a pole of any order of a function f, then

$$(1) \qquad\qquad \lim_{z \to z_0} f(z) = \infty;$$

that is, for each positive number ε there exists a positive number δ such that

$$(2) \qquad\qquad |f(z)| > \frac{1}{\varepsilon} \qquad \text{whenever} \qquad 0 < |z - z_0| < \delta.$$

As a consequence, there is always some neighborhood of a pole containing no zeros of the function f.

Since poles are isolated singular points, it follows that *if z_0 is a pole of a function f, there is a neighborhood of z_0 which contains neither a zero of f nor a singular point of f other than z_0 itself.*

According to Sec. 58, if z_0 is a zero of order m of an analytic function $f(z)$, then z_0 is a pole of order m of the reciprocal function $1/f(z)$. The converse is easily established. If z_0 is a pole of order m of a function $f(z)$, we know from Exercise 11, Sec. 58, that

$$(3) \qquad\qquad f(z) = \frac{\phi(z)}{(z - z_0)^m}$$

where $\phi(z)$ is analytic at z_0 and $\phi(z_0) \neq 0$. Now the function $g(z) = 1/\phi(z)$ is analytic at z_0, and $g(z_0) \neq 0$. Since

$$(4) \qquad\qquad \frac{1}{f(z)} = (z - z_0)^m g(z),$$

this shows that (see Sec. 53) *if z_0 is a pole of order m of a function $f(z)$, then it is a zero of order m of the reciprocal function $1/f(z)$.*

In contrast to condition (2), suppose f is bounded and analytic in a domain $0 < |z - z_0| < \delta$. Then the following theorem due to Riemann applies.

Theorem. *If a function f is bounded and analytic throughout a domain $0 < |z - z_0| < \delta$, then either f is analytic at z_0 or else z_0 is a removable singular point of that function.*

To prove this, observe that $f(z)$ is represented by its Laurent series in the given domain about z_0. If C denotes a positively oriented circle $|z - z_0| = \rho$, where $\rho < \delta$, the coefficients b_n of $1/(z - z_0)^n$ in that series are (Sec. 46)

$$b_n = \frac{1}{2\pi i} \int_C \frac{f(z)\,dz}{(z - z_0)^{-n+1}} = \frac{\rho^n}{2\pi} \int_0^{2\pi} f(z_0 + \rho e^{i\theta}) e^{in\theta}\,d\theta,$$

where $n = 1, 2, \ldots$. Since f is bounded, there is a positive real number M such that

$$|f(z)| < M \qquad\qquad (0 < |z - z_0| < \delta);$$

hence

$$|b_n| < M\rho^n.$$

But the coefficients are constants; and, since ρ can be chosen arbitrarily small, $b_n = 0$ ($n = 1, 2, \ldots$). Thus the Laurent series for $f(z)$ reduces to a power series

$$f(z) = \sum_{n=0}^{\infty} a_n(z - z_0)^n \qquad\qquad (0 < |z - z_0| < \delta).$$

If $f(z_0)$ is defined as the number a_0, it follows that f is analytic at z_0. This proves the theorem.

107. Essential Singular Points

The behavior of a function near an essential singular point is quite irregular. This was pointed out in Sec. 56, where Picard's theorem was stated; namely, *in each neighborhood of an essential singular point the function assumes every finite value, with one possible exception, an infinite number of times.* We also illustrated Picard's theorem in Example 4 there by showing that the function $\exp(1/z)$, which has an essential singular point at the origin, assumes the value -1 an infinite number of times in any given neighborhood of that singular point. We do not prove Picard's theorem, but do prove a related theorem due to Weierstrass. This theorem shows that the value of a function is arbitrarily close to any prescribed number c at certain points arbitrarily near an essential singular point of that function.

Theorem. *Let z_0 be an essential singular point of a function f, and let c be any given complex number. Then for each positive number ε, however small, the inequality*

(1) $$|f(z) - c| < \varepsilon$$

is satisfied at some point z different from z_0 in each neighborhood of z_0.

To prove the theorem, suppose that condition (1) is not satisfied at any point in a neighborhood $|z - z_0| < \delta$, where δ is small enough that f is analytic in the domain $0 < |z - z_0| < \delta$. Then $|f(z) - c| \geq \varepsilon$ for all points in that domain, and the function

(2) $$g(z) = \frac{1}{f(z) - c} \qquad\qquad (0 < |z - z_0| < \delta)$$

is analytic and bounded there. According to Riemann's theorem (Sec. 106), z_0 is a removable singular point of g. Let $g(z_0)$ be defined so that g is analytic at z_0. Since f cannot be a constant function, neither can g; and, in view of the Taylor series for g at z_0, either $g(z_0) \neq 0$ or else g has a zero of some finite order at z_0. Consequently, its reciprocal,

$$\frac{1}{g(z)} = f(z) - c,$$

is either analytic at z_0 or has a pole there. But this contradicts the hypothesis that z_0 is an essential singular point of f. Hence condition (1) must be satisfied at some point in the given neighborhood.

108. Number of Zeros and Poles

Properties of the logarithmic derivative found in Exercises 13 and 14, Sec. 58, can be generalized.

Let a function f be analytic inside and on a simple closed contour C except for at most a finite number of poles interior to C. Also, let f have no zeros on C and at most a finite number of zeros interior to C. Then, if C is described in the positive sense,

(1) $$\frac{1}{2\pi i} \int_C \frac{f'(z)}{f(z)}\, dz = N - P$$

where N is the total number of zeros of f inside C and P is the total number of poles of f there. A zero of order m_0 is to be counted m_0 times, and a pole of order m_p is to be counted m_p times.

To prove statement (1), we show that the integer $N - P$ is the sum of the residues of the function $f'(z)/f(z)$ at its singular points inside the simple closed contour C. Those singular points are, of course, the zeros and poles of f interior to C.

Let z_0 be a zero of f of order m_0. In some neighborhood of z_0 we can write (Sec. 53)

(2) $$f(z) = (z - z_0)^{m_0} g(z),$$

where g is analytic in that neighborhood and $g(z_0) \neq 0$. Hence

$$f'(z) = m_0(z - z_0)^{m_0 - 1} g(z) + (z - z_0)^{m_0} g'(z),$$

or

$$\frac{f'(z)}{f(z)} = \frac{m_0}{z - z_0} + \frac{g'(z)}{g(z)}.$$

Since $g'(z)/g(z)$ is analytic at z_0, the function $f'(z)/f(z)$ has a simple pole at z_0, with residue m_0. The sum of the residues of $f'(z)/f(z)$ at all the zeros of f inside C is therefore the integer N.

If z_p is a pole of f of order m_p, the function

(3) $$\phi(z) = (z - z_p)^{m_p} f(z) \qquad\qquad (z \neq z_p)$$

can be defined at z_p so that ϕ is analytic there and $\phi(z_p) \neq 0$ (see Exercise 11, Sec. 58). Thus in some neighborhood of z_p, except at the point $z = z_p$ itself,

(4) $$f(z) = (z - z_p)^{-m_p} \phi(z);$$

and this means that

$$f'(z) = -m_p(z - z_p)^{-m_p-1} \phi(z) + (z - z_p)^{-m_p} \phi'(z).$$

Therefore

$$\frac{f'(z)}{f(z)} = -\frac{m_p}{z - z_p} + \frac{\phi'(z)}{\phi(z)},$$

and we see that $f'(z)/f(z)$ has a simple pole at z_p, with residue $-m_p$. Hence the sum of the residues of $f'(z)/f(z)$ at all the poles of f inside C is the integer $-P$. Statement (1) is now verified.

One form of what is known as the *Bolzano-Weierstrass theorem* can be stated as follows.* *A set of infinitely many points each of which lies in a closed bounded region has at least one accumulation point in that region.* The proof can be made by selecting any infinite sequence $z_1, z_2, \ldots, z_n, \ldots$ of points from the set and applying to that sequence the process for subdividing squares used in Exercise 12, Sec. 37.

In view of that theorem, the condition that the number of zeros and poles inside C be finite which was used in proving statement (1) can be removed. For the number of zeros and poles within the simple closed contour C must necessarily be finite if the function f is to be analytic within and on C, except possibly for poles within C, because zeros and poles are isolated. The full argument is left as an exercise.

109. Argument Principle

Let C be a simple closed contour having positive orientation in the z plane, and let f be a function which is analytic within and on C, except possibly for poles interior to C. Also, let f have no zeros on C. The image Γ of C under the transformation $w = f(z)$ is a closed contour in the w plane (Fig. 123). As a point z traverses C in the positive direction, its image w traverses Γ in a particular direction which determines the orientation of Γ.

*See, for example, A. E. Taylor and W. R. Mann, "Advanced Calculus," 3d ed., pp. 517 and 521, 1983.

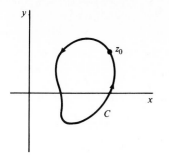
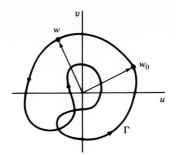

Figure 123. $w = f(z)$.

Since f has no zeros on C, the contour Γ does not pass through the origin in the w plane. Let w_0 be a fixed point on Γ, and let ϕ_0 be a value of arg w_0. Then let arg w vary continuously, starting with the value ϕ_0, as the point w begins at w_0 and traverses Γ once in the direction of orientation assigned to it by the mapping $w = f(z)$. When w returns to the starting point w_0, arg w assumes a particular value of arg w_0 which we denote by ϕ_1. Thus the change in arg w as w describes Γ once in its direction of orientation is $\phi_1 - \phi_0$. Note that this change is independent of the particular point w_0 chosen to determine it.

The number $\phi_1 - \phi_0$ is also the change in the argument of $f(z)$ as z describes C once in the positive direction, and we write

$$(1) \qquad\qquad \Delta_C \arg f(z) = \phi_1 - \phi_0.$$

The value of $\Delta_C \arg f(z)$ is a multiple of 2π, and the integer

$$\frac{1}{2\pi} \Delta_C \arg f(z)$$

represents the number of times the point w winds around the origin in the w plane as z describes C once in the positive direction. If, for instance, this integer is -1, then Γ winds around the origin once in the clockwise direction. In Fig. 123 the value of $\Delta_C \arg f(z)$ is zero. The value of $\Delta_C \arg f(z)$ is always zero when the contour Γ does not enclose the origin; the verification of this fact for a special case is left to the exercises.

The value of $\Delta_C \arg f(z)$ can be determined from the number of zeros and poles of f interior to C.

Theorem 1. *Let C be a simple closed contour, described in the positive sense, and let f be a function which is analytic inside and on C, except possibly for poles interior to C. Also, let f have no zeros on C. Then*

$$(2) \qquad\qquad \frac{1}{2\pi} \Delta_C \arg f(z) = N - P$$

where N and P are the number of zeros and the number of poles of f, counting multiplicities, interior to C.

Our proof of this result, known as the *argument principle,* is based on the statement

$$(3) \qquad \frac{1}{2\pi i} \int_C \frac{f'(z)}{f(z)} \, dz = N - P,$$

obtained in Sec. 108. If C is expressed parametrically as $z = z(t)$ $(a \leq t \leq b)$, a parametric representation for its image Γ under the transformation $w = f(z)$ is then

$$w = w(t) = f[z(t)] \qquad\qquad (a \leq t \leq b).$$

Now according to Exercise 11, Sec. 30,

$$w'(t) = f'[z(t)]z'(t)$$

along each of the smooth arcs making up the contour Γ. Since $z'(t)$ and $w'(t)$ are piecewise continuous on the interval $a \leq t \leq b$, we can write

$$\int_a^b \frac{f'[z(t)]}{f[z(t)]} z'(t) \, dt = \int_a^b \frac{w'(t)}{w(t)} \, dt.$$

That is,

$$\int_C \frac{f'(z)}{f(z)} \, dz = \int_\Gamma \frac{dw}{w}.$$

Equation (3) therefore becomes

$$(4) \qquad \frac{1}{2\pi i} \int_\Gamma \frac{dw}{w} = N - P.$$

Since Γ never passes through the origin in the w plane, we can express each point on that contour in exponential form as $w = \rho \exp(i\phi)$. If we then represent Γ in terms of some parameter τ as

$$w = w(\tau) = \rho(\tau) \exp[i\phi(\tau)] \qquad\qquad (c \leq \tau \leq d),$$

we obtain the equation

$$w'(\tau) = \rho'(\tau) \exp[i\phi(\tau)] + \rho(\tau) \exp[i\phi(\tau)]i\phi'(\tau),$$

where $\rho'(\tau)$ and $\phi'(\tau)$ are piecewise continuous on the interval $c \leq \tau \leq d$. Thus we can write

$$\int_\Gamma \frac{dw}{w} = \int_c^d \frac{w'(\tau)}{w(\tau)} \, d\tau = \int_c^d \frac{\rho'(\tau)}{\rho(\tau)} \, d\tau + i \int_c^d \phi'(\tau) \, d\tau,$$

or

$$\int_\Gamma \frac{dw}{w} = \text{Log } \rho(\tau) \Big]_c^d + i\phi(\tau) \Big]_c^d.$$

But $\rho(d) = \rho(c)$ and

$$\phi(d) - \phi(c) = \phi_1 - \phi_0 = \Delta_C \arg f(z).$$

Hence

(5)
$$\int_\Gamma \frac{dw}{w} = i\Delta_C \arg f(z).$$

Expression (2) now follows immediately from equations (4) and (5).

We next present a useful consequence of the argument principle, known as *Rouché's theorem*.

> **Theorem 2.** *Let f and g be functions which are analytic inside and on a positively oriented simple closed contour C. If $|f(z)| > |g(z)|$ at each point z on C, the functions $f(z)$ and $f(z) + g(z)$ have the same number of zeros, counting multiplicities, inside C.*

To prove this, we note first that since $|f(z)| > |g(z)| \geq 0$ on C, $f(z)$ has no zeros on C. Moreover,

$$|f(z) + g(z)| \geq |f(z)| - |g(z)| > 0$$

on C, and so the function $f(z) + g(z)$ has no zeros on C either. Now

(6)
$$\frac{1}{2\pi}\Delta_C \arg f(z) = N_f$$

and

(7)
$$\frac{1}{2\pi}\Delta_C \arg[f(z) + g(z)] = N_{f+g},$$

where N_f is the number of zeros of $f(z)$ interior to C and N_{f+g} is the number of zeros of $f(z) + g(z)$ there. Equations (6) and (7) follow from the argument principle and the fact that the functions $f(z)$ and $f(z) + g(z)$ have no poles inside C. Observe that

$$\Delta_C \arg[f(z) + g(z)] = \Delta_C \arg\left\{f(z)\left[1 + \frac{g(z)}{f(z)}\right]\right\}$$

$$= \Delta_C \arg f(z) + \Delta_C \arg\left[1 + \frac{g(z)}{f(z)}\right].$$

Under the transformation $w = 1 + [g(z)/f(z)]$, the image Γ of C lies inside the circle $|w - 1| = 1$ because $|w - 1| = |g(z)/f(z)| < 1$ on C. Hence the point $w = 0$ is not enclosed by the curve Γ, and the value of $\Delta_C \arg\{1 + [g(z)/f(z)]\}$ is zero. Thus $\Delta_C \arg [f(z) + g(z)] = \Delta_C \arg f(z)$; and $f(z) + g(z)$ has the same number of zeros as $f(z)$ interior to C, according to equations (6) and (7).

Example. For an application of Rouché's theorem, let us determine the number of roots of the equation $z^7 - 4z^3 + z - 1 = 0$ interior to the circle $|z| = 1$. Write $f(z) = -4z^3$ and $g(z) = z^7 + z - 1$, and observe that $|f(z)| = 4$ and $|g(z)| \leq 3$ when $|z| = 1$. The conditions in Rouché's theorem are thus satisfied. Consequently, since $f(z)$ has three zeros, counting multiplicities, interior to the circle $|z| = 1$, so does $f(z) + g(z)$. That is, the equation $z^7 - 4z^3 + z - 1 = 0$ has three roots interior to the circle $|z| = 1$.

EXERCISES

1. Let c be a fixed complex number different from zero. Show that the function $\exp(1/z)$, which has an essential singular point at $z = 0$, assumes the value c an infinite number of times in any neighborhood of the origin.

   Suggestion: Write $c = c_0 \exp(i\gamma)$, where $c_0 > 0$, and show that $\exp(1/z)$ assumes the value c at the points $z = r \exp(i\theta)$ where r and θ satisfy the equations

   $$r^2 = \frac{1}{\gamma^2 + (\text{Log } c_0)^2}, \quad \sin\theta = \frac{-\gamma}{\sqrt{\gamma^2 + (\text{Log } c_0)^2}}, \quad \cos\theta = \frac{\text{Log } c_0}{\sqrt{\gamma^2 + (\text{Log } c_0)^2}}.$$

   Note that r can be made arbitrarily small by adding integral multiples of 2π to the angle γ, leaving c unaltered.

2. *If a function f is analytic in a domain $0 < |z - z_0| < R_0$ and if z_0 is an accumulation point of zeros of the function, then z_0 is an essential singular point of f or else $f(z)$ is identically zero.* Prove this theorem with the aid of results found in Secs. 53 and 106.

3. Examine the set of zeros of the function $z^2 \sin(1/z)$, and apply the theorem stated in Exercise 2 to show that the origin is an essential singular point of the function. Note that this conclusion also follows from the nature of the Laurent series that represents the function in the domain $0 < |z| < \infty$.

4. Let C be a simple closed contour in the z plane, described in the positive sense, and let w_0 be any given complex number. Let a function g be analytic within and on C, and suppose that $g'(z) \neq 0$ at any point z interior to C. If $g(z) \neq w_0$ at any point z on C, then

   $$\frac{1}{2\pi i} \int_C \frac{g'(z)}{g(z) - w_0}\, dz = N$$

   where the integer N is the number of points z interior to C at which $g(z) = w_0$. Show how this result follows from those given in Sec. 108. (Compare the result with that found in Exercise 12, Sec. 91.)

5. Complete the argument (Sec. 108), based on the Bolzano-Weierstrass theorem, that if a function f is analytic within and on a simple closed contour C, except possibly for poles inside C, and if $f(z) \neq 0$ at any point z on C, then the zeros and poles of f inside C are finite in number and statement (1) of Sec. 108 is valid.

6. Let f be a function which is analytic inside and on a simple closed contour C, and suppose that $f(z)$ is never zero on C. Let the image of C under the transformation $w = f(z)$ be the closed contour Γ shown in Fig. 124. Using the contour Γ, determine the value of $\Delta_C \arg f(z)$. Also, determine the number of zeros of f interior to C.

   Ans. $6\pi; 3.$

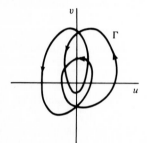

Figure 124

7. Let C denote the unit circle $|z| = 1$, described in the positive sense. Determine the value of $\Delta_C \arg f(z)$ for the function

(a) $f(z) = z^2$;   (b) $f(z) = \dfrac{z^3 + 2}{z}$.

Also, for each of the transformations $w = f(z)$ defined by these functions, state the number of times the image point w winds around the origin in the w plane as the point z describes C once in the positive direction.

Ans.   (a) $4\pi, 2$;   (b) $-2\pi, -1$.

8. Using the notation of Sec. 109, show that when Γ does not enclose the point $w = 0$ and there is a ray from that point which does not intersect Γ, then $\Delta_C \arg f(z) = 0$.

   Suggestion: Note that the change in the value of $\arg f(z)$ must be less than 2π in absolute value when a point z makes one cycle around C. Then use the fact that $\Delta_C \arg f(z)$ is an integral multiple of 2π.

9. Determine the number of zeros of the polynomial

(a) $z^6 - 5z^4 + z^3 - 2z$;   (b) $2z^4 - 2z^3 + 2z^2 - 2z + 9$

inside the circle $|z| = 1$.

Ans.   (a) 4;   (b) 0.

10. Determine the number of roots of the equation $2z^5 - 6z^2 + z + 1 = 0$ in the region $1 \leqq |z| < 2$.

Ans.   3.

11. Show that if c is a complex number such that $|c| > e$, the equation $cz^n = e^z$ has n roots inside the circle $|z| = 1$.

12. Using Rouché's theorem, prove that any polynomial

$$P(z) = a_0 + a_1 z + \cdots + a_{n-1} z^{n-1} + a_n z^n \qquad (a_n \neq 0),$$

where $n \geqq 1$, has precisely n zeros. Thus give an alternative proof of the fundamental theorem of algebra (Sec. 42).

   Suggestion: Note that it is sufficient to let a_n be unity. Then, by writing

$$f(z) = z^n \qquad \text{and} \qquad g(z) = a_0 + a_1 z + \cdots + a_{n-1} z^{n-1},$$

show that $P(z)$ has n zeros inside a circle $|z| = R$ where R is larger than either of the two numbers 1 and $|a_0| + |a_1| + \cdots + |a_{n-1}|$. To show that $P(z)$ has no other zeros, show that

$$|z^n + g(z)| \geqq |z|^n - |g(z)| > 0$$

when $|z| \geqq R$.

C. RIEMANN SURFACES

A *Riemann surface* is a generalization of the complex plane to a surface of more than one sheet such that a multiple-valued function has only one value corresponding to each point on that surface. Once such a surface is devised for a given function, the function is single-valued on the surface and the theory of single-valued functions applies there. Complexities arising because the function is multiple-valued are thus relieved by a geometric device. However, the description of those surfaces and the arrangement of proper connections between the sheets can become quite involved. We limit our attention to fairly simple examples.

110. A Surface for $\log z$

Corresponding to each nonzero number z, the multiple-valued function

$$\log z = \mathrm{Log}\, r + i\theta$$

has infinitely many values. To describe $\log z$ as a single-valued function, we replace the z plane, with the origin deleted, by a surface on which a new point is located whenever the argument of the number z is increased or decreased by 2π, or an integral multiple of 2π.

Consider the z plane, with the origin deleted, as a thin sheet R_0 which is cut along the positive half of the real axis. On that sheet let θ range from 0 to 2π. Let a second sheet R_1 be cut in the same way and placed in front of the sheet R_0. The lower edge of the slit in R_0 is then joined to the upper edge of the slit in R_1. On R_1 the angle θ ranges from 2π to 4π; so when z is represented by a point on R_1, the imaginary component of $\log z$ ranges from 2π to 4π.

A sheet R_2 is then cut in the same way and placed in front of R_1. The lower edge of the slit in R_1 is joined to the upper edge of the slit in this new sheet, and similarly for sheets $R_3, R_4, \ldots$. A sheet R_{-1} on which θ varies from 0 to -2π is cut and placed behind R_0, with the lower edge of its slit connected to the upper edge of the slit in R_0; the sheets $R_{-2}, R_{-3}, \ldots$ are constructed in like manner. The coordinates r and θ of a point on any sheet can be considered as polar coordinates of the projection of the point onto the original z plane, the angular coordinate θ being restricted to a definite range of 2π radians on each sheet.

Consider any continuous curve on this connected surface of infinitely many sheets. As a point z describes that curve, the values of $\log z$ vary continuously since θ, in addition to r, now varies continuously; and $\log z$ assumes just one value corresponding to each point on the curve. For example, as the point makes a complete cycle around the origin on the sheet R_0 over the path indicated in Fig. 125, the angle changes from 0 to 2π. As it moves across the ray $\theta = 2\pi$, the point passes to the sheet R_1 of the surface. As the point completes a cycle in R_1, the angle θ varies from 2π to 4π; and as it crosses the ray $\theta = 4\pi$, the point passes to the sheet R_2.

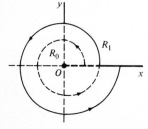

Figure 125

The surface described here is a Riemann surface for log z. It is a connected surface of infinitely many sheets arranged so that log z is a single-valued function of points on it.

The transformation $w = \log z$ maps the whole Riemann surface in a one to one manner onto the entire w plane. The image of the sheet R_0 is the strip $0 \leq v \leq 2\pi$. As a point z moves onto the sheet R_1 over the arc shown in Fig. 126, its image w moves upward across the line $v = 2\pi$, as indicated in the figure.

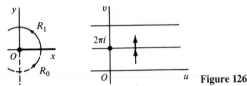

Figure 126

Note that log z, defined on the sheet R_1, represents the analytic continuation of the single-valued analytic function

$$\text{Log } r + i\theta \qquad\qquad (0 < \theta < 2\pi)$$

upward across the positive real axis. In this sense, log z is not only a single-valued function of all points z on the Riemann surface but also an *analytic* function at all points there.

The sheets could, of course, be cut along the negative real axis, or along any other ray from the origin, and properly joined along the slits to form other Riemann surfaces for log z.

111. A Surface for $z^{1/2}$

Corresponding to each point in the z plane other than the origin, the function

$$z^{1/2} = \sqrt{r}\, e^{i\theta/2}$$

has two values. A Riemann surface for $z^{1/2}$ is obtained by replacing the z plane with a surface made up of two sheets R_0 and R_1, each cut along the positive real axis with R_1 placed in front of R_0. The lower edge of the slit in R_0 is joined to the upper edge of the slit in R_1, and the lower edge of the slit in R_1 is joined to the upper edge of the slit in R_0.

As a point z starts from the upper edge of the slit in R_0 and describes a continuous circuit around the origin in the counterclockwise direction (Fig. 127), the angle θ

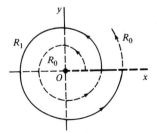

Figure 127

increases from 0 to 2π. The point then passes from the sheet R_0 to the sheet R_1, where θ increases from 2π to 4π. As the point moves still further, it passes back to the sheet R_0, where the values of θ can vary from 4π to 6π or from 0 to 2π, a choice that does not affect the value of $z^{1/2}$, etc. Note that the value of $z^{1/2}$ at a point where the circuit passes from the sheet R_0 to the sheet R_1 is different from the value of $z^{1/2}$ at a point where the circuit passes from the sheet R_1 to the sheet R_0.

We have thus constructed a Riemann surface on which $z^{1/2}$ is single-valued for each nonzero z. In that construction, the edges of the sheets R_0 and R_1 are joined in pairs in such a way that the resulting surface is closed and connected. The points where two of the edges are joined are distinct from the points where the other two edges are joined. Thus it is physically impossible to build a model of that Riemann surface. In visualizing a Riemann surface, it is important to understand how we are to proceed when we arrive at an edge of a slit.

The origin is a special point on this Riemann surface. It is common to both sheets, and a curve around the origin on the surface must wind around it twice in order to be a closed curve. A point of this kind on a Riemann surface is called a *branch point*.

The image of the sheet R_0 under the transformation $w = z^{1/2}$ is the upper half of the w plane since the argument of w is $\theta/2$, where $0 \leq \theta/2 \leq \pi$, on R_0. Likewise, the image of the sheet R_1 is the lower half of the w plane. As defined on either sheet, the function is the analytic continuation, across the cut, of the function defined on the other sheet. In this respect, the single-valued function $z^{1/2}$ of points on the Riemann surface is analytic at all points except the origin.

112. Surfaces for Related Functions

Example 1. Let us describe a Riemann surface for the double-valued function

$$(1) \qquad f(z) = (z^2 - 1)^{1/2} = \sqrt{r_1 r_2} \, \exp \frac{i(\theta_1 + \theta_2)}{2}$$

where $z - 1 = r_1 \exp(i\theta_1)$ and $z + 1 = r_2 \exp(i\theta_2)$, as shown in Fig. 128. A branch of this function, with the line segment $P_1 P_2$ between the branch points $z = \pm 1$ as a branch cut, was described in Example 2, Sec. 69. That branch is as written above, with the restrictions $r_k > 0$, $0 \leq \theta_k < 2\pi$ $(k = 1, 2)$ and $r_1 + r_2 > 2$. The branch is not defined on the segment $P_1 P_2$.

A Riemann surface for the double-valued function (1) must consist of two sheets R_0 and R_1. Let both sheets be cut along the segment $P_1 P_2$. The lower edge of the slit in R_0 is then joined to the upper edge of the slit in R_1, and the lower edge in R_1 is joined to the upper edge in R_0.

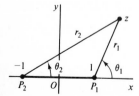

Figure 128

On the sheet R_0, let the angles θ_1 and θ_2 range from 0 to 2π. If a point on the sheet R_0 describes a simple closed curve which encloses the segment P_1P_2 once in the counterclockwise direction, then both θ_1 and θ_2 change by the amount 2π upon the return of the point to its original position. The change in $(\theta_1 + \theta_2)/2$ is also 2π, and the value of f is unchanged. If a point starting on the sheet R_0 describes a path which passes twice around just the branch point $z = 1$, it crosses from the sheet R_0 onto the sheet R_1 and then back onto the sheet R_0 before it returns to its original position. In this case, the value of θ_1 changes by the amount 4π, while the value of θ_2 does not change at all. Similarly, for a circuit twice around the point $z = -1$, the value of θ_2 changes by 4π, while the value of θ_1 remains unchanged. Again, the change in $(\theta_1 + \theta_2)/2$ is 2π; and the value of f is unchanged. Thus on the sheet R_0 the range of the angles θ_1 and θ_2 may be extended by changing both θ_1 and θ_2 by the same integral multiple of 2π or by changing just one of the angles by a multiple of 4π. In either case the total change in both angles is an even integral multiple of 2π.

To obtain the range of values for θ_1 and θ_2 on the sheet R_1, we note that if a point starts on the sheet R_0 and describes a path around just one of the branch points once, it crosses onto the sheet R_1 and does not return to the sheet R_0. In this case, the value of one of the angles is changed by 2π while the value of the other remains unchanged. Hence on the sheet R_1 one angle can range from 2π to 4π while the other ranges from 0 to 2π. Their sum then ranges from 2π to 4π, and the value of $(\theta_1 + \theta_2)/2$, the argument of $f(z)$, ranges from π to 2π. Again, the range of the angles is extended by changing the value of just one of the angles by an integral multiple of 4π or by changing the value of both angles by the same integral multiple of 2π.

The double-valued function (1) may now be considered as a single-valued function of the points on the Riemann surface just constructed. The transformation $w = f(z)$ maps each of the sheets used in the construction of that surface onto the entire w plane.

Example 2. Consider the double-valued function

(2) $$f(z) = [z(z^2 - 1)]^{1/2} = \sqrt{rr_1r_2}\, \exp \frac{i(\theta + \theta_1 + \theta_2)}{2}$$

(Fig. 129). The points $z = 0, \pm 1$ are branch points of this function. We note that if the point z describes a circuit that includes all three of those points, the argument of $f(z)$ changes by the angle 3π and the value of the function thus changes. Consequently, a branch cut must run from one of those branch points to the point at infinity in order to describe a single-valued branch of f. Hence the point at infinity is also a branch point, as one can show by noting that the function $f(1/z)$ has a branch point at $z = 0$.

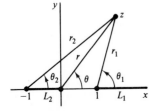

Figure 129

Let two sheets be cut along the line segment L_2 from $z = -1$ to $z = 0$ and along the part L_1 of the real axis to the right of the point $z = 1$. We specify that each of the three angles θ, θ_1, and θ_2 may range from 0 to 2π on the sheet R_0 and from 2π to 4π on the sheet R_1. We also specify that the angles corresponding to a point on either sheet may be changed by integral multiples of 2π in such a way that the sum of the three angles changes by an integral multiple of 4π; the value of the function f is therefore unaltered.

A Riemann surface for the double-valued function (2) is obtained by joining the lower edges in R_0 of the slits along L_1 and L_2 to the upper edges in R_1 of the slits along L_1 and L_2, respectively. The lower edges in R_1 of the slits along L_1 and L_2 are then joined to the upper edges in R_0 of the slits along L_1 and L_2, respectively. It is readily verified with the aid of Fig. 129 that one branch of the function is represented by its values at points on R_0 and the other branch at points on R_1.

EXERCISES

1. Describe a Riemann surface for the triple-valued function $w = (z - 1)^{1/3}$, and point out which third of the w plane represents the image of each sheet of that surface.

2. Describe the Riemann surface for $\log z$ obtained by cutting the z plane along the negative real axis. Compare this Riemann surface for $\log z$ with the one obtained in Sec. 110.

3. Determine the image under the transformation $w = \log z$ of the sheet R_n, where n is an arbitrary integer, of the Riemann surface for $\log z$ given in Sec. 110.

4. Verify that under the transformation $w = z^{1/2}$, the sheet R_1 of the Riemann surface for $z^{1/2}$ given in Sec. 111 is mapped onto the lower half of the w plane.

5. Describe the curve, on a Riemann surface for $z^{1/2}$, whose image is the entire circle $|w| = 1$ under the transformation $w = z^{1/2}$.

6. Corresponding to each point on the Riemann surface described in Example 2, Sec. 112, for the function $w = f(z)$ in that example, there is just one value of w. Show that corresponding to each value of w, there are, in general, three points on the surface.

7. Describe a Riemann surface for the multiple-valued function

$$f(z) = \left(\frac{z - 1}{z}\right)^{1/2}.$$

8. Let C denote the positively oriented circle $|z - 2| = 1$ on the Riemann surface described in Sec. 111 for $z^{1/2}$, where the upper half of that circle lies on the sheet R_0 and the lower half on R_1. Note that for each point z on C one can write

$$z^{1/2} = \sqrt{r}\, e^{i\theta/2} \qquad \text{where} \qquad 4\pi - \frac{\pi}{2} < \theta < 4\pi + \frac{\pi}{2}.$$

State why it follows that

$$\int_C z^{1/2}\, dz = 0.$$

Generalize this result to fit the case of the other simple closed curves that cross from one sheet to another without enclosing the branch points. Generalize to other functions, thus extending the Cauchy-Goursat theorem to integrals of multiple-valued functions.

9. Note that the Riemann surface described in Example 1, Sec. 112, for $(z^2 - 1)^{1/2}$ is also a Riemann surface for the function

$$g(z) = z + (z^2 - 1)^{1/2}.$$

Let f_0 denote the branch of $(z^2 - 1)^{1/2}$ defined on the sheet R_0, and show that the branches g_0 and g_1 of g on the two sheets are given by the equations

$$g_0(z) = \frac{1}{g_1(z)} = z + f_0(z).$$

10. In Exercise 9 the branch f_0 of $(z^2 - 1)^{1/2}$ can be described by the equation

$$f_0(z) = \sqrt{r_1 r_2}\left(\exp\frac{i\theta_1}{2}\right)\left(\exp\frac{i\theta_2}{2}\right),$$

where θ_1 and θ_2 range from 0 to 2π and

$$z - 1 = r_1 \exp(i\theta_1), \qquad z + 1 = r_2 \exp(i\theta_2).$$

Note that $2z = r_1 \exp(i\theta_1) + r_2 \exp(i\theta_2)$, and show that the branch g_0 of the function $g(z) = z + (z^2 - 1)^{1/2}$ can be written in the form

$$g_0(z) = \frac{1}{2}\left(\sqrt{r_1}\,\exp\frac{i\theta_1}{2} + \sqrt{r_2}\,\exp\frac{i\theta_2}{2}\right)^2.$$

Find $g_0(z)\overline{g_0(z)}$, and note that $r_1 + r_2 \geq 2$ and $\cos[(\theta_1 - \theta_2)/2] \geq 0$ for all z, to prove that $|g_0(z)| \geq 1$. Then show that the transformation $w = z + (z^2 - 1)^{1/2}$ maps the sheet R_0 of the Riemann surface onto the region $|w| \geq 1$, the sheet R_1 onto the region $|w| \leq 1$, and the branch cut between the points $z = \pm 1$ onto the circle $|w| = 1$. Note that the transformation used here is an inverse of the transformation

$$z = \frac{1}{2}\left(w + \frac{1}{w}\right),$$

and compare the results obtained with the results in Exercise 5(a), Sec. 12.

BIBLIOGRAPHY

The following list of supplementary books is far from exhaustive. Further references can be found in many of the books listed here.

Theory

Ahlfors, L. V.: "Complex Analysis," 3d ed., McGraw-Hill Book Company, Inc., New York, 1979.

Bieberbach, L.: "Conformal Mapping," Chelsea Publishing Co., New York, 1953.

Carathéodory, C.: "Conformal Representation," Cambridge University Press, London, 1952.

———: "Theory of Functions of a Complex Variable," vols. 1 and 2, Chelsea Publishing Co., New York, 1954.

Copson, E. T.: "Theory of Functions of a Complex Variable," Oxford University Press, London, 1962.

Evans, G. C.: "The Logarithmic Potential," American Mathematical Society, Providence, R. I., 1927.

Flanigan, F. J.: "Complex Variables: Harmonic and Analytic Functions," Allyn and Bacon, Inc., Boston, 1972.

Grove, E. A., and G. Ladas: "Introduction to Complex Variables," Houghton Mifflin Company, Boston, 1974.

Hille, E.: "Analytic Function Theory," vols. 1 and 2, 2d ed., Chelsea Publishing Co., New York, 1973.

Kaplan, W.: "Advanced Calculus," 3d ed., Addison-Wesley Publishing Company, Inc., Reading, Mass., 1984.

———: "Advanced Mathematics for Engineers," Addison-Wesley Publishing Company, Inc., Reading, Mass., 1981.

Kellogg, O. D.: "Foundations of Potential Theory," Dover Publications, Inc., New York, 1953.

Knopp, K.: "Elements of the Theory of Functions," Dover Publications, Inc., New York, 1952.

Krzyż, J. G.: "Problems in Complex Variable Theory," American Elsevier Publishing Company, Inc., New York, 1971.

Levinson, N., and R. M. Redheffer: "Complex Variables," Holden-Day, Inc., San Francisco, 1970.

Markushevich, A. I.: "Theory of Functions of a Complex Variable," 3 vols. in one, Chelsea Publishing Co., New York, 1977.

Marsden, J. E.: "Basic Complex Analysis," W. H. Freeman and Company, San Francisco, 1973.

Mitrinović, D. S.: "Calculus of Residues," P. Noordhoff, Ltd., Groningen, 1966.

Nehari, Z.: "Conformal Mapping," Dover Publications, Inc., New York, 1975.

Newman, M. H. A.: "Elements of the Topology of Plane Sets of Points," Cambridge University Press, London, 1964.

Pennisi, L. L.: "Elements of Complex Variables," Holt, Rinehart and Winston, Inc., New York, 1963.

Saff, E. B., and A. D. Snider: "Fundamentals of Complex Analysis for Mathematics, Science, and Engineering," Prentice-Hall, Inc., Englewood Cliffs, N. J., 1976.

Silverman, R. A.: "Complex Analysis with Applications," Prentice-Hall, Inc., Englewood Cliffs, N. J., 1974.

Springer, G.: "Introduction to Riemann Surfaces," 2d ed., Chelsea Publishing Co., New York, 1981.

Taylor, A. E., and W. R. Mann: "Advanced Calculus," 3d ed., John Wiley & Sons, Inc., New York, 1983.

Thron, W. J.: "Introduction to the Theory of Functions of a Complex Variable," John Wiley & Sons, Inc., New York, 1953.

Titchmarsh, E. C.: "Theory of Functions," Oxford University Press, London, 1939.

Whittaker, E. T., and G. N. Watson: "A Course of Modern Analysis," 4th ed., Cambridge University Press, London, 1963.

Applications

Bowman, F.: "Introduction to Elliptic Functions, with Applications," English Universities Press, London, 1953.

Brown, G. H., C. N. Hoyler, and R. A. Bierwirth: "Theory and Application of Radio-Frequency Heating," D. Van Nostrand Company, Inc., New York, 1947.

Churchill, R. V.: "Operational Mathematics," 3d ed., McGraw-Hill Book Company, New York, 1972.

——— and J. W. Brown: "Fourier Series and Boundary Value Problems," 3d ed., McGraw-Hill Book Company, New York, 1978.

Hayt, W. H., Jr.: "Engineering Electromagnetics," 4th ed., McGraw-Hill Book Company, New York, 1981.

Henrici, P.: "Applied and Computational Complex Analysis," vols. 1 and 2, John Wiley & Sons, Inc., New York, 1974, 1977.

Kober, H.: "Dictionary of Conformal Representations," Dover Publications, Inc., New York, 1952.

Lamb, H.: "Hydrodynamics," 6th ed., Dover Publications, Inc., New York, 1945.

Lebedev, N. N.: "Special Functions and Their Applications," rev. ed., Dover Publications, Inc., New York, 1972.

Milne-Thomson, L. M.: "Theoretical Hydrodynamics," Macmillan & Co., Ltd., London, 1955.

Oberhettinger, F., and W. Magnus: "Anwendung der elliptischen Funktionen in Physik und Technik," Springer-Verlag OHG, Berlin, 1949.

Rothe, R., F. Ollendorff, and K. Pohlhausen: "Theory of Functions as Applied to Engineering Problems," Technology Press, Massachusetts Institute of Technology, Cambridge, Mass., 1948.

Sokolnikoff, I. S.: "Mathematical Theory of Elasticity," 2d ed., McGraw-Hill Book Company, New York, 1956.

Streeter, V. L., and E. B. Wylie: "Fluid Mechanics," 7th ed., McGraw-Hill Book Company, New York, 1979.

Timoshenko, S. P., and J. N. Goodier: "Theory of Elasticity," 3d ed., McGraw-Hill Book Company, New York, 1970.

TABLE OF TRANSFORMATIONS OF REGIONS
(See Sec. **73**)

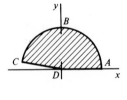

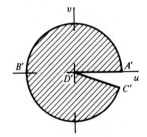

Figure 1. $w = z^2$.

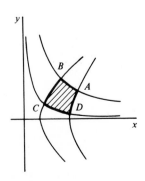

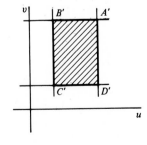

Figure 2. $w = z^2$.

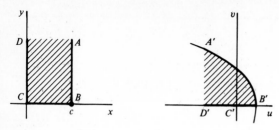

Figure 3. $w = z^2$; $A'B'$ on parabola $v^2 = -4c^2(u - c^2)$.

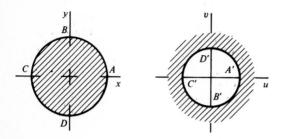

Figure 4. $w = \dfrac{1}{z}$.

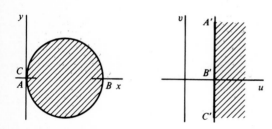

Figure 5. $w = \dfrac{1}{z}$.

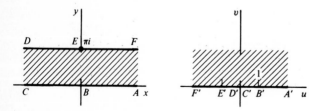

Figure 6. $w = \exp z$.

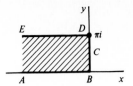
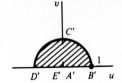

Figure 7. $w = \exp z$.

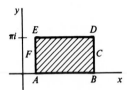
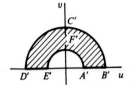

Figure 8. $w = \exp z$.

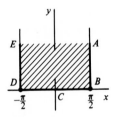
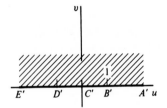

Figure 9. $w = \sin z$.

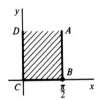
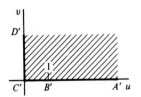

Figure 10. $w = \sin z$.

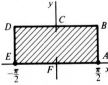
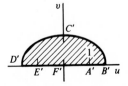

Figure 11. $w = \sin z$; BCD on line $y = b$ $(b > 0)$, $B'C'D'$ on ellipse $\dfrac{u^2}{\cosh^2 b} + \dfrac{v^2}{\sinh^2 b} = 1$.

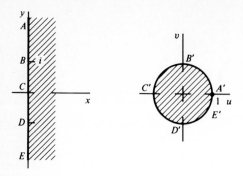

Figure 12. $w = \dfrac{z-1}{z+1}$.

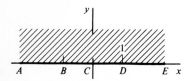

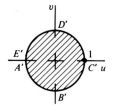

Figure 13. $w = \dfrac{i-z}{i+z}$.

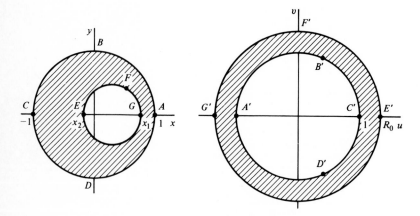

Figure 14. $w = \dfrac{z-a}{az-1}$; $a = \dfrac{1 + x_1 x_2 + \sqrt{(1 - x_1^2)(1 - x_2^2)}}{x_1 + x_2}$,

$$R_0 = \frac{1 - x_1 x_2 + \sqrt{(1 - x_1^2)(1 - x_2^2)}}{x_1 - x_2} \quad (a > 1 \text{ and } R_0 > 1 \text{ when } -1 < x_2 < x_1 < 1).$$

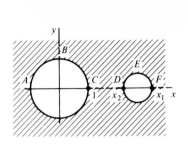
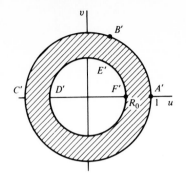

Figure 15. $w = \dfrac{z-a}{az-1}$; $a = \dfrac{1 + x_1 x_2 + \sqrt{(x_1^2 - 1)(x_2^2 - 1)}}{x_1 + x_2}$,

$$R_0 = \dfrac{x_1 x_2 - 1 - \sqrt{(x_1^2 - 1)(x_2^2 - 1)}}{x_1 - x_2}\,(x_2 < a < x_1 \text{ and } 0 < R_0 < 1 \text{ when } 1 < x_2 < x_1).$$

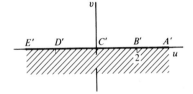

Figure 16. $w = z + \dfrac{1}{z}$.

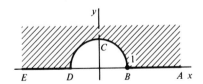
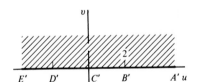

Figure 17. $w = z + \dfrac{1}{z}$.

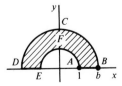
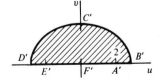

Figure 18. $w = z + \dfrac{1}{z}$; $B'C'D'$ on ellipse $\dfrac{u^2}{(b + 1/b)^2} + \dfrac{v^2}{(b - 1/b)^2} = 1$.

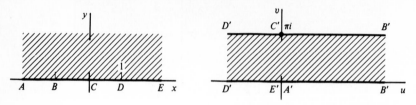

Figure 19. $w = \text{Log}\dfrac{z-1}{z+1}$; $z = -\coth\dfrac{w}{2}$.

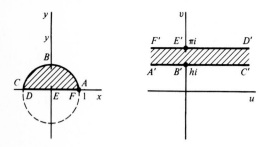

Figure 20. $w = \text{Log}\dfrac{z-1}{z+1}$; ABC on circle $x^2 + (y + \cot h)^2 = \csc^2 h \quad (0 < h < \pi)$.

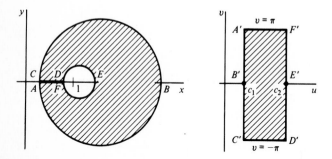

Figure 21. $w = \text{Log}\dfrac{z+1}{z-1}$; centers of circles at $z = \coth c_n$, radii: $\text{csch } c_n \quad (n = 1, 2)$.

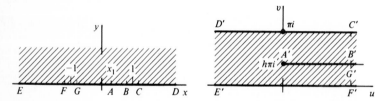

Figure 22. $w = h\,\text{Log}\dfrac{h}{1-h} + \text{Log } 2(1-h) + i\pi - h\,\text{Log}(z+1) - (1-h)\,\text{Log}(z-1)$;

$x_1 = 2h - 1$.

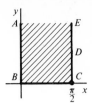
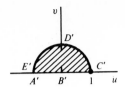

Figure 23. $w = \left(\tan \dfrac{z}{2}\right)^2 = \dfrac{1 - \cos z}{1 + \cos z}$.

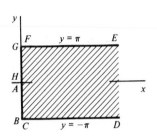
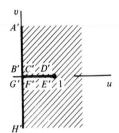

Figure 24. $w = \coth \dfrac{z}{2} = \dfrac{e^z + 1}{e^z - 1}$.

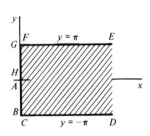
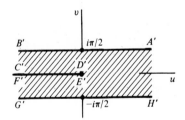

Figure 25. $w = \text{Log}\left(\coth \dfrac{z}{2}\right)$.

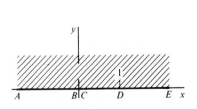
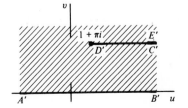

Figure 26. $w = \pi i + z - \text{Log } z$.

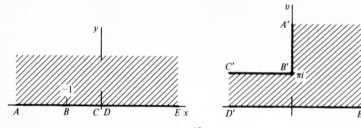

Figure 27. $w = 2(z + 1)^{1/2} + \mathrm{Log}\, \dfrac{(z + 1)^{1/2} - 1}{(z + 1)^{1/2} + 1}.$

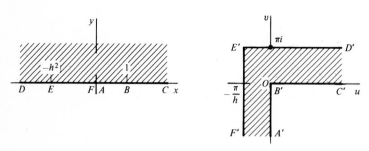

Figure 28. $w = \dfrac{i}{h}\,\mathrm{Log}\,\dfrac{1 + iht}{1 - iht} + \mathrm{Log}\,\dfrac{1 + t}{1 - t}; \quad t = \left(\dfrac{z - 1}{z + h^2}\right)^{1/2}.$

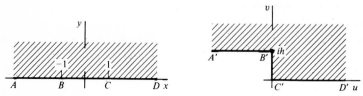

Figure 29. $w = \dfrac{h}{\pi}[(z^2 - 1)^{1/2} + \cosh^{-1} z].$*

*See Exercise 3, Sec. 94.

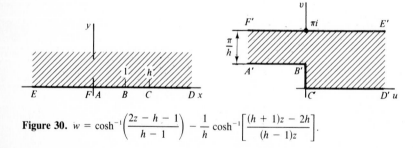

Figure 30. $w = \cosh^{-1}\left(\dfrac{2z - h - 1}{h - 1}\right) - \dfrac{1}{h}\cosh^{-1}\left[\dfrac{(h + 1)z - 2h}{(h - 1)z}\right].$

INDEX

Absolute convergence, 137
Absolute value, 7
Accumulation point, 23
Aerodynamics, 249
Analytic continuation, 298–306
Analytic function(s), 50–52
   derivative of, 111–114
   quotients of, 51, 165–167
   zeros of, 152–153, 169
Angle of inclination, 82, 217
Angle of rotation, 218
Antiderivative, 102–106
Arc, 80
   differentiable, 81
   Jordan, 81
   simple, 81
   smooth, 82
Argument, 12
Argument principle, 310–313

Bernoulli's equation, 250
Bessel function, 146n.
Beta function, 183, 264
Bibliography, 323
Bilinear transformation, 192
Binomial formula, 6
Boas, R. P., Jr., 119n.
Bolzano-Weierstrass theorem, 310

Boundary conditions, 226–230
   transformations of, 228–230
Boundary point, 22
Boundary value problem, 226–228, 281
Bounded function, 39, 308
Bounded set, 23
Branch of function, 68
   principal, 68, 74, 200
Branch cut, 69, 200–205, 318
   integration through, 179
Branch point, 69

Cartesian coordinates, 6
Cauchy, A. L., 45
Cauchy-Goursat theorem, 94–95, 100
   converse of, 114
   proof of, 97–99
Cauchy integral formula, 109–111
   for half plane, 289
Cauchy principal value, 170
Cauchy product, 149
Cauchy-Riemann equations, 43–46
   in polar form, 48
   sufficiency of, 46–47
Cauchy's inequality, 118
Christoffel, E. B., 262
Circle of convergence, 138

333